CAMBRIDGE TRACTS IN
MATHEMATICS

General Editors

B. BOLLOBAS, H. HALBERSTAM, C. T. C. WALL

95 Nonlinear superposition operators

JÜRGEN APPELL

Universität Würzburg, Fakultät für Mathematik, Am Hubland,
D-8700 Würzburg, WEST GERMANY

PETR P. ZABREJKO

Belgosuniversitet, Matematicheskij Fakul'tet, Pl. Lenina 4,
SU-220080 Minsk, SOVIET UNION

Nonlinear superposition operators

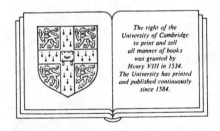

The right of the
University of Cambridge
to print and sell
all manner of books
was granted by
Henry VIII in 1534.
The University has printed
and published continuously
since 1584.

CAMBRIDGE UNIVERSITY PRESS

Cambridge
New York   Port Chester
Melbourne   Sydney

CAMBRIDGE UNIVERSITY PRESS
Cambridge, New York, Melbourne, Madrid, Cape Town, Singapore, São Paulo, Delhi

Cambridge University Press
The Edinburgh Building, Cambridge CB2 8RU, UK

Published in the United States of America by Cambridge University Press, New York

www.cambridge.org
Information on this title: www.cambridge.org/9780521361026

First published 1990
This digitally printed version 2008

A catalogue record for this publication is available from the British Library

ISBN 978-0-521-36102-6 hardback
ISBN 978-0-521-09093-3 paperback

Contents

Preface

The present monograph is concerned with a thorough study of the nonlinear operator

$$Fx(s) = f(s, x(s)). \tag{1}$$

Here $f = f(s, u)$ is a given function which is defined on the Cartesian product of some set Ω, which in most cases is either a metric space or a measure space or both, with the set $\mathbb{R}$ of real or the set $\mathbb{C}$ of complex numbers, and takes values in $\mathbb{R}$ or $\mathbb{C}$, respectively. By definition, the operator F associates to each real (or complex) function $x(s)$ on Ω the real (or complex) function $f(s, x(s))$ on Ω; therefore F is usually called a *superposition operator* (sometimes also *composition operator*, *substitution operator*, or *Nemytskij operator*).

In an implicit form, the superposition operator (1) can be found in the first pages of any calculus textbook (in the old terminology, as "composite function", "function of a function", etc.), where some of its elementary properties are described. Typical examples of such properties are the continuity of the superposition of continuous functions, the differentiability of the superposition of differentiable functions, and similar statements. Many other results of this type are scattered, mostly as lemmas or auxiliary results, in a vast literature on *mathematical analysis*, *functional analysis*, *differential* and *integral equations*, *probability theory* and *statistics*, *variational calculus*, *optimization theory*, and other fields of contemporary mathematics – the superposition operator occurs everywhere.

In many situations, the investigation of the basic properties of the operator (1) is quite straightforward and does not involve any particular difficulties. But this is not always so. In fact, at the beginning of nonlinear analysis it was often tacitly assumed that "nice" properties of the function f carry over to the corresponding operator F; this turned out to be false even in well-known classical function spaces. A typical example of this phenomenon is the behaviour of the superposition operator in Lebesgue spaces. For instance, the smoothness (and even the analyticity) of the function f does not imply the smoothness of F, considered as an operator between two Lebesgue spaces. Moreover, just the fact that F acts from the Lebesgue space L_p into the Lebesgue space L_q, say, leads to the very restrictive growth condition $f(s, u) = O(|u|^{p/q})$. Further, if F is (Fréchet-) differentiable between L_p and L_q, and the partial derivative f'_u of f with respect to u exists, then necessarily $f'_u(s, u) = O(|u|^{(p-q)/q})$ if $p \geq q$, and $f'_u(s, u) \equiv 0$ if $p < q$. Finally, if F is analytic between L_p and L_q, then the function f reduces to a polynomial in u (of degree at most p/q). All these facts are rather surprising; they show that many of the important properties of the function f do not imply analogous properties of the operator F, or vice versa.

Classical mathematical analysis mainly dealt with *spaces of continuous* or *differentiable functions*; already *Lebesgue spaces* arose only in special fields, e.g. Fourier series, approximation theory, probability theory. In modern nonlinear analysis, however, the arsenal of available function spaces has been considerably enlarged. In this connection, one should mention *Sobolev spaces* and their generalizations which are simply indispensable for the study of partial differential equations, *Orlicz spaces* which are the natural tool in the theory of both linear and nonlinear integral equations, *Hölder spaces* and their generalizations which are basic for the investigation of singular integral equations, *Lorentz* and *Marcinkiewicz spaces* which are widely used in interpolation theory for linear operators, and special classes of *spaces of differentiable* or *smooth functions* which frequently occur in the theory of ordinary or partial differential equations and variational calculus. The usefulness of all these spaces in various fields of mathematical analysis emphasizes the need for a systematic study of the superposition operator (1), considered as an operator from one such space into another.

In this connection, there are still many open problems. In particular, for many of these spaces one does not even know *acting conditions* for F, by means of conditions on f, which are both necessary and sufficient (sufficient conditions are often easily formulated). On the other hand, many special facts regarding the elementary properties of F, such as *continuity, boundedness*, or *compactness*, are well-known in, say, Orlicz spaces, Hölder spaces, or Sobolev spaces. Unfortunately, all these results are scattered in research papers and special monographs. We therefore conclude that it would be useful to collect the basic facts on the superposition operator, to present the main ideas which have been shown to be useful in studying its properties, and to provide a comparison of its behaviour in different spaces. This is the purpose of the present monograph.

Here the key problem is, as already mentioned, to find conditions on the function f which imply certain properties of the corresponding operator F. In this connection, the main properties we are interested in are: *boundedness* and *compactness* on certain subsets, *continuity* and *differentiability* at single points, *continuity* and *continuous differentiability* on open subsets, *special continuity properties* (such as *Lipschitz, uniform*, or *weak continuity*), *analyticity*, and related properties. These are just the properties which occur most frequently in the application of methods of nonlinear analysis, such as *fixed-point principles, degree theory, bifurcation methods, variational techniques*, to nonlinear equations involving superposition operators. Thus, the reader may typically find answers to questions of the following type: what are necessary and sufficient conditions for the function f such that the corresponding operator F maps the Lebesgue space L_p into the Lebesgue space L_q, or is continuous between two Orlicz spaces L_M and L_N, or differentiable between two Hölder spaces H_ϕ and H_ψ, or bounded between two Sobolev spaces W_p^k and W_q^m, or Lipschitz continuous in the space BV?

When preparing the material for this monograph, we intentionally confined ourselves to the *scalar case*. The vector case, i.e. when the superposition operator F maps $\mathbb{R}^m$- (or $\mathbb{C}^m$-) valued functions on Ω into $\mathbb{R}^n$- (or $\mathbb{C}^n$-) valued functions on Ω (and f is defined, of course, on $\Omega \times \mathbb{R}^m$ or $\Omega \times \mathbb{C}^m$ with values in $\mathbb{R}^n$ or $\mathbb{C}^n$, respectively), is at least as important as the scalar case. However, much less is known in this case, and the development of a "higher-dimensional" theory would be beyond the scope of the present work and would probably require us to increase the size of this survey at least twofold. In large parts of the monograph, Ω may also be the set of all natural numbers, equipped with the counting measure; consequently, our results cover superposition operators in sequence spaces as well. The main emphasis is put, however, on "usual" functions, i.e. the case when Ω is some domain in Euclidean space.

Apart from the superposition operator (1), the related operator

$$\Phi x(s) = x(\phi(s)) \qquad (2)$$

is sometimes also called superposition operator in the literature, where ϕ is some bijection of Ω onto itself; more precisely, operators of this type should be called *"inner" superposition operators*, in contrast to the "outer" superposition operator (1). In spite of the similar structure of the operators (1) and (2), their properties are quite different; this is clear, for instance, from the fact that the operator (2) is *linear*, while the major difficulty in the study of the operator (1) lies in its *nonlinearity*. Throughout this monograph, we shall be concerned only with the outer superposition operator (1).

Another operator which is closely related to the operator (1) is the *integral functional*

$$\Phi x = \int_\Omega f(s, x(s)) ds \,, \qquad (3)$$

which is of fundamental importance, for example, in *variational problems* of nonlinear analysis. We shall be concerned with the operator (3) only marginally and refer to the vast literature on variational methods.

The monograph consists of nine chapters. Each chapter is divided into a number of sections and provides a self-contained systematic study of the superposition operator in some class of function (or sequence) spaces. We have tried to make the exposition as complete and explicit as possible, including proofs, examples, and counterexamples. The last section of each chapter is devoted to possible generalizations, special cases, open problems, related fields and detailed bibliographical references. Each theorem, lemma, or formula is indexed within the corresponding chapter; thus, for example, Lemma 1.2 is the second lemma of the first chapter. By $\Rightarrow$ and $\Leftarrow$ we denote the beginning and the end, respectively, of a proof.

The contents of the monograph go as follows. The first chapter is entirely devoted to the study of the superposition operator in the *space* $S = S(\Omega)$ *of measurable functions* on Ω, where Ω is an arbitrary nonempty set with measure. Here a basic problem is that of finding conditions on the function f which ensure that the operator F maps measurable functions into measurable functions. Surprisingly enough, this turns out to be a highly nontrivial problem.

As a matter of fact, the space S is a complete metric linear space, but not normable. Most fundamental principles of linear and nonlinear functional analysis are formulated, however, in a Banach space setting. Consequently, it is desirable to study the properties of the superposition operator not only in S, but also in normed subspaces of S. It turns out that the most appropriate class of Banach spaces of measurable functions is that of so-called *ideal spaces* (or *Banach lattices*), which were considered by many authors for different purposes. General properties of the superposition operator in ideal spaces are described in detail in the second chapter.

The third and fourth chapters are concerned with the superposition operator in *Lebesgue* and *Orlicz spaces*, respectively. Here the theory is most complete and advanced, and one can characterize all basic properties of the operator F (in particular, acting conditions) in terms of the generating function f.

Some other classes of ideal spaces which include, for example, the classical *Lorentz* and *Marcinkiewicz spaces* are dealt with in the fifth chapter. In this connection, only very few elementary results are presently known.

The sixth chapter is devoted to the superposition operator in the *space* $C = C(\Omega)$ *of continuous functions* on Ω, where Ω is a compact domain without isolated points in Euclidean space. Here the basic facts are well-known "folklore"; however, we shall also discuss some special problems which have not been studied yet. Moreover, we briefly discuss the superposition operator in the *space BV of functions of bounded variation*.

In the seventh chapter we shall present a systematic study of the superposition operator in *Hölder-type spaces*. It turns out that the behaviour of the superposition operator in such spaces is quite different from that in spaces of measurable functions.

The eighth chapter will be concerned with the superposition operator in spaces of functions which are characterized by certain differentiability or smoothness properties. Moreover, we shall consider the operator F in various *spaces of finitely* or *infinitely differentiable functions*, including *Roumieu, Beurling* and *Gevrey classes*.

Some results on the superposition operator in *Sobolev spaces* are given in the ninth chapter. Unfortunately, in spite of the importance of these spaces in the theory of distributions and partial differential equations, they have been given very little attention in the literature.

Some remarks on the bibliography are in order. We hope to present a rather exhaustive list of references on the superposition operator in function and sequence spaces. The bibliography at the end covers the period from 1918 to 1988 and contains about 400 items, half of them in Russian; thus, it may also serve as a guide to the Soviet literature. For the reader's convenience, we have added English translations (if there are any) of Russian books and major journal papers, and, beginning with 1960, the corresponding review numbers of Zentralblatt für Mathematik (Zbl.), Referativnyj Zhurnal Matematika (R.Zh.), and Mathematical Reviews (M.R.). We are indebted to Nguyễn Hồng Thái, Heinz-Willi Kröger and Reiner Welk for computer-aided help in finding many review numbers.

It is a great pleasure to thank all colleagues and friends who sent us reprints, preprints, and unpublished manuscripts, and helped us to make the list of references more complete by suggesting new (or simply forgotten) entries. In particular, we are indebted to Nguyễn Hồng Thái , Jevgenij M. Semjonov, and Marek Z. Berkolajko for several helpful discussions on Chapter 4, Chapter 5, and Chapter 7, respectively. Moreover, we are grateful to the publishers, especially to David Tranah and Mark Hendy, for fruitful collaboration and useful advice. Last but not least, our special thanks go to Fritzi Stegmüller for her excellent typing of the manuscript with extraordinary patience, and for never grumbling at a lot of changes.

This book could not have been realized without the possibility of travels and meetings in both Germany and the Soviet Union, generously supported by the Deutscher Akademischer Austauschdienst (Bonn) and the Ministry of Higher Education (Moscow). The first author acknowledges the hospitality of the Belorusskij Gosudarstvennyj Universitet at Minsk, the second author that of the Universität Augsburg and the Ruhruniversität Bochum.

Spring 1989   Jürgen Appell, Petr P. Zabrejko

Chapter 1

The superposition operator
in the space S

In this chapter we study the superposition operator $Fx(s) = f(s, x(s))$ in the complete metric space S of measurable functions over some measure space Ω. First, we consider some classes of functions f which generate a superposition operator F from S into S; a classical example is the class of Carathéodory functions, a more general class that of Shragin functions.

As a matter of fact, there exist functions f, called "monsters", which generate the zero operator $Fx \equiv \theta$, but are not measurable on $\Omega \times \mathbb{R}$, and hence are not Carathéodory functions; this disproves the old-standing Nemytskij conjecture. On the other hand, we show that a function which generates a continuous superposition operator (in measure) is "almost" a Carathéodory function.

We give a necessary and sufficient condition for the function f to generate a bounded superposition operator F in the space S. In particular, this conditions holds always if f is a Carathéodory function. On the other hand, we show that the superposition operator F is "never" compact in the space S, except for the trivial case when F is constant.

Finally, we consider superposition operators which are generated by functions f with special properties (e.g. monotonicity), and characterize the points of discontinuity of such operators.

1.1 The space S

Let Ω be an arbitrary set, $\mathcal{M}$ some σ-algebra of subsets of Ω (which will be called measurable in what follows), and μ a countably additive and σ-finite measure on $\mathcal{M}$. By λ we denote some normalized ("probability") measure on $\mathcal{M}$ which is equivalent to μ (i.e. has the same null sets); one possible choice of λ could be, for instance,

$$\lambda(D) = \int_D n(s)d\mu,$$

where n is any positive function on Ω with

$$\int_\Omega n(s)d\mu = 1\,.$$

In most examples, we shall deal with either some bounded perfect set Ω with nonempty interior in some finite dimensional space, together with the algebra $\mathcal{M}$ of Borel- or Lebesgue-measurable subsets and the Lebesgue measure μ, or the set of natural numbers, together with the algebra of all subsets and the counting measure. More complicated examples, of course, are also possible: Ω being an arbitrary Lebesgue or Borel subset, and μ the Lebesgue or Borel measure, or Ω being a "nice" subset of a finite dimensional manifold, together with a suitable algebra $\mathcal{M}$ of subsets and some measure μ. Such examples will be considered only in quite exceptional cases. We point out that we do not suppose the σ-algebra $\mathcal{M}$ to be complete with respect to the measure μ.

Recall (Saks' lemma) that the set Ω can be divided, uniquely up to null sets, into two parts Ω_c and Ω_d such that μ is *atomic-free* ("continuous") on Ω_c (i.e. any subset of Ω_c can be divided into two parts of equal measure), and μ is *purely atomic* ("discrete") on Ω_d, i.e. Ω_d is a finite or countable union of atoms of positive measure. In "natural" examples, one of the sets Ω_c or Ω_d is usually empty, and thus one deals with real "function spaces" or "sequence spaces".

As usual, we denote by $S = S(\Omega, \mathcal{M}, \mu)$ the set of all (real or complex-valued) almost everywhere finite μ-measurable functions on Ω; more precisely, S consists of equivalence classes of such functions, where two functions x and y are called *equivalent* if they coincide almost everywhere on Ω. The set S can be equipped with the usual algebraic operations, where the zero element is the function $\theta(s) = 0$ almost everywhere, as well as with the metric $\rho(x, y) = [x - y]$, where

$$[z] = \inf_{0 < h < \infty} \{h + \lambda(\{s : s \in \Omega, |z(s)| > h\})\} \qquad (1.1)$$

or

$$[z] = \int_\Omega \frac{|z(s)|}{1 + |z(s)|}d\lambda\,. \qquad (1.2)$$

With respect to this metric, S becomes a *complete metric space*, and convergence $\rho(x_n, x) \to 0$ is equivalent to *convergence of x_n in measure* to x, i.e. $\lambda(\{s : |x_n(s) - x(s)| > h\}) \to 0$ as $n \to \infty$, for any $h > 0$.

It is convenient to introduce also a *partial ordering* in the space S: we write $x \le y$ $(x, y \in S)$ if $x(s) \le y(s)$ for almost all $s \in \Omega$. In this way, S

becomes an ordered linear space, i.e. $x \leq y$ implies that $x + z \leq y + z$ for $z \in S$, and that $\lambda x \leq \lambda y$ for $\lambda \geq 0$; moreover, if x_n and y_n are two sequences in S which converge to $x \in S$ and $y \in S$, respectively, then $x_n \leq y_n$ implies that also $x \leq y$. Finally, S is a K-*space* (in the sense of L.V.Kantorovich), which means that any set which is bounded from above (respectively below) admits a least upper bound (respectively greatest lower bound), where these notions are defined as usual.

As in every ordered linear space, one can consider convergence with respect to the above ordering in S. A sequence x_n in S is *order convergent* to $x \in S$ if $\varliminf\limits_{n \to \infty} x_n = \varlimsup\limits_{n \to \infty} x_n = x$, where

$$\varliminf_{n \to \infty} x_n = \sup_k \inf_{m \geq k} x_m\,, \quad \varlimsup_{n \to \infty} x_n = \inf_k \sup_{m \geq k} x_m\,.$$

In the space S, this type of convergence coincides with *convergence almost everywhere*. It is well known (Lebesgue's theorem) that convergence almost everywhere implies convergence in measure; the converse is true only if the measure μ is discrete (i.e. $\Omega_c = \emptyset$). Nevertheless (Riesz' theorem), each sequence which is convergent in measure admits a subsequence which converges almost everywhere (to the same limit, of course). We still mention a well known fact about convergence almost everywhere (Jegorov's theorem): if x_n converges almost everywhere to $x \in S$, then x_n converges uniformly outside some set $D \in \mathcal{M}$ of arbitrarily small λ-measure.

In the sequel, we shall denote by χ_D $(D \in \mathcal{M})$ the *characteristic function* of D,

$$\chi_D(s) = \begin{cases} 1 & s \in D, \\ 0 & s \notin D, \end{cases}$$

and by P_D the *multiplication operator* by χ_D

$$P_D x(s) = \chi_D(s) x(s)\,. \tag{1.3}$$

The functions

$$x(s) = \sum_{j=1}^{m} c_j \chi_{D_j}(s) \quad (D_j \in \mathcal{M}; \, j = 1, \cdots, m) \tag{1.4}$$

are usually called *simple functions*. It is not hard to see that the linear space S_0 of simple functions in S is dense in S; this implies, in particular, that the separability of S is equivalent to the separability of the metric space $(\mathcal{M}, d)$, where $d(A, B) = \lambda(A \triangle B)$ is the λ-measure of the "symmetric difference" of

A and B. Moreover (Mikusiński's theorem), every nonnegative function in S is the limit of a monotonically increasing sequence of nonnegative simple functions.

In many situations, the set Ω is a complete metric space, and the algebra $\mathcal{M}$ includes the subalgebra $\mathcal{B}(\Omega)$ of all Borel subsets of Ω. In this case the measure μ is usually supposed to be *regular*, i.e. the following compatibility condition holds between the metric spaces Ω and $(\mathcal{M}, d)$: given $D \in \mathcal{M}$ and $\varepsilon > 0$, there exists a compact subset D_ε of Ω such that $d(D, D_\varepsilon) < \varepsilon$.

We suppose that the reader is familiar with the construction and the basic properties of the (Lebesgue) integral. In what follows, we shall denote by L the set of all (Lebesgue) integrable functions over Ω, equipped with the norm

$$\|x\| = \int_\Omega |x(s)| d\mu(s) . \tag{1.5}$$

If the measure μ under consideration is fixed, we shall write simply ds instead of $d\mu(s)$.

1.2 The superposition operator

Let $f = f(s, u)$ be a function defined on $\Omega \times \mathbb{R}$ (or $\Omega \times \mathbb{C}$), and taking values in $\mathbb{R}$ (respectively $\mathbb{C}$). Given a function $x = x(s)$ on Ω, by applying f we get another function $y = y(s)$ on Ω, defined by $y(s) = f(s, x(s))$. In this way, the function f generates an operator

$$Fx(s) = f(s, x(s)) \tag{1.6}$$

which is usually called *superposition operator* (also outer superposition operator, composition operator, substitution operator, or Nemytskij operator).

In this chapter, and in most other chapters, unless otherwise stated, we shall consider f as function from $\Omega \times \mathbb{R}$ into $\mathbb{R}$.

The superposition operator (1.6) has some remarkable properties. One "algebraic" property which is called the *local determination* of F is described in the following:

Lemma 1.1   *The superposition operator F has the following three (equivalent) properties:*
(a) For $D \subseteq \Omega$,

$$FP_D - P_D F = P_{\Omega \setminus D} F\theta , \tag{1.7}$$

where θ is the almost everywhere zero function.

(b) For $D \subseteq \Omega$,

$$P_D F P_D x = P_D F x, \quad P_{\Omega \setminus D} F P_D x = P_{\Omega \setminus D} F \theta.$$

(c) If two functions x_1 and x_2 coincide on $D \subseteq \Omega$, then the functions $F x_1$ and $F x_2$ also coincide on D.

$\Rightarrow$ The validity of the three conditions follows immediately from the definition of the superposition operator; therefore we shall show only their equivalence (for any operator F). The equivalence of (a) and (b) follows from the equality $F P_D = P_D F P_D + P_{\Omega \setminus D} F P_D$. Now, if (a) holds and x_1 and x_2 coincide on $D \subseteq \Omega$, we have $P_D x_1 = P_D x_2$ and hence, by (1.7), $P_D F x_1 - P_D F x_2 = F P_D x_1 - F P_D x_2 = \theta$, i.e. $F x_1$ and $F x_2$ coincide on D. Finally, suppose that (c) holds, x is some function on Ω, and $D \subset \Omega$. Then $F x$ and $F P_D x$ coincide on D, since x and $P_D x$ do so. On the other hand, $P_D x$ and θ coincide on $\Omega \setminus D$, and hence also $F P_D x$ and $F \theta$. This shows that (b) holds. $\Leftarrow$

If, in particular, $F \theta = \theta$ (which means that the function f satisfies

$$f(s, 0) = 0 \tag{1.8}$$

for almost all $s \in \Omega$), all these conditions are equivalent to the fact that F commutes with any of the multiplication operators (1.3), i.e.

$$F P_D = P_D F. \tag{1.9}$$

In this case, the superposition operator is *disjointly additive*; this means that

$$F(x_1 + x_2) = F x_1 + F x_2, \tag{1.10}$$

whenever the functions x_1 and x_2 are *disjoint* (i.e. their supports supp $x_j = \{s : s \in \Omega, x_j(s) \neq 0\}$ $(j = 1, 2)$ are disjoint). In fact, from (1.9) we get $F(x_1 + x_2) = F(P_{D_1 \cup D_2} x) = P_{D_1 \cup D_2} F x = P_{D_1} F x + P_{D_2} F x = F P_{D_1} x + F P_{D_2} x = F x_1 + F x_2$, where $x = x_1 + x_2$ and $D_j = \text{supp } x_j$ $(j = 1, 2)$. We remark that an analogous partial additivity holds also for countably many functions.

Observe that the condition (1.8) is not really restrictive in many cases, because one can often pass from the superposition operator (1.6) to the superposition operator $\tilde{F} x(s) = \tilde{f}(s, x(s))$ generated by the function

$$\tilde{f}(s, u) = f(s, x_0(s) + u) - f(s, x_0(s)), \tag{1.11}$$

where x_0 is any fixed function on Ω (for example $x_0 = \theta$).

1.3 Sup-measurable functions

Lemma 1.1 implies, in particular, that the superposition operator (1.6) maps
equivalent functions on Ω into equivalent ones, i.e. acts actually on classes.
Therefore it is natural to ask for *acting conditions* for F in S, i.e. conditions
on the function f which guarantee that the corresponding superposition
operator F maps all (equivalence classes of) functions in S into such. Surprisingly enough, this problem turns out to be very hard, and a large part of
this chapter is actually devoted to the discussion of the known results in this
direction. We shall call a function f *superpositionally measurable*, or *sup-
measurable*, for short, if the corresponding superposition operator F maps
the space S into itself. It is natural to try to characterize sup-measurability
by means of (possibly simple) intrinsic properties of f; one basic difficulty
in this connection lies in the fact that a sup-measurable function f is by
no means uniquely determined by the corresponding operator F, since two
functions f_1 and f_2, although generating the same superposition operator,
may be "essentially different".

At this point, we introduce special relations between functions of two
variables which allow us to formulate most of our results very easily and
precisely. Given two functions f_1 and f_2 on $\Omega \times \mathbb{R}$ and some subset Δ of
$\Omega \times \mathbb{R}$, we shall write

$$f_1(s,u) \preceq f_2(s,u) \quad ((s,u) \in \Delta) \tag{1.12}$$

if, whenever $x \in S$ has its graph in Δ, we have

$$f_1(s,x(s)) \leq f_2(s,x(s))$$

for almost all $s \in \Omega$ (in case $\Delta = \Omega \times \mathbb{R}$ we drop the condition on the
right-hand side of (1.12)). Moreover, the notation

$$f_1(s,u) \simeq f_2(s,u) \quad ((s,u) \in \Delta) \tag{1.13}$$

means that both $f_1(s,u) \preceq f_2(s,u)$ and $f_2(s,u) \preceq f_1(s,u)$, i.e. the operators
F_1 and F_2 coincide on the set of all $x \in S$ whose graphs are contained in Δ.
In this case we shall call the functions f_1 and f_2 *superpositionally equivalent*,
or *sup-equivalent*, for short, on Δ. Note that sup-measurability is then
invariant under the equivalence relation (1.13), i.e. if f is sup-measurable
then so is every function $\tilde{f}$ which is sup-equivalent to f.

It is very striking that even functions f which are sup-equivalent to the
zero function $\tilde{f}(s,u) \equiv 0$ may exhibit a very pathological behaviour; in the

literature such functions are called *monsters*. We give now two examples of such monsters which are both constructed on the interval $\Omega = [0,1]$ and essentially build on the validity of the continuum hypothesis.

The first one (which we call the *Russian monster*) was invented by M.A. Krasnosel'skij and A.V. Pokrovskij. Recall first that both the set Ω and the space S (more precisely, a complete representation system of pairwise non-equivalent functions) have the cardinality of the continuum, and therefore their elements ($s_\alpha \in \Omega$ and $x_\beta \in S$, say) can be indexed by the ordinals from 1 to ω, the first uncountable ordinal. Let

$$f(s_\alpha, u) = \begin{cases} 0 & \text{if } u = x_\beta(s_\alpha) \text{ for some } \beta < \alpha, \\ 1 & \text{otherwise.} \end{cases} \qquad (1.14)$$

This function is sup-measurable, but not measurable! In fact, for any $x \in S$, the function Fx is different from zero at most on a countable subset of Ω. To see this, fix $x = x_\beta \in S$. By (1.14), we have $f(s, x_\beta(s)) \neq 0$ only for $s = s_\alpha$ with $\alpha \leq \beta$, and these points s form only a finite or countably infinite set (it is here that one uses the continuum hypothesis). This means that $f(s, u) \simeq 0$; in other words, f generates the zero operator, and hence is certainly sup-measurable.

To prove that f is not measurable on the product $\Omega \times \mathbb{R}$, we remark that, for any fixed $s_0 \in \Omega$, the function $f(s_0, \cdot)$ vanishes at most on a countable subset of $\mathbb{R}$. Therefore the set $Q = \{(s, u) : f(s, u) = 1\}$ meets any horizontal line $u = u_0$ in at most countably many points, but contains all vertical lines $s = s_0$ except for at most countably many points. This shows that Q is a non-measurable subset of $\Omega \times \mathbb{R}$ (for example, by Fubini's theorem) and hence f, being the characteristic function of Q, is not measurable either.

The second example (which we call the *Polish monster*) was given by Z. Grande and J. Lipiński and is also quite "exotic". Recall that both the class of all closed subsets of $\Omega \times \mathbb{R}$ of positive measure and the set of all Borel functions on Ω have the cardinality of the continuum, and can therefore again be indexed by the ordinals from 1 to ω (M_α and x_β, say). Further, by transfinite induction one can construct a sequence of points $(s_\alpha, u_\alpha) \in \Omega \times \mathbb{R}$ such that

$$(s_\alpha, u_\alpha) \in M_\alpha \setminus \bigcup_{\beta < \alpha} \Gamma_\beta, \quad (s_\alpha \neq s_\beta \text{ for } \beta < \alpha),$$

where M_α is a closed subset of $\Omega \times \mathbb{R}$ of positive measure, and Γ_β is the graph of the Borel function x_β. The possibility of choosing (s_α, u_α) in such a way follows from the fact that the set of all ordinals β "preceding" α is at most countable, the measure of M_α is positive, and the graph Γ_β of a Borel function x_β is a null set in $\Omega \times \mathbb{R}$. The Polish monster

$$f(s,u) = \begin{cases} 1 & \text{if } (s,u) = (s_\alpha, u_\alpha) \text{ for some } \alpha, \\ 0 & \text{otherwise,} \end{cases} \qquad (1.15)$$

is then again sup-measurable, but not measurable, as can be seen as follows: if x is any measurable function and x_β is a Borel function which is equivalent to x, the equality $f(s, x_\beta(s)) \neq 0$ holds only for those $s_\alpha \in \Omega$ which satisfy $(s_\alpha, u_\alpha) \in \Gamma_\beta$. But these points form only a countable set, and hence the Polish monster again generates the zero operator, and thus is trivially sup-measurable.

To show that f is non-measurable one can again consider the set $Q = \{(s,u) : f(s,u) = 1\}$; in fact, if Q were measurable, Q would have measure zero (by Fubini's theorem). But in this case one would have $Q \cap M_\alpha = \emptyset$ for some closed set M_α of positive measure, which is impossible since Q contains at least one point of each set M_α.

The above examples show that the sup-measurability of f does not imply its measurability. The converse is not true either; this is easier to see: given an arbitrary non-measurable function z on Ω, just consider

$$f(s,u) = \begin{cases} z(s) & \text{if } u = s, \\ 0 & \text{if } u \neq s; \end{cases} \qquad (1.16)$$

this function f is obviously measurable on $\Omega \times \mathbb{R}$, but the corresponding operator F maps the measurable function $x(s) = s$ into the non-measurable function z.

To make the following arguments more transparent, we still introduce some terminology. Suppose that N is some subset of functions of S, and assume that f_1 and f_2 are two sup-measurable functions on $\Omega \times \mathbb{R}$ with the property that the operators F_1 and F_2 take the same values on N. We shall call the set N *thick* if this implies that $f_1(s,u) \simeq f_2(s,u)$, i.e. F_1 and F_2 take the same values on the whole space S. In other words, a function set N is thick if every superposition operator F admits a unique extension from N to S.

A necessary and sufficient condition for "thickness" is provided by the following:

Lemma 1.2   *A subset N of S is thick if and only if, given $x \in S$, one can find a sequence x_n in N and a countable partition $\{D_1, D_2, \cdots\}$ of Ω ($D_n \in \mathcal{M}$) such that $x(s) = x_n(s)$ for almost all $s \in D_n$.*

⇒ The proof is quite simple: in fact, if two superposition operators F_1 and F_2 coincide on N then, by Lemma 1.1, they also coincide on the set N^* of all functions of the form

$$x(s) = \sum_{n=1}^{\infty} P_{D_n} x_n(s) \qquad (x_n \in N),$$

where $\{D_1, D_2, \cdots\}$ is an arbitrary countable partition of Ω (and only on such functions!). But Lemma 1.2 just states that N is thick if and only if $N^* = S$. ⇐

It is also possible to give other simple (sufficient) conditions for the "thickness" of some set of measurable functions. Given $N \subset S$, we call the set $\mathfrak{lu}(N)$ of all functions $x \in S$ such that, for each $\varepsilon > 0$, there exists $x_\varepsilon \in N$ with $\lambda(\{s : x(s) \neq x_\varepsilon(s)\}) < \varepsilon$, the *Luzin hull* of N. Obviously, $N \subset S$ is thick if $\mathfrak{lu}(N) = S$.

For example, if Ω is a bounded domain in Euclidean space, Luzin's theorem (see Section 6.1) states that $\mathfrak{lu}(C) = S$, where C is the set of continuous functions on Ω. It is rather surprising that, if C is replaced by the set C^1 of continuously differentiable functions on Ω, one has $\mathfrak{lu}(C^1) \neq S$; in fact, C^1 is not thick in S!

To conclude this section, we shall prove an auxiliary result which will be used several times in the sequel.

Lemma 1.3   *Let a be a nonnegative sup-measurable function on $\Omega \times \mathbb{R}$, and suppose that*

$$\int_\Omega a(s, x(s)) ds \leq c < \infty$$

for all $x \in S$. Then there exists a function $\bar{a} \in L$ such that $a(s, u) \preceq \bar{a}(s)$ and

$$\int_\Omega \bar{a}(s) ds \leq c. \qquad (1.17)$$

⇒ By Kantorovich's theorem, the function set $H = \{\arctan Ax : x \in S\}$ admits a least upper bound z_* in S, where A is the superposition operator generated by the function a. Moreover, one can find a sequence x_n in S such that $\sup H = \sup_n \arctan Ax_n$. By induction, we construct a sequence z_n in S putting $z_1 = x_1$ and

$$z_n(s) = \begin{cases} z_{n-1}(s) & \text{if } a(s, z_{n-1}(s)) \geq a(s, x_n(s)), \\ x_n(s) & \text{if } a(s, z_{n-1}(s)) \leq a(s, x_n(s)). \end{cases}$$

Obviously, z_n has the property that arctan Az_n converges monotonically to sup H. By Levi's theorem, the function $\bar{a}(s) = \lim\limits_{n \to \infty} Az_n(s)$ is integrable and satisfies (1.17). $\Leftarrow$

1.4 Carathéodory and Shragin functions

Already in 1918, K. Carathéodory gave the following sufficient condition for sup-measurability: a function $f = f(s, u)$ is sup-measurable if $f(s, \cdot)$ is continuous on $\mathbb{R}$ for almost all $s \in \Omega$, and $f(\cdot, u)$ is measurable on Ω for all $u \in \mathbb{R}$. Such functions are now called *Carathéodory functions* (or functions which satisfy a Carathéodory condition).

To prove the sup-measurability of a Carathéodory function f, note first that the corresponding operator F maps any simple function into a measurable function; in fact, if x has the form (1.4), all functions $f(\cdot, c_j)$ $(j = 1, \cdots, m)$ are measurable, and Lemma 1.1 shows that

$$Fx(s) = \sum_{j=1}^{m} P_{D_j} f(s, c_j),$$

which is clearly a measurable function. Now, if $x \in S$ is arbitrary, and $x_n \in S_0$ are simple functions which converge to x almost everywhere on Ω, we have, by the continuity of $f(s, \cdot)$, that $Fx(s) = \lim\limits_{n \to \infty} Fx_n(s)$ for almost all $s \in \Omega$. By Lebesgue's theorem, Fx is then also measurable.

In the sequel we shall need the following obvious lemma:

Lemma 1.4   Let f_n be a sequence of sup-measurable functions, and suppose that $f_n(s, u)$ converges to $f(s, u)$ for almost all $s \in \Omega$ and all $u \in \mathbb{R}$. Then f is also sup-measurable.

$\Rightarrow$ The assertion follows easily from Lebesgue's theorem and the fact that, for any $x \in S$, the sequence $f_n(s, x(s))$ converges almost everywhere to $f(s, x(s))$. $\Leftarrow$

Lemma 1.4 allows us to enlarge the class of sup-measurable functions, by adopting an analogous construction to that leading to the known Baire classes: let B_0 denote the class of all Carathéodory functions, and B_α (α a countable ordinal number) the class of all functions f which admit a representation

$$f(s, u) = \lim_{n \to \infty} f_n(s, u) \qquad (s \in \Omega \setminus D_0, \ u \in \mathbb{R}), \qquad (1.18)$$

where the functions f_n $(n = 1, 2, \cdots)$ belong to classes B_{α_n} with $\alpha_n < \alpha$, and $D_0 \in \mathcal{M}$ is some null set. We call the classes B_α $(0 \leq \alpha < \omega)$ *Baire–Carathéodory classes*, and their elements *Baire–Carathéodory functions*. Roughly speaking, all functions f which occur in applications and are not too "exotic" belong to some Baire-Carathéodory class. For example, the functions $f_1(s, u) = \text{sgn } u$ and $f_2(s, u) = \text{sgn } (s-u)$ $(\Omega = [0, 1])$ are of Baire-Carathéodory class B_1, since $f_1(s, u) = \lim\limits_{n \to \infty} \frac{2}{\pi} \arctan nu$ and $f_2(s, u) = \lim\limits_{n \to \infty} \frac{2}{\pi} \arctan n(s - u)$. Another important class of sup-measurable functions was introduced and studied by I.V. Shragin in 1976 under the name "standard functions"; we shall call them "Shragin functions" in what follows.

To define this class, recall that a set $D \in \mathcal{M}$ is called *negligible* if any subset D' of D also belongs to $\mathcal{M}$. If the measure μ is purely atomic, every set $D \in \mathcal{M}$ is negligible; roughly speaking, this is the reason why almost all the following considerations are rather trivial in the case of a purely atomic measure, but complicated and significant only in the case of an atomic-free measure. Observe that, if the algebra $\mathcal{M}$ is complete with respect to the measure μ, the negligible sets in $\mathcal{M}$ are just the null sets.

A function $f = f(s, u)$ is called *Shragin function* (or function which satisfies a Shragin condition) if there exists a negligible set $D_0 \in \mathcal{M}$ with the property that, for any $C \in \mathcal{B}$ (the Borel subsets of $\mathbb{R}$), the set $f^{-1}(C) \backslash (D_0 \times \mathbb{R})$ belongs to $\mathcal{M} \otimes \mathcal{B}$, the minimal σ-algebra which contains all products $M \times B$ with $M \in \mathcal{M}$ and $B \in \mathcal{B}$.

The definition shows that the class of Shragin functions is somewhat larger than the class of $(\mathcal{M} \otimes \mathcal{B}, \mathcal{B})$-measurable functions.

The Shragin functions have some important properties: for instance, they form a linear space, and if a sequence of Shragin functions converges to some function f in the sense (1.18), the limit function f is also a Shragin function (this is in contrast to Carathéodory functions and shows that it does not make sense to define "Baire–Shragin classes").

Let us now show that every Shragin function is sup-measurable. Let $D_0 \in \mathcal{M}$ be the negligible set in the definition of a Shragin function f, and let x be measurable. Given a Borel set $C \subseteq \mathbb{R}$, we have

$$\{s : s \in \Omega, \ f(s, x(s)) \in C\}$$
$$= \hat{x}^{-1}[f^{-1}(C) \backslash (D_0 \times \mathbb{R})] \cup \{s : s \in D_0, \ f(s, x(s)) \in C\},$$

where $\hat{x}(s) = (s, x(s))$. Since the function $\hat{x}$ is obviously $(\mathcal{M}, \mathcal{M} \otimes \mathcal{B})$-measurable and f is a Shragin function, the set $\hat{x}^{-1}[f^{-1}(C) \backslash (D_0 \times \mathbb{R})]$ is measurable. On the other hand, the set $\{s : s \in D_0, \ f(s, x(s)) \in C\}$ is also measurable, as a subset of the negligible set D_0. Since $C \in \mathcal{B}$ was arbitrary, Fx is a measurable function as claimed.

As already mentioned above, the Baire–Carathéodory classes are fairly large and embrace all "natural" functions. It turns out that the class of Shragin functions is even larger, in the sense that it contains the union of all Baire–Carathéodory classes. To see this, it suffices to show that every Carathéodory function is a Shragin function, because Shragin functions are "stable" under the limit (1.18). So, let f be a Carathéodory function, and suppose, without loss of generality, that $f(s, \cdot)$ is continuous on $\mathbb{R}$ even for all $s \in \Omega$. Consider the auxiliary functions

$$f^0(s, u) = \lim_{\delta \to 0} f^\delta(s, u) \quad \text{and} \quad f_0(s, u) = \lim_{\delta \to 0} f_\delta(s, u),$$

where

$$f^\delta(s, u) = \sup\{f(s, v) : v \in \mathbb{Q}, |u - v| < \delta\},$$
$$f_\delta(s, u) = \inf\{f(s, v) : v \in \mathbb{Q}, |u - v| < \delta\}.$$

Obviously, $f_0(s, u) \le f(s, u) \le f^0(s, u)$; by the continuity of $f(s, \cdot)$, we even have $f_0(s, u) = f(s, u) = f^0(s, u)$. It is therefore sufficient to prove that all functions f^δ and f_δ are Shragin functions. The function f^δ, for instance, satisfies the equality

$$\{(s, u) : f^\delta(s, u) > c\} = \bigcup_{v \in \mathbb{Q}} [\{s : f(s, v) > c\} \times (v - \delta, v + \delta)],$$

which implies that the set $(f^\delta)^{-1}((c, \infty))$ belongs to the σ-algebra $\mathcal{M} \otimes \mathcal{B}$. Since the intervals (c, ∞) generate all Borel sets in $\mathbb{R}$, the functions f^δ are Shragin functions, and hence the function f as well.

As mentioned in Section 1.3, any function which is sup-equivalent to a sup-measurable function is sup-measurable itself. Consequently, any function which is sup-equivalent to a Shragin function is a fortiori sup-measurable; such a function, however, need not be a Shragin function itself, as the example of the monsters in Section 1.3 shows.

For further reference, we collect all results proved so far in the following:

Theorem 1.1 *Every Carathéodory function is a Baire-Carathéodory function, every Baire-Carathéodory function is a Shragin function, and every Shragin function is sup-measurable.*

We point out that sup-measurability can be regarded as usual measurability with respect to a special σ-algebra. In fact, given a σ-algebra $\mathcal{M}$, denote by $\mathcal{M}(\mathcal{B})$ the σ-algebra of all subsets $\Delta \subseteq \Omega \times \mathbb{R}$ such that

$\{s : s \in \Omega, \ \hat{x}(s) \in \Delta\} \in \mathcal{M}$ for all $x \in S$, where $\hat{x}(s) = (s, x(s))$ as above. It is not hard to see that a function f is sup-measurable if and only if f is $(\mathcal{M}(\mathcal{B}), \mathcal{B})$-measurable. Unfortunately, almost nothing is known about the σ-algebra $\mathcal{M}(\mathcal{B})$; in particular, it is not at all clear what its elements should look like. Observe that the sup-measurability of Shragin functions can be stated simply as $\mathcal{M} \otimes \mathcal{B} \subseteq \mathcal{M}(\mathcal{B})$, where the inclusion is, in general, strict.

To conclude, let us once more return to the ordering defined in (1.12). Obviously, a sufficient condition for (1.12) in case $\Delta = \Omega \times \mathbb{R}$ is that $f_1(s, u) \leq f_2(s, u)$ for almost all $s \in \Omega$ and all $u \in \mathbb{R}$. As the monsters show, this is not necessary for (1.12). If we restrict ourselves to Shragin functions, however, the situation is different.

Lemma 1.5 *If f_1 and f_2 are Shragin functions, the two conditions $f_1 \preceq f_2$ and $f_1 \leq f_2$ are equivalent.*

$\Rightarrow$ Suppose that f is a Shragin function and $f(s, u) \succeq 0$; we have to show that $f(s, u)$ is almost everywhere nonnegative. If we put $\Delta = \{(s, u) : s \in \Omega, \ u \in \mathbb{R}, \ f(s, u) < 0\}$ the set $\Delta \setminus (D_0 \times \mathbb{R})$ belongs to $\mathcal{M} \otimes \mathcal{B}$ for some negligible set $D_0 \in \mathcal{M}$. Consequently, by Sainte-Beuve's selection theorem, the projection D_* of $\Delta \setminus (D_0 \times \mathbb{R})$ is measurable, and there exists some measurable function x_* on D_* whose graph lies entirely in Δ. Extending x_* equal to zero outside D_*, we get $Fx_* \in S$ and $Fx_* \geq \theta$, since $f(s, u) \succeq 0$ by assumption. But $Fx_*(s) < 0$ for $s \in D_*$, by definition of the set Δ, and hence D_* is a negligible set. Altogether, this means that $f(s, u) \geq 0$ for $s \in \Omega \setminus (D_0 \cup D_*)$ and $u \in \mathbb{R}$, and so we are done. $\Leftarrow$

Observe that Lemma 1.5 is trivial for purely atomic measures, since the orderings $\preceq$ and $\leq$ are always equivalent on Ω_d.

1.5 Boundedness conditions

Recall that a set $N \subset S$ is called *bounded* if N can be absorbed by any neighbourhood of zero in S. In other words, N is bounded if, given $\varepsilon > 0$, one can find $k > 0$ such that $[x/k] \leq \varepsilon$ for all $x \in N$, where $[z]$ is defined as in (1.1) or (1.2).

One can easily give other (equivalent) definitions of boundedness in the space S. For sake of simplicity, let us introduce the *distribution functions*

$$\begin{aligned}
\lambda(x, h) &= \lambda(\{s : s \in \Omega, \ |x(s)| > h\}), \\
\lambda_c(x, h) &= \lambda(\{s : s \in \Omega_c, \ |x(s)| > h\})
\end{aligned} \tag{1.19}$$

for every $x \in S$ (recall that Ω_c is that part of Ω on which the measure μ is atomic-free).

For convenience, we mention some boundedness criteria which are completely elementary to prove:

Lemma 1.6   *For $N \subset S$, the following four conditions are equivalent:*
(a) N is bounded.
(b) The equality $\lim\limits_{h \to \infty} \sup\limits_{x \in N} \lambda(x, h) = 0$ holds.
(c) The equality $\lim\limits_{h \to \infty} \sup\limits_{x \in N} \lambda_c(x, h) = 0$ holds, and there exists a function
$u \in S$ *such that $|x(s)| \leq u(s)$ for $s \in \Omega_d$.*
(d) If x_n is a sequence in N and δ_n is a sequence of real numbers converging to 0, the sequence $\delta_n x_n$ converges to θ in S.

The main boundedness result for the superposition operator is the following:

Theorem 1.2   *Let f be a sup-measurable function. Then the superposition operator F generated by f is bounded on bounded sets in S if and only if for each $r > 0$ there exists a function $c_r \in S$ such that*

$$|f(s, u)| \preceq c_r(s) \quad ((s, u) \in \Omega \times [-r, r]), \qquad (1.20)$$

where the ordering $\preceq$ is defined in (1.12). In particular, if f is a Carathéodory function, the operator F is always bounded on bounded sets in S.

$\Rightarrow$ First we prove the sufficiency of (1.20). Let $N \subset S$ be bounded. Let x_n be any sequence in N, and let δ_n be a sequence in $\mathbb{R}$ tending to zero. Given $\varepsilon > 0$, by Lemma 1.6 we can find a number $h_\varepsilon > 0$ such that $\lambda(x_n, h_\varepsilon) \leq \varepsilon$ ($n = 1, 2, \cdots$), with $\lambda(x, h)$ as in (1.19). Let $\tilde{x}_n(s) = \min\{|x_n(s)|, h_\varepsilon\}\mathrm{sgn}\, x_n(s)$. By (1.20), we have $|\delta_n F\tilde{x}_n(s)| \leq |\delta_n|c_\varepsilon(s)$ almost everywhere in Ω, where c_ε is the function corresponding to $r = h_\varepsilon$ in (1.20). Consequently,

$$\lim_{n \to \infty} \lambda(\delta_n F\tilde{x}_n, h) = 0 \quad (0 < h < \infty).$$

Since, by construction, the set of all $s \in \Omega$ for which $Fx_n(s) \neq F\tilde{x}_n(s)$ has λ-measure at most ε, we get

$$\overline{\lim_{n \to \infty}} \, \lambda(\delta_n Fx_n, h) \leq \varepsilon \quad (0 < h < \infty);$$

this means that $\delta_n Fx_n$ converges to θ in S.

To prove the necessity of (1.20), suppose that F is bounded in S, and let $r > 0$. By Kantorovich's theorem, the function set $H_r = \{\arctan |Fx| : x \in S, |x| \leq r\}$ admits a least upper bound z_* in S. Moreover, one can find a sequence x_n in S such that $|x_n(s)| \leq r$ and $\sup_n \arctan |Fx_n| = \sup H_r$. By induction, we construct a sequence z_n putting $z_1 = x_1$ and

$$z_n(s) = \begin{cases} z_{n-1}(s) & \text{if } |f(s, z_{n-1}(s))| \geq |f(s, x_n(s))|, \\ x_n(s) & \text{if } |f(s, z_{n-1}(s))| \leq |f(s, x_n(s))|. \end{cases}$$

Obviously, $|z_n(s)| \leq r$, $z_n \in S$, and the sequence $\arctan |Fz_n|$ converges monotonically to $\sup H_r$. By hypothesis, the set $\{Fz_n : n = 1, 2, \cdots\}$ is bounded in S. Consequently, the limit $c_r(s) = \lim_{n \to \infty} Fz_n(s)$ belongs also to S. Since $c_r = \tan \sup H_r$, we have $|f(s, x(s))| \leq c_r(s)$ for any $x \in S$ with $|x(s)| \leq r$, and this is precisely (1.20). This proves the necessity of (1.20). If f is a Carathéodory function, by Lemma 1.5 condition (1.20) is equivalent to

$$|f(s, u)| \leq c_r(s) \quad ((s, u) \in \Omega \times [-r, r]), \tag{1.21}$$

and this is always fulfilled, since $f(s, \cdot)$ is continuous. $\Leftarrow$

Note the analogy of the preceding proof to that of Lemma 1.3.

We point out that one can prove the following refinement of Theorem 1.2: if the operator F is bounded on bounded sets in S there exists a sup-equivalent function $\tilde{f}$ such that (1.20) holds for $\tilde{f}(s, u)$ with $\preceq$ replaced by $\leq$. In fact, it suffices to consider the function

$$\tilde{f}(s, u) = \min\{|f(s, u)|, c_n(s)\}\operatorname{sgn} f(s, u) \quad (s \in \Omega, \ n - 1 < |u| \leq n),$$

which generates the same superposition operator F and has the required properties.

One could also study somewhat different boundedness properties for the superposition operator. For instance, one could ask for conditions on f which guarantee that F is bounded on some fixed set $N \subset S$ which need not even be bounded itself (for example, a set of the form $\{x : u \leq x \leq v\}$ with two not necessarily measurable or not necessarily finite functions u and v). It is evident that one can easily find trivial sufficient conditions; general conditions, however, are not known.

Further, we remark that in the case $\Omega_c \neq \emptyset$ the *convex hull* coN of a bounded set N in S need not be bounded (coN may even be dense in S if $\Omega_d = \emptyset$). We call a set $N \subset S$ *co-bounded* if its convex hull is also bounded. Thus, one could ask for conditions on the function f which ensure that the corresponding superposition operator F maps co-bounded sets into co-bounded sets in S. Also, this problem has not been studied at all up to the present.

1.6 Continuity conditions

In this section we shall be concerned with the following problem: given a sup-measurable function f, what conditions ensure the continuity of the corresponding superposition operator F in the space S (i.e. with respect to the metric ρ defined in (1.1) or (1.2))?

Surprisingly enough, one can give a complete answer to this problem. We shall give the corresponding results in two parts: first, we shall show that a Shragin function which generates a continuous superposition operator F is actually a Carathéodory function; afterwards, we shall give a complete characterization of the sup-measurable functions f which generate a continuous superposition operator F.

Theorem 1.3 *Let f be a Shragin function, and suppose that the superposition operator F generated by f is continuous in the space S. Then f is a Carathéodory function.*

$\Rightarrow$ It is sufficient to show that, for all $s \in \Omega$, the function $f(s, \cdot)$ is uniformly continuous on any bounded interval $[-c, c] \subset \mathbb{R}$.

Suppose that this is not true for some interval $[-c, c]$; then the set

$$D = \{s : s \in \Omega; \ \varlimsup_{\substack{|u-v| \to 0 \\ |u|, |v| \leq c}} |f(s, u) - f(s, v)| \geq \delta\}$$

has positive measure for some $\delta > 0$. Consider the sets $(n = 1, 2, \cdots)$

$$D_n = \{(s, u, v) : s \in \Omega; |u|, |v| \leq c; |u - v| \leq \frac{1}{n}; |f(s, u) - f(s, v)| \geq \delta\}.$$

Since f is a Shragin function, the sets D_n belong to the σ-algebra $\mathcal{M} \otimes \mathcal{B}(\mathbb{R}^2)$, where $\mathcal{B}(\mathbb{R}^2)$ denotes the σ-algebra of all Borel subsets of $\mathbb{R}^2$. By Sainte-Beuve's theorem, one can construct two sequences of functions u_n and v_n in S such that $|u_n(s)|, |v_n(s)| \leq c$, $|u_n(s) - v_n(s)| \leq \frac{1}{n}$ and $|f(s, u_n(s)) - f(s, v_n(s))| \geq \delta$. Let D_* denote the set of all pairs (s, u) such that $u = \lim_{k \to \infty} u_{n_k}(s)$ for some sequence n_k of natural numbers. Then

$$D_* = \bigcap_{m \geq 1} \bigcup_{n \geq m} \bigcup_{r \in \mathbb{Q}} \{(s, u) : |u_n(s) - r|, |u - r| < \frac{1}{m}\},$$

hence $D_* \in \mathcal{M} \otimes \mathcal{B}$. Again by Sainte-Beuve's theorem, there exists a measurable function x_* on D whose graph lies in D_* for almost all $s \in D$. The sequence of functions $\tilde{u}_n = P_D u_n$ and $\tilde{v}_n = P_D v_n$ (P_D as in (1.3)) converges in S to the function $\tilde{x}_* = P_D x_*$.

By hypothesis, the operator F is continuous in S, and hence the sequence $F\tilde{u}_n - F\tilde{v}_n$ converges in measure to zero. On the other hand, the inequality $|F\tilde{u}_n(s) - F\tilde{v}_n(s)| \geq \delta$ holds on the set D of positive measure, a contradiction. $\Leftarrow$

To state our main continuity result, suppose, for simplicity, that $f(s,0) = 0$, and consider the functional

$$\Phi(x, D) = \int_D f(s, x(s)) ds. \qquad (1.22)$$

Obviously, $\Phi(\theta, D) = 0$ for all $D \in \mathcal{M}$, and $P_D x = P_D y$ implies that $\Phi(x, D) = \Phi(y, D)$.

The following lemma will be very important in what follows.

Lemma 1.7 *Let f be a sup-measurable function, and suppose that the functional $\Phi(\cdot, D)$ is continuous on S for any $D \in \mathcal{M}$, i.e.*

$$\int_D f(s, x_n(s)) ds \to \int_D f(s, x(s)) ds \qquad (1.23)$$

as $x_n \to x$ in measure. Then f is sup-equivalent to some Carathéodory function.

$\Rightarrow$ Given $\eta > 0$, let Φ_η be defined by

$$\Phi_\eta(x, D) = \inf_{w \in S} \left\{ \Phi(w, D) + \eta \int_D |w(s) - x(s)| ds \right\}; \qquad (1.24)$$

obviously, $\Phi_\eta(x, D) \leq \Phi(x, D)$ for each $\eta > 0$. Moreover, it follows from the assumption (1.23) that $\lim_{\eta \to \infty} \Phi_\eta(x, D) = \Phi(x, D)$. We claim that the functional Φ_η satisfies a "Lipschitz-type" condition

$$|\Phi_\eta(x, D) - \Phi_\eta(y, D)| \leq \eta \int_D |x(s) - y(s)| ds \quad (D \in \mathcal{M}). \qquad (1.25)$$

On the one hand, for $w \in S$ we get, by (1.24)

$$\Phi_\eta(x, D) \leq \Phi(w, D) + \eta \int_D |x(s) - w(s)| ds \quad (D \in \mathcal{M}). \qquad (1.26)$$

On the other hand, given $\delta > 0$, we find $w_\delta \in S$ such that

$$\Phi_\eta(y, D) \geq \Phi(w_\delta, D) + \eta \int_D |w_\delta(s) - y(s)| ds - \delta.$$

Putting $w = w_\delta$ in (1.26) yields

$$\Phi_\eta(x, D) - \Phi_\eta(y, D) \leq \eta \int_D |x(s) - w_\delta(s)| ds - \eta \int_D |w_\delta(s) - y(s)| ds + \delta$$

$$\leq \eta \int_D |x(s) - y(s)| ds + \delta.$$

Since $\delta > 0$ is arbitrary and the last expression is symmetric in x and y, we have (1.25).

Consider now the restriction $\Psi_\eta(\cdot, D)$ of $\Phi_\eta(\cdot, D)$ to $\mathbb{R}$ (i.e. we identify the real number u with the constant function $x_u(s) \equiv u$). Condition (1.25) reads then

$$|\Psi_\eta(u, D) - \Psi_\eta(v, D)| \leq \eta\mu(D)|u - v|. \tag{1.27}$$

By the Radon–Nikodým theorem, the measure $\Psi_\eta(u, \cdot)$ can be represented as integral

$$\Psi_\eta(u, D) = \int_D g_\eta(s, u) ds,$$

where the function $g_\eta(\cdot, u)$ is integrable over Ω. Note that this function is defined only for $s \in \Omega \setminus D_u$, where D_u is some null set depending, in general, on u. The union D_0 of all sets D_u ($u \in \mathbb{Q}$) is also a null set, and the function g_η extends to a function $\tilde{g}_\eta$ whose domain of definition $(\Omega \setminus D_0) \times \mathbb{R}$ does not depend any more on u. By the continuity of $\Psi_\eta(\cdot, D)$ on $\mathbb{R}$ and by Lebesgue's theorem we have

$$\Psi_\eta(u, D) = \int_D \tilde{g}_\eta(s, u) ds, \tag{1.28}$$

hence $|\int_D [\tilde{g}_\eta(s, u) - \tilde{g}_\eta(s, v)] ds| \leq \eta\mu(D)|u - v|$, by (1.27). Since $D \in \mathcal{M}$ is arbitrary, this yields $|\tilde{g}_\eta(s, u) - \tilde{g}_\eta(s, v)| \leq \eta|u - v|$; consequently, $\tilde{g}_\eta$ is a Carathéodory function for each $\eta > 0$. The equality (1.28) shows that

$$\Phi_\eta(x, D) = \int_D \tilde{g}_\eta(s, x(s)) ds \tag{1.29}$$

for any constant function $x = x_u$ ($u \in \mathbb{R}$). Since superposition operators satisfying $F\theta = \theta$ commute with the multiplication operator (1.3), (1.29) holds also for all simple functions x, and hence just for $x \in S$, since the simple functions are dense in S, and both sides of (1.29) are continuous functionals of x.

Now let $\tilde{f}(s,u) = \lim\limits_{\eta \to \infty} \tilde{g}_\eta(s,u)$. Clearly, the function $\tilde{f}$ is measurable, and by Levi's theorem we have

$$\Phi(x, D) = \lim_{\eta \to \infty} \Phi_\eta(x, D) = \lim_{\eta \to \infty} \int_D \tilde{g}_\eta(s, x(s)) ds$$

$$= \int_D \lim_{\eta \to \infty} \tilde{g}_\eta(s, x(s)) ds = \int_D \tilde{f}(s, x(s)) ds,$$

hence

$$\int_D f(s, x(s)) ds = \int_D \tilde{f}(s, x(s)) ds$$

which means that $f(s,u) \simeq \tilde{f}(s,u)$, since $D \in \mathcal{M}$ is arbitrary.

It remains to show that $\tilde{f}$ is a Carathéodory function. Since all $\tilde{g}_\eta$ are Carathéodory functions, $\tilde{f}$ is a Shragin function, by Theorem 1.1, and hence a Carathéodory function, by Theorem 1.3. $\Leftarrow$

We are now in a position to state the main continuity result for the superposition operator in the space S.

Theorem 1.4 *Let f be a sup-measurable function. Then the superposition operator F generated by f is continuous in the space S if and only if f is sup-equivalent to some Carathéodory function.*

$\Rightarrow$ The "if" part of the assertion is not hard to prove: first, observe that, if x_n is a sequence in S which converges almost everywhere on Ω to $x \in S$, the sequence Fx_n converges also almost everywhere on Ω to Fx if f is (equivalent to) some Carathéodory function. Suppose now that F is not continuous at some $x_0 \in S$. Given $\varepsilon > 0$, choose a sequence x_n in S such that

$$\rho(Fx_n, Fx_0) \geq \varepsilon \quad (n = 1, 2, \cdots). \tag{1.30}$$

By Riesz' theorem, we can extract a subsequence x_{n_k} which converges almost everywhere to x_0; by the above remark, Fx_{n_k} converges almost everywhere to Fx_0, i.e. $\lim\limits_{k \to \infty} \rho(Fx_{n_k}, Fx_0) = 0$, contradicting (1.30). This proves the sufficiency.

The "only if" part follows from Lemma 1.7: in fact, if F is continuous in the space S, the functional $\Phi(\cdot, D)$ is continuous on the set of all $x \in S$ for which $Fx \in L$ for each $D \in \mathcal{M}$, and thus f is sup-equivalent to a Carathéodory function. $\Leftarrow$

Theorem 1.4 is global, since it gives a necessary and sufficient continuity condition for F on the whole space S. Sometimes it is also important, however, to have a condition for F to be continuous at some point $x_0 \in S$. Here one cannot give a definite answer as in Theorem 1.4, but the following holds:

Theorem 1.5   *Let f be a Shragin function and $x_0 \in S$. Then the superposition operator F generated by f is continuous at x_0 if and only if, for almost all $s \in \Omega$, the function $f(s, \cdot)$ is continuous at $u = x_0(s)$.*

$\Rightarrow$ The "if" part can be proved by the same reasoning as in Theorem 1.4. To prove the "only if" part, consider the sequence of measurable functions

$$\sigma_n(s) = \sup\{|f(s,u) - f(s,x_0(s))| : |u - x_0(s)| < \frac{1}{n}\}.$$

Evidently, it is sufficient to show that the sequence σ_n converges almost everywhere to zero. The set D of all $s \in \Omega$ with $\overline{\lim}_{n \to \infty} \sigma_n(s) > 0$ is measurable, since the functions σ_n are measurable. If D would have positive measure, we could find a set $D_0 \in \mathcal{M}$ of positive measure and measurable functions x_n on D_0 such that

$$|x_n(s) - x_0(s)| < \frac{1}{n}, \quad |f(s, x_n(s)) - f(s, x(s))| \geq \delta \quad (s \in D) \qquad (1.31)$$

for some $\delta > 0$. Extending x_n to be equal to x_0 outside D_0, we get a sequence x_n in S which converges to x_0. On the other hand, by (1.31) the sequence Fx_n does not converge to Fx_0 in S, contradicting the continuity of F at x_0. $\Leftarrow$

Theorem 1.5 gives a necessary and sufficient local continuity condition for superposition operators F which are generated by Shragin functions. For arbitrary sup-measurable functions, only sufficient conditions are known.

From Theorem 1.5 it follows, for example, that the superposition operator F generated by the function $f(s,u) = \text{sgn}\ (u-w(s))$ ($w \in S$ fixed) is continuous at each $x_0 \in S$ for which the set $\{s : x_0(s) = w(s)\}$ has measure zero. Analogously, if the discontinuity points of some sup-measurable function f form a countable family of "curves" $u = w_\alpha(s)$ $(\alpha \in A)$ (here the functions w_α need even not be measurable!), the corresponding superposition operator F is continuous at each $x_0 \in S$ for which the set $\{s : x_0(s) = w_\alpha(s)\}$ $(\alpha \in A)$ has measure zero.

In Section 1.8 we shall obtain more precise information about the measurability and continuity behaviour of special functions. As mentioned in Section 1.1, there is another type of convergence in the space S which is important in applications, namely convergence almost everywhere. The following analogue to Theorem 1.5 deals with "mixed" continuity properties of the superposition operator in the space S.

Theorem 1.6 *Let f be a sup-measurable function. Suppose that the superposition operator F generated by f maps every sequence which converges almost everywhere on Ω into a sequence which converges in measure. Then f is sup-equivalent to some Carathéodory function. The same holds if F maps almost everywhere convergent sequences into almost everywhere convergent sequences. On the other hand, suppose that F maps every sequence which converges in measure into a sequence which converges almost everywhere on Ω. Then the function $f(s, \cdot)$ is constant for $s \in \Omega_c$ and continuous for $s \in \Omega_d$.*

$\Rightarrow$ The first two statements follow from Theorem 1.4, since every superposition operator which maps almost everywhere convergent sequences into measure-convergent sequences is continuous in S, and every sequence which converges in measure contains an almost everywhere convergent subsequence.

To prove the last statement, suppose that the assertion is not true; in this case we can find a $u \neq 0$ such that $f(s, u) \neq 0$ for $s \in D$, where $D \subset \Omega_c$ is some set of positive measure. In the usual way, considering partitions $\{D_1^n, D_2^n, \cdots, D_{2^n}^n\}$ of D ($n = 1, 2, \cdots$) of mesh tending to zero, one obtains a sequence χ_n of characteristic functions which tend to zero in measure, but not almost everywhere on D. By hypothesis, the sequence $z_n(s) = f(s, u\chi_n(s))$ tends to zero almost everywhere, contradicting the fact that $z_n(s) = f(s, u)$ on a subset of D of positive measure. The continuity of $f(s, \cdot)$ for $s \in \Omega_d$ follows from the fact that convergence almost everywhere on Ω_d coincides with both pointwise convergence and convergence in measure. $\Leftarrow$

To conclude this section, we present a result on the uniform continuity of the superposition operator in the space S. It is well known that a nonlinear operator may be continuous in an infinite-dimensional metric space without being uniformly continuous on bounded subsets of this space. A remarkable exception is the following:

Theorem 1.7 *If the superposition operator F is continuous in the space S, F is also uniformly continuous on bounded sets.*

$\Rightarrow$ In view of Theorem 1.4, we may assume that f is a Carathéodory function. Let N be a bounded subset of S, and let $x_n, y_n \in N$ such that the sequence $x_n - y_n$ converges to θ in S. We have to show that the sequence

$Fx_n - Fy_n$ converges to θ in S as well. Without loss of generality, we may assume that $x_n - y_n$ converges to θ almost everywhere.

Let $\varepsilon > 0$. By Lemma 1.6, we have $\lambda(x_n, h_\varepsilon) \leq \varepsilon$ and $\lambda(y_n, h_\varepsilon) \leq \varepsilon$ ($n = 1, 2, \cdots$) for some $h_\varepsilon > 0$. Let

$$\tilde{x}_n(s) = \min\{|x_n(s)|, h_\varepsilon\}\mathrm{sgn}\, x_n(s)\,,$$
$$\tilde{y}_n(s) = \min\{|y_n(s)|, h_\varepsilon\}\mathrm{sgn}\, y_n(s)\,.$$

Obviously, the sequence $\tilde{x}_n - \tilde{y}_n$ converges also almost everywhere to θ. Since $|\tilde{x}_n(s)| \leq h_\varepsilon$ and $|\tilde{y}_n(s)| \leq h_\varepsilon$, and because the function $f(s, \cdot)$ is, for almost all $s \in \Omega$, uniformly continuous on $[-h_\varepsilon, h_\varepsilon]$, by Cantor's theorem, the sequence $F\tilde{x}_n - F\tilde{y}_n$ converges almost everywhere to θ. Consequently, $F\tilde{x}_n - F\tilde{y}_n$ converges also in measure to θ, i.e.

$$\lim_{n\to\infty} \lambda(F\tilde{x}_n - F\tilde{y}_n, h) = 0 \quad (0 < h < \infty)\,.$$

But, by construction, the set of all $s \in \Omega$ for which $Fx_n(s) - Fy_n(s) \neq F\tilde{x}_n(s) - F\tilde{y}_n(s)$ has λ-measure at most 2ε, hence

$$\varlimsup_{n\to\infty} \lambda(Fx_n - Fy_n, h) \leq 2\varepsilon \quad (0 < h < \infty)\,,$$

and the statement follows, since $\varepsilon > 0$ is arbitrary. $\Leftarrow$

1.7 Compactness conditions

Recall that a set $N \subset S$ is called *precompact* (or totally bounded) if any sequence in N has a convergent subsequence. Equivalently, N is precompact if N admits for any $\varepsilon > 0$ a *finite ε-net* (i.e. a finite set C such that $\mathrm{dist}(x, C) \leq \varepsilon$ for all $x \in N$).

A set N is called *compact* if N is precompact and closed. The following compactness criterion holds in the space S:

Lemma 1.8 *A set $N \subset S$ is precompact if and only if for each $\varepsilon > 0$ there exist a number $h > 0$ and a partition $\{D_1, \cdots, D_m\}$ of Ω ($D_j \in \mathcal{M}$) such that, for each $x \in N$, $|x(s)| \leq h$ ($s \in \Omega \setminus D_x$) and $|x(s) - x(t)| \leq \varepsilon$ ($s, t \in D_j \setminus D_x$; $j = 1, \cdots, m$), where $D_x \in \mathcal{M}$ is a subset of Ω (in general, depending on x) of λ-measure at most ε.*

Unfortunately, this compactness criterion is very cumbersome. Only in the case $\Omega_c = \emptyset$ the conditions of Lemma 1.8 are easy to verify, but in this case compactness simply reduces to boundedness.

For applications it would be very useful if the superposition operator F would be compact (i.e. maps bounded sets in S into precompact sets). It turns out, however, that the class of compact superposition operators in S is highly degenerate: in the case $\Omega_d = \emptyset$ it consists just of constant operators!

Theorem 1.8   *Let f be a sup-measurable function. Then the superposition operator F generated by f is compact if and only if the following two conditions hold:*
(a) For almost all $s \in \Omega_c$, the function $f(s, \cdot)$ does not depend on u.
(b) For all $s \in \Omega_d$, the function $f(s, \cdot)$ is bounded on each bounded interval in $\mathbb{R}$.

$\Rightarrow$ The sufficiency of the two conditions is obvious. For proving the necessity, we can restrict ourselves to the separate cases $\Omega_d = \emptyset$ for (a), and $\Omega_c = \emptyset$ for (b).

Let $\Omega_d = \emptyset$, suppose that F maps each bounded set in S into a precompact set, and let x_1 and x_2 be two functions in S for which $Fx_1 \neq Fx_2$. This means that there exists a set $D \subseteq \Omega$ of positive measure such that $|f(s, x_1(s)) - f(s, x_2(s))| \geq \gamma$ $(s \in D)$ for some $\gamma > 0$. By induction, we construct a sequence of partitions of D into sets $D(\varepsilon_1, \cdots, \varepsilon_n)$ $(\varepsilon_i \in \{0, 1\})$ as follows: first, let $\{D(0), D(1)\}$ be a partition of D such that $\mu(D(0)) = \mu(D(1)) = \frac{1}{2}\mu(D)$ (recall that $D \subseteq \Omega_c$). Further, if $\{D(\varepsilon_1, \cdots, \varepsilon_n) : \varepsilon_i \in \{0, 1\}\}$ is the n-th partition of D, divide each $D(\varepsilon_1, \cdots, \varepsilon_n)$ into two parts $D(\varepsilon_1, \cdots, \varepsilon_n, \varepsilon_{n+1})$ such that $\mu(D(\varepsilon_1, \cdots, \varepsilon_n, 0)) = \mu(D(\varepsilon_1, \cdots, \varepsilon_n, 1)) = \frac{1}{2}\mu(D(\varepsilon_1, \cdots\cdots, \varepsilon_n))$; this defines the $(n+1)$-st partition $\{D(\varepsilon_1, \cdots, \varepsilon_{n+1}) : \varepsilon_i \in \{0, 1\}\}$. Now set

$$
\begin{aligned}
D_n^0 &= \bigcup_{\varepsilon_i \in \{0,1\}} D(\varepsilon_1, \cdots, \varepsilon_n, 0) \,, \\
D_n^1 &= \bigcup_{\varepsilon_i \in \{0,1\}} D(\varepsilon_1, \cdots, \varepsilon_n, 1) \,,
\end{aligned}
\tag{1.32}
$$

and observe that $\mu(D_n^0) = \mu(D_n^1) = \frac{1}{2}\mu(D)$ and $D_n^0 \cap D_n^1 = \emptyset$. Let z_n be the sequence of functions defined by $z_n(s) = P_{D_n^0} x_1(s) + P_{D_n^1} x_2(s)$. Obviously, the sequence z_n is bounded in S, and for $i < j$ we have

$$
|f(s, z_i(s)) - f(s, z_j(s))| = |f(s, x_1(s)) - f(s, x_2(s))| \geq \gamma \quad (s \in D_{i,j}) \,,
$$

where the set

$$
D_{i,j} = \bigcup_{\varepsilon_i \neq \varepsilon_j} D(\varepsilon_1, \cdots, \varepsilon_i, \cdots, \varepsilon_j)
$$

has measure $\mu(D_{i,j}) = \frac{1}{2}\mu(D) > 0$. But this shows that the sequence Fz_n cannot contain a convergent subsequence.

The proof of (b) in the case $\Omega_c = \emptyset$ is trivial, since every compact operator is bounded, and boundedness of F implies boundedness of $f(s, \cdot)$ for $s \in \Omega_d$. $\Leftarrow$

At this point, one can make analogous remarks as at the end of the preceding section. For instance, one could ask for conditions under which F is compact on some fixed set N; for example, in the case $N \supseteq \{x : u \leq x \leq v\}$, the function $f(s, \cdot)$ should not depend on u for $s \in \Omega_c$, and the set $N(s) = \{x(s) : x \in N\}$ should be bounded in $\mathbb{R}$ for $s \in \Omega_d$.

In the case $\Omega_c \neq \emptyset$ the convex hull $co\,N$ of a precompact set N need not be precompact; it may even happen that N is precompact but $co\,N$ is unbounded! Here is an example: let $\{D_{k,n} : k = 1, \cdots, n\}$ be a partition of Ω_c into n subsets of equal λ-measure, and let $x_{k,n} = n^2 \chi_{D_{k,n}}$ ($k = 1, \cdots, n$; $n = 1, 2, \cdots$); then the set N of all functions $x_{k,n}$ is precompact in S, as Lemma 1.8 shows, but the set of the convex combinations $z_n = \frac{1}{n}(x_{1,n} + \cdots + x_{n,n}) \equiv n$ is unbounded in S!

As before, we shall call a set $N \subset S$ co-precompact if its convex hull $co\,N$ is precompact. Theorem 1.8 shows that, whenever F is compact, F is even "co-compact" (in the sense that F maps bounded sets into co-precompact sets); this is a nontrivial statement only in the case $\Omega_c \neq \emptyset$, since in the case $\Omega_c = \emptyset$ the notions of precompactness and co-precompactness coincide.

1.8 Special classes of functions

In this section we shall deal with superposition operators F whose generating functions f have special properties with respect to u, such as monotonicity, convexity, semi-additivity, etc.; for such operators much more can be said about their analytical and topological properties. We begin with *monotone* functions; more precisely, we suppose that $f(s, \cdot)$ is increasing for almost all $s \in \Omega$. Even for such functions, measurability on $\Omega \times \mathbb{R}$ does not imply sup-measurability. For example, let $\Omega = [0, 1]$, $D \subset \Omega$ be a non-measurable subset, and

$$f(s, u) = \begin{cases} 1 & \text{if } u > s, \text{ or } u = s \text{ and } s \in D, \\ 0 & \text{if } u < s, \text{ or } u = s \text{ and } s \notin D. \end{cases} \tag{1.33}$$

Then f is obviously measurable on $\Omega \times \mathbb{R}$, $f(\cdot, u)$ is measurable on Ω for all $u \in \mathbb{R}$, $f(s, \cdot)$ is increasing for all $s \in \Omega$, and $f(s, \cdot)$ is even continuous except for the point $u = s$ (roughly speaking, f is "almost" a Carathéodory function). Nevertheless, f is not sup-measurable, since F maps the function $x_0(s) = s$ into the function $Fx_0(s) = \chi_D(s)$.

Lemma 1.9   *Suppose that $f(s, \cdot)$ is increasing for almost all $s \in \Omega$ and $f(\cdot, u)$ is measurable for all $u \in \mathbb{R}$. For $h \in \mathbb{R}$, let*

$$g(s, h) = \sup\{u : f(s, u) < h\}. \tag{1.34}$$

Then the function $g(\cdot, h)$ is measurable for each $h \in \mathbb{R}$, and f is $(\overline{\mathcal{M} \otimes \mathcal{B}}, \mathcal{B})$-measurable, where $\overline{\mathcal{M} \otimes \mathcal{B}}$ denotes the completion of the σ-algebra $\mathcal{M} \otimes \mathcal{B}$.

$\Rightarrow$ It is easy to see that

$$g(s, h) = \sup_{u \in \mathbb{Q}} u \chi_{D(u,h)}(s),$$

where $D(u, h) = \{s : s \in \Omega, f(s, u) < h\}$; hence $g(\cdot, h)$ is measurable for each $h \in \mathbb{R}$.

By definition (1.34), the inclusions

$$\{(s, u) : g(s, h) > u\} \subseteq \{(s, u) : f(s, u) < h\} \subseteq \{(s, u) : g(s, h) \geq u\} \tag{1.35}$$

hold. As a matter of fact, the left-hand and right-hand sides in (1.35) (i.e. the "open" and "closed" subgraph of g) are measurable sets. Moreover, the set $\{(s, u) : g(s, h) = u\}$ (i.e. the "difference" of these subgraphs) is a null set. Consequently, the Lebesgue set $\{(s, u) : f(s, u) < h\}$ of f is the union of the $(\mathcal{M} \otimes \mathcal{B})$-measurable set $\{(s, u) : g(s, h) > u\}$ and of a subset of some $(\mathcal{M} \otimes \mathcal{B})$-null set. But this means that this set belongs to the completion $\overline{\mathcal{M} \otimes \mathcal{B}}$. $\Leftarrow$

It follows from this reasoning that, under the hypotheses of Lemma 1.9, the function f is $(\overline{\mathcal{M} \otimes \mathcal{B}}, \mathcal{B})$-measurable if and only if the set $D(h) = \{s : f(s, g(s, h)) < h\}$ belongs to $\mathcal{M}$ for all $h \in \mathbb{R}$. This in turn is the case if f is sup-measurable, by the measurability of $g(\cdot, h)$. Thus, in contrast to the counterexample (1.33), here sup-measurability implies measurability; in particular, "monotone monsters" (of any nationality) cannot exist!

In case Ω is a complete metric space and μ is regular (see Section 1.1), one can show that, if $f(s, \cdot)$ is increasing and f is sup-measurable, one can find a null set $D_0 \subset \Omega$ and a Borel function $\tilde{f}$ (i.e. $\tilde{f}$ is $(\mathcal{B}(\Omega) \otimes \mathcal{B}, \mathcal{B})$-measurable) such that $f(s, u) = \tilde{f}(s, u)$ $(s \in \Omega \setminus D_0, u \in \mathbb{R})$. In fact, in this case every set $D \in \mathcal{M}$ can be represented as union of a null set and a Borel set, and hence f can be re-defined on a null set (with respect to s) in such a way that the Lebesgue sets $\{(s, u) : f(s, u) < h\}$ $(h \in \mathbb{Q})$ become Borel subsets of $\Omega \times \mathbb{R}$. This gives a function $\tilde{f}$ with the required properties.

We summarize what we have shown so far in the following:

Theorem 1.9   *Suppose that $f(\cdot, u)$ is measurable for all $u \in \mathbb{R}$, and $f(s, \cdot)$ is increasing for almost all $s \in \Omega$. Then f is sup-measurable if and only if f is a Shragin function. Moreover, if Ω is a complete metric space and the measure μ is regular, f is sup-measurable if and only if f is equivalent to some Borel function.*

We now pass to the continuity properties of a superposition operator F which is generated by a monotone function f. If $f(s, \cdot)$ is increasing, there are only countably many discontinuities of the first kind (i.e. jump discontinuities). By Theorem 1.5, the corresponding superposition operator F is continuous at $x \in S$ if and only if

$$f_+(s, x(s)) = f_-(s, x(s)), \qquad (1.36)$$

where $f_+(s, u) = \lim_{v \downarrow u} f(s, v)$ and $f_-(s, u) = \lim_{v \uparrow u} f(s, v)$. This simple observation allows us to describe the set of discontinuity points of F rather precisely. First, we claim that (1.48) is true if

$$\lambda(\{s : s \in \Omega,\, x(s) = g(s, u)\}) = 0 \quad (u \in \mathbb{Q}), \qquad (1.37)$$

with g given by (1.34). In fact, if (1.36) fails to hold on a set D_1 of positive measure, then on a set $D_2 \subseteq D_1$, also of positive measure, we have $f_-(s, x(s)) < u < f_+(s, x(s))$ for some $u \in \mathbb{Q}$. But this implies that $x(s) = g(s, u)$ for $s \in D_2$, contradicting (1.37).

Observe that the set $\mathcal{M}(D, u)$ of all functions $x \in S$ for which $\{s : s \in \Omega,\, x(s) = g(s, u)\} = D$ is nowhere dense in S $(D \in \mathcal{M},\, u \in \mathbb{Q})$. Consequently, if the measure μ is *separable* (i.e. the corresponding space S is separable, see Section 1.1), the set of all discontinuity points of F can be represented as countable union of nowhere dense sets in S, and thus is a set of first category. This result extends to a larger class of superposition operators:

Theorem 1.10   *Suppose that f is sup-measurable, and $f(s, \cdot)$ is of bounded variation on each bounded interval of $\mathbb{R}$ for almost all $s \in \Omega$. Then the set of discontinuity points of the superposition operator F generated by f is of first category.*

$\Rightarrow$ It suffices to show that f can be represented in the form $f(s, u) = f_\oplus(s, u) - f_\ominus(s, u)$, where $f_\oplus$ and $f_\ominus$ are two sup-measurable functions such that $f_\oplus(s, \cdot)$ and $f_\ominus(s, \cdot)$ are increasing for almost all $s \in \Omega$. Such functions can be defined, for instance, by

$$f_\oplus(s,u) = \frac{1}{2}[\check{f}(s,u) + f(s,u)], \quad f_\ominus(s,u) = \frac{1}{2}[\check{f}(s,u) - f(s,u)],$$

where

$$\check{f}(s,u) = \begin{cases} \displaystyle\lim_{\substack{r\downarrow u \\ r\in\mathbb{Q}}} \text{var } (f(s,\cdot);0,r) & \text{if } u \geq 0, \\[2ex] \displaystyle-\lim_{\substack{r\uparrow u \\ r\in\mathbb{Q}}} \text{var } (f(s,\cdot);r,0) & \text{if } u < 0. \end{cases}$$

and var $(\varphi; a, b)$ denotes the *total variation* of the function φ on $[a, b]$. Obviously, $f_\oplus$ and $f_\ominus$ are increasing in u. They are also sup-measurable, since the function $\check{f}$ can also be defined by

$$\check{f}(s,u) = \lim_{\substack{r\downarrow u \\ r\in\mathbb{Q}}} \sup\left\{ \sum_{j=1}^{m} |f(s,r_j) - f(s,r_{j-1})| : 0 \right.$$

$$\left. = r_0 < r_1 < \cdots < r_m = r, \, r_j \in \mathbb{Q} \right\}$$

for $u \geq 0$ (similarly for $u < 0$). This proves the theorem. $\Leftarrow$

It would be interesting to study the properties of superposition operators F which are generated by functions f with other special properties (convexity, semi-additivity etc.). For some of these properties, this is very simple. For instance, if $f(s, \cdot)$ is convex for almost all $s \in \Omega$, then f is sup-measurable if and only if $f(\cdot, u)$ is measurable on Ω for all $u \in \mathbb{R}$; moreover, f is always a Carathéodory function in this case. The same holds if $f(s, \cdot)$ is *semi-additive* for almost all $s \in \Omega$, and

$$\lim_{u\to 0} f(s,u) = f(s,0) = 0. \tag{1.38}$$

But already if $f(s, \cdot)$ is merely semi-additive, and (1.38) fails to hold, nothing can be said about the sup-measurability of f.

1.9 Notes, remarks and references

1. Detailed information on the space S and its properties, as well as all measure-theoretical results discussed in Section 1.1, can be found in the books [104] and [156]. The basic results on the space S carry over almost without changes to *vector-valued functions*, and even to functions which take values in arbitrary metric spaces, equipped with the σ-algebra of Borel sets, see e.g. the book [194]. As pointed out in Section 1.1, most "natural" examples of sets Ω lead to a separable space $S = S(\Omega)$. However, non-separable measures occur sometimes in probability theory (see e.g. [194]). For instance, uncountable products of probability spaces give non-separable (and hence non-metrizable) measure spaces.

2. In large parts of the text, we restrict ourselves to superposition operators generated by scalar functions f. Many results, however, carry over to functions from some product $\Omega \times U$ into some set V, where U and V are arbitrary measure spaces. The case when U and V are metric spaces, with the σ-algebra of Borel subsets, is particularly important. All results of Section 1.1 – 1.8 are valid in case $U = \mathbb{R}^m$, $V = \mathbb{R}^n$ as well.

The first results on the superposition operator were obtained by K. Cara-théodory, M.A.Krasnosel'skij, V.V. Nemytskij and M.M. Vajnberg [78], [164], [165], [238–241], [341–346]; in the last mentioned paper, the superposition operator is called "Nemytskij operator" for the first time.

The property of *local determination* or (what is equivalent in case $F\theta = \theta$) *disjoint additivity* of the superposition operator was implicitly used by many authors, and systematically studied first by P.P. Zabrejko. In this connection, the question arises if the local determination (1.9) characterizes the superposition operator, or if there are other operators which are locally determined. It turns out that, under the continuum hypothesis, the super-position operator (1.6) is the only locally determined operator on the space S (over a separable measure): in fact, since the space S has the cardinality of the continuum, its elements can be indexed by the ordinals from 1 to ω, the first uncountable ordinal. Given a locally determined operator F in S, define a function f by $f(s, u) = Fx_\alpha(s)$, where α is the first ordinal such that $x_\alpha(s) = u$. The function f is then determined by the values of F on classes in S, and generates the superposition operator F [168].

Locally determined and similar operators were also studied in the papers [158], [159], [161–163], [269], [270], [287], [288], [312], [314], [362], and [366]. G. Buttazzo [71], [72] constructed explicitly an example of a locally determined operator which is generated by a non-measurable function. The papers [1–3] are concerned with the problem of characterizing those locally determined operators F which are *multiplication operators* $Fx(s) = a(s)x(s)$ by some measurable function a.

3. The notion of *sup-measurability* was introduced in the book [182]. Various sufficient (and partially necessary) conditions for sup-measurability which are often very delicate, were studied by Z. Grande (see [123–128],

[130–132], and, in particular, the monograph [129] and the bibliography therein), M.A. Krasnosel'skij and A.V. Pokrovskij [170–174], I.V. Shragin [304–306], [308–310], [316], [317], [319], and others [91], [149], [191], [265–267].

The problem of whether or not sup-measurability implies measurability was raised by Z. Grande at the end of his monograph [129]. The monsters (which give a negative answer to this problem) were introduced by M.A. Krasnosel'skij and A.V. Pokrovskij [171], [173] and, independently, by Z. Grande and J.S. Lipiński [133]. The notion of *sup-equivalence* was introduced and studied in the papers [35] and [36].

The monsters and the example (1.16) show that sup-measurability and (Lebesgue) measurability are independent concepts. If we replace Lebesgue measurability by Borel measurability, however, the situation is slightly different: sup-measurability does not imply measurability (just choose f equal to the characteristic function of $D \times \mathbb{R}$, where D is a Lebesgue measurable non-Borel set [306]), but measurability implies sup-measurability in this case. Some sufficient conditions for sup-measurability which are weaker than those in [129] are given in [325]. Sup-measurable functions between abstract spaces are studied in [99], generalizing [306] and [310]. Measurability of f on the product $\Omega \times \mathbb{R}$ is dealt with in many papers, e.g. [38].

The notion of *thickness* of a set of measurable functions is due to P.P. Zabrejko. Lemma 1.2 was proved (in another terminology) in [171]. The fact that the set C^1 of continuously differentiable functions is not thick in S is also mentioned in the latter paper. To see this, it is sufficient to construct a real function φ (on $\Omega = [0, 1]$, say) with the property that the projection (onto Ω) of the intersection points of the graph $\Gamma(\varphi)$ of φ and the graph $\Gamma(\psi)$ of any C^1-function ψ is a null-set; the characteristic function $f = \chi_{\Gamma(\varphi)}$ has then the required properties [174]. The fact that the Luzin hull of the space C^1 is a strict subset of S is implicitly contained in the paper [45]; more general information on this topic may be found in [192]. The problem of recovering the function f if the values of the corresponding superposition operator F are given on certain classes of "test functions" is important also in view of applications, see e.g. [6].

Lemma 1.3 is new; the facts about the partial ordering in the space S used there may be found in [156] or [157].

4. The functions which we call "Carathéodory functions" were introduced by K. Carathéodory in 1918 [78] in connection with a generalized Peano theorem on the solvability of the Cauchy problem for an ordinary differential equation with discontinuous (in s) right-hand side. Actually, Carathéodory also proved the sup-measurability of such functions. [340] seems to be the first paper where it is observed that the Carathéodory conditions imply the measurability of f on $\Omega \times \mathbb{R}$, see also [79] and [139]. The "Baire-type" classification was first proposed in the book [182].

The functions which we call "Shragin functions" were introduced by I.V. Shragin in [306] under the name *standard functions*. The fact that these functions are "stable" under pointwise limits, in contrast to Carathéodory functions, was also established in [306], the sup-measurability of Shragin functions in [306], [310]. Observe that, if the function f does not depend on s, f is a Shragin function if and only if f is Borel measurable. Thus, in general a sup-measurable function is not equivalent to a Shragin function. We do not know whether or not there are Shragin functions which are not Baire-Carathéodory functions.

The σ-algebra $\mathcal{M}(\mathcal{B})$ was introduced in [90], where it is also shown that the inclusion $\mathcal{M} \otimes \mathcal{B} \subset \mathcal{M}(\mathcal{B})$ is in general strict. Sainte-Beuve's selection theorem which we used several times can be found in [281].

5. Theorem 1.2 on the boundedness of the superposition operator in the space S is contained in [35]. The class of "co-bounded" superposition operators has not been studied yet. One can show that a set $N \subset S$ is co-bounded if and only if there exists a positive function $u_0 \in S$ such that

$$\sup_{x \in N} \int_{\Omega} |x(s)| u_0(s) d\lambda < \infty.$$

6. The fact that a Carathéodory function f generates a superposition operator F which is continuous in measure was actually already known to Carathéodory himself, and was proved in full generality, i.e. without superfluous assumptions, by V.V. Nemytskij. The converse was an open question for many years, called the "Nemytskij conjecture" by some people [267]; this conjecture was disproved by the existence of monsters. The fact that a function f which generates a continuous superposition operator is supequivalent to a Carathéodory function was first proved by I.Vrkoč in the case $\Omega = [0, 1]$ (see [362]). The same result for general Ω, and even for the case when f maps $\Omega \times U$ into V, where U and V are two metric spaces, is due to A.V. Ponosov [265–267]. Both Vrkoč's and Ponosov's proofs, however, are rather technical and cumbersome. Our proof is taken from [36] and builds essentially on the approximation scheme given in Lemma 1.7; this approximation scheme, in turn, was used in a different context in [72], see also [71].

Theorem 1.3 shows that the Nemytskij conjecture is true within the class of Shragin functions. This was first observed in [316], [317].

Theorem 1.4 is contained in [36], as well as various continuity properties of the *integral functional*

$$\Phi x = \int_\Omega f(s, x(s)) ds \qquad (1.39)$$

which is obviously closely related to the superposition operator $Fx(s) = f(s, x(s))$. Much general information on the functional (1.39), which is of course a special case of (1.22), can be found, in particular, in the recent survey article [71].

Conditions for the continuity of F at a single point $x_0 \in S$ which are both necessary and sufficient are not known; the only result presently known is our Theorem 1.5 which refers to Shragin functions (compare, however, the remarks below). General conditions for the convergence of sequences $f_m(s, x_n(s))$, as $m, n \to \infty$, can be found in [315]. In [264] the author studies the "random continuity" of superposition operators between probabilistic topological spaces.

The fact that "continuity implies uniform continuity" (Theorem 1.7) in its general form is given here for the first time. We point out that an analogous result in well-known Banach spaces of measurable functions (Lebesgue spaces, Orlicz spaces etc.) is not true, see the counterexample preceding Theorem 2.10, and also [58].

7. On the other hand, the fact that a compact superposition operator "degenerates" (Theorem 1.8) has analogues in various normed function spaces. In case of Lebesgue spaces (with $\Omega = \Omega_c$) this was first observed by M.A. Krasnosel'skij [166]. The class of "co-precompact" sets has not beed studied yet; the importance of such sets can be seen from the fact that Schauder's fixed point theorem may actually be formulated in terms of such sets in metric linear spaces [248].

8. Superposition operators which are generated by monotone functions were studied only in the last years; they seem to be important in the investigation of nonlinearities of hysteresis type [170–174]. As already stated at the beginning of Section 1.8, results on superposition operators F which are generated by monotone functions (in u) are much more precise. For example, in this case one can give a precise characterization of sup-measurability (Theorem 1.9, see also [306]). Moreover, if f is sup-measurable and monotone in u, the corresponding operator F is continuous at $x_0 \in S$ if and only if

$$\lim_{u \to 0} \lambda(\{s : s \in \Omega, \, |f[s, x_0(s) + u] - f[s, x_0(s)]| > h\}) = 0$$

for every $h > 0$. Theorem 1.10 was known in the case of monotone functions ([172], see also [174]); in the form given here it is new.

9. As already observed, in some cases it is necessary to consider the superposition operator F not on the whole space S, but only on a subset G

of S. Here many additional difficulties may occur which are due to the "disposition" of the set G in the space S. Of course, in this case one should only assume that F maps functions from G into S; in general, this does not imply that the function f is sup-measurable. This is clear if the set G is, say, a "conical" interval $\{x : x \in S, u \leq x \leq v\}$ or $\{x : x \in S, u < x < v\}$, where u and v are fixed functions. A similar phenomenon occurs if G is a sufficiently "thin" set of functions like the space C^1 of continuously differentiable functions (over a compact domain Ω in Euclidean space). Superposition operators between the space C of continuous functions and the space S of measurable functions will be dealt with in Section 6.4 (see also [292], [308], [319]).

Chapter 2

The superposition operator
in ideal spaces

In this chapter we are concerned with the basic properties of the superposition operator in so-called ideal spaces which are, roughly speaking, Banach spaces of measurable functions with monotone norm. To formulate our results in a sufficiently general framework, we must introduce a large number of auxiliary notions which will be justified by the results in concrete function spaces given in subsequent chapters; we request the reader's indulgence until then.

First, we give conditions for the local and global boundedness of the superposition operator F between ideal spaces X and Y which are typically ensured by special properties of the "source space" X. Second, special properties, such as absolute boundedness and compactness, are treated. Afterwards, we give conditions for the continuity and uniform continuity of F which are now typically ensured by special properties of the "target space" Y. For example, F is "always" continuous if Y is regular, and "never" continuous if Y is completely irregular (see the definitions below).

Weak continuity of F between ideal spaces is also considered; here we mention the surprising fact that, loosely speaking, only linear superposition operators are weakly continuous.

Next, we give necessary and sufficient conditions under which F satisfies a Lipschitz or Darbo condition. It turns out that in many spaces these two conditions are in fact equivalent.

Finally, the last part of this chapter is concerned with differentiability conditions for the superposition operator between ideal spaces. Various existence and degeneracy results for the derivative of F at a single point, the asymptotic linearity, and higher derivatives are given. Moreover, we are concerned with analyticity properties of the superposition operator. Surprisingly enough, it turns out that any analytic superposition operator between two ideal spaces X and Y reduces to a polynomial if X belongs to a certain class of spaces to be defined below. On the other hand, given a

specific nonlinearity f, we provide a "recipe" to construct two ideal spaces
such that the corresponding operator F becomes analytic between them.

2.1 Ideal spaces

The first chapter was devoted to a rather detailed and complete description of the superposition operator in the space S of measurable functions.
Unfortunately, although S is a complete metric space, S is not normable,
and even not locally convex. This makes it rather difficult to study operator
equations in this space, since the main principles of (both linear and nonlinear) functional analysis are formulated in complete normed spaces (Banach
spaces) or complete locally convex spaces. Thus it became necessary to
consider the superposition operator in important normed spaces such as
Lebesgue spaces, Orlicz spaces, Lorentz and Marcinkiewicz spaces, spaces
of continuous and differentiable functions, and others. It turned out, however, that the basic properties of the superposition operator, e.g. continuity,
non-compactness, boundedness, etc., are not connected with special properties of the norms in these spaces, but only with very few general properties
which the above mentioned spaces have in common. Only the conditions on
the function f which ensure that the operator F acts between two concrete
spaces are directly related to the specific properties of these spaces. Therefore, before considering special acting conditions for F in particular spaces,
it seems natural to study first the general properties of the superposition
operator in large classes of spaces, where acting is a priori assumed.

This is the purpose of the present chapter. We shall study the superposition operator in so-called ideal spaces of measurable functions (also
called Banach lattices) which embrace all spaces mentioned above, except
for spaces of continuous or differentiable functions. The theory of the superposition operator in ideal spaces is rather advanced and complete.

As in the first chapter, let Ω be an arbitrary set, $\mathcal{M}$ some σ-algebra of
subsets of Ω, μ a countably additive and σ-finite measure on $\mathcal{M}$, λ an equivalent normalized measure, and S the corresponding space of real measurable
functions on Ω. A Banach space X of measurable functions over Ω is called
ideal space if the relations $|x| \leq |y|$, $x \in S$ and $y \in X$ imply that also $x \in X$
and $\|x\|_X \leq \|y\|_X$.

An important property of ideal spaces X is that of being continuously
imbedded into the space S; hence every sequence which converges in the
norm of X is also convergent in measure, or, equivalently, every ball $B_r(X)$
is a bounded set in S. (Here and in what follows, $B_r(X)$ denotes the set
of all $x \in X$ such that $\|x\|_X \leq r$; when the context is clear, we shall write
simply B_r instead of $B_r(X)$ and $\|x\|$ instead of $\|x\|_X$).

Recall that the *support* supp x of a function x is the set of all $s \in \Omega$ such that $x(s) \neq 0$ (which is uniquely defined up to null sets). If N is any set of measurable functions we define the support of N by

$$\text{supp } N = \text{supp} \left[\sup\{ \chi_{\text{supp } x} : x \in N \} \right].$$

Thus, every function $x \in N$ vanishes outside supp N, and for any set $D \subseteq$ supp N of positive measure one can find a subset $D_0 \subseteq D$, also of positive measure, and a function $x_0 \in N$ for which $D_0 \subseteq \text{supp } x_0$.

In every ideal space X there exist nonnegative functions u_0 for which supp $u_0 = $ supp X; such functions will be called *units* of X in the sequel.

An element $x \in X$ is said to have an *absolutely continuous norm* if

$$\lim_{\lambda(D) \to 0} \| P_D x \| = 0, \tag{2.1}$$

where P_D is the multiplication operator (1.3). The set X^0 of all functions with an absolutely continuous norm in X is a closed (ideal) subspace of X; we call X^0 the *regular part* of X.

The regular part X^0 of X has many remarkable properties. In particular, convergent sequences in X^0 admit an easy characterization: a sequence x_n in X^0 converges to $x \in X$ (actually, $x \in X^0$) if and only if x_n converges to x in S and

$$\lim_{\lambda(D) \to 0} \sup_n \| P_D x_n \| = 0, \tag{2.2}$$

i.e. the elements x_n have *uniformly absolutely continuous norms*. Further, the space X^0 is separable if and only if the underlying space S is separable, and the space X itself is separable if and only if S is separable and $X = X^0$. Unfortunately, the subspace X^0 can be much "smaller" than the whole space X. An ideal space X is called *regular* if $X = X^0$, and *quasi-regular* if supp $X = $ supp X^0; the latter condition means that X^0 is dense in X with respect to convergence in measure.

We still need some other notions. An ideal space X is called *almost perfect* if, given a sequence x_n in X which converges in measure to $x \in X$, one has

$$\| x \| \leq \varliminf_{n \to \infty} \| x_n \|, \tag{2.3}$$

i.e. the norm has the *Fatou property*; X is called *perfect* if, given a sequence x_n in X which converges in measure to $x \in S$, one has $x \in X$ and (2.3) holds. It is easy to see that a space X is almost perfect (resp. perfect) if and only if its unit ball $B_1(X)$ is a closed set in X (resp. in S) with respect to convergence in measure.

Every regular space is almost perfect; the converse is false. An ideal space which is both regular and perfect will be called *completely regular*.

Although we do not discuss concrete examples of ideal spaces in this chapter, we just mention two special spaces which are of particular interest. Given a nonnegative function $u_0 \in S$, the space $L(u_0)$ is, by definition, the set of all $x \in S$, vanishing outside supp u_0, for which the norm

$$\|x\|_{L(u_0)} = \int_\Omega |x(s)| u_0(s) ds \qquad (2.4)$$

is finite. Similarly, the space $M(u_0)$ is the set of all $x \in S$, vanishing outside supp u_0, for which the norm

$$\|x\|_{M(u_0)} = \inf\{\lambda : |x| \le \lambda u_0\} \qquad (2.5)$$

is finite.

The importance of these spaces in the general theory of ideal spaces follows, for example, from the fact that $X = \bigcup \{M(u_0) : M(u_0) \subseteq X\}$ for any space X, and $X = \bigcap \{L(u_0) : L(u_0) \supseteq X\}$ for perfect X. These equalities show that, roughly speaking, $L(u_0)$ and $M(u_0)$ play the role of "maximal" and "minimal" spaces, respectively.

Sometimes it suffices to choose $u_0(s) \equiv 1$; in this case the spaces $L(u_0)$ and $M(u_0)$ are usually denoted by L (or L_1) and M (or L_∞), respectively (see (1.5)). If Ω is a domain in Euclidean space, $\mathcal{M}$ is the system of all Lebesgue measurable subsets of Ω, and μ is the Lebesgue measure, the spaces L and M are the classical space L_1 of integrable functions, and the classical space L_∞ of essentially bounded functions, respectively. Likewise, if Ω is the set of natural numbers, $\mathcal{M}$ is the system of all subsets of Ω, and μ is the counting measure, the spaces L and M are the classical spaces ℓ_1 and ℓ_∞ of summable respectively bounded sequences. All these spaces will be studied in detail in Chapter 3.

Let X be an ideal space. By $\tilde{X}$ we denote the set of all functions $y \in S$, vanishing outside supp X, for which

$$| < x, y > | < \infty \quad (x \in X), \qquad (2.6)$$

where

$$< x, y > = \int_\Omega x(s) y(s) ds. \qquad (2.7)$$

Equipped with the usual algebraic operations and the norm

$$\|y\|_{\widetilde{X}} = \sup\{< x, y >: \|x\|_X \le 1\}, \tag{2.8}$$

the set $\widetilde{X}$ becomes an ideal space which is always perfect; we call $\widetilde{X}$ the
associate space to X. For example, a straightforward computation shows
that $\widetilde{L}(u_0) = M(u_0)$ and $\widetilde{M}(u_0) = L(u_0)$.
It follows from the definition that X is always imbedded into the associate
space $\widetilde{\widetilde{X}}$ of $\widetilde{X}$, i.e.

$$\|x\|_{\widetilde{\widetilde{X}}} \le \|x\|_X \quad (x \in X);$$

in general, $\widetilde{\widetilde{X}}$ and X do not coincide. One can show that $\widetilde{\widetilde{X}}$ and X coincide,
and are even isometrically isomorphic, if and only if X is perfect.

Obviously, the adjoint space $\widetilde{X}$ is closely related to the usual *dual space*
X^* of continuous linear functionals on X. In fact, any $a \in \widetilde{X}$ defines a
continuous linear functional

$$f_a(x) = < x, a > \quad (x \in X) \tag{2.9}$$

on X, where $\|f_a\|_{X^*} = \|a\|_{\widetilde{X}}$, and hence $\widetilde{X}$ is a closed (in general, strict)
subspace of X^*. One can show that $\widetilde{X}$ coincides with X^* if and only if X
coincides with X^0, i.e. X is regular. In general, functionals in $\widetilde{X}$ can be
characterized by special continuity properties: a functional $f \in X^*$ belongs
to $\widetilde{X}$ if and only if the relation

$$\lim_{\lambda(D)\to 0} \sup_{|x|\le z} |f(P_D x)| = 0$$

holds for any nonnegative function $z \in X$; in this case, the element $a \in \widetilde{X}$
which corresponds to f_a by formula (2.9) belongs to $(\widetilde{X})^0$ if and only if f_a
maps any sequence x_n which is bounded in X and convergent in S into a
convergent sequence in $\mathbb{R}$.

The dual space X^* of X is, of course, very well studied, and a lot of
information can be found in any textbook on functional analysis. From the
viewpoint of ideal spaces, however, the adjoint space $\widetilde{X}$ is more natural.
Some information on X^* can also be obtained from information on $\widetilde{X}$, and
vice versa. Thus, a space X is reflexive if and only if both X and $\widetilde{X}$
are completely regular; in particular, the spaces $L(u_0)$ and $M(u_0)$ are not
reflexive.

2.2 The domain of definition of the superposition operator

We continue the study of the superposition operator (1.6); to this end, we shall suppose throughout this chapter that F acts between two ideal spaces X and Y, where, without loss of generality, supp $X =$ supp $Y = \Omega$. The condition supp $X = \Omega$ means, in particular, that the space X is a thick set (see Section 1.3) in S. Consequently, if F maps X into S, F extends to all of S, and hence the function f is sup-measurable.

The aim of the present section is to describe the "natural" domain of definition of the superposition operator, viewed as an operator between two ideal spaces. To this end, we need an important notioh. Given a subset N of X, we denote by $\Sigma(N)$ the set of all $x \in X$ for which there exist a finite partition $\{D_1, \cdots, D_m\}$ of Ω and functions $x_j \in N$ $(j = 1, \cdots, m)$ such that $x(s) = x_j(s)$ for $s \in D_j$ $(j = 1, \cdots, m)$. In other words, $\Sigma(N)$ is the set of all functions of the form

$$x = \sum_{j=1}^{m} P_{D_j} x_j, \qquad (2.10)$$

where $x_j \in N$ and $D_j \in \mathcal{M}$ $(j = 1, \cdots, m)$. In the sequel, we shall call $\Sigma(N)$ the Σ-*hull* of N. Roughly speaking, one could say that this definition plays the same role in ideal spaces as the definition of thick sets in the space S; note, however, that we consider only finite partitions $\{D_1, \cdots, D_m\}$ here.

The following assertion follows immediately from the definition of the Σ-hull and from the disjoint additivity of the operator F (see Lemma 1.1):

Lemma 2.1   *If F maps some set $N \subseteq X$ into Y, then F maps also the Σ-hull $\Sigma(N)$ of N into Y.*

Generally speaking, the importance of the Σ-hull consists in the fact that many "nice" properties which the operator F has on some set N (boundedness, continuity, differentiability, analyticity etc.) carry over to the set $\Sigma(N)$. If, for example, N is the unit ball $B_1(L(u_0))$ in the space $L(u_0)$, then $\Sigma(N)$ coincides with the whole space $L(u_0)$. On the other hand, $\Sigma(N)$ is in some sense the "maximal" set on which F can be defined; for a precise formulation see Theorem 2.1.

For general N, the set $\Sigma(N)$ may be rather complicated. We consider now the special case when N is the unit ball $B_1(X)$ in X; in this case we denote the Σ-hull $\Sigma(B_1(X))$ simply by Σ.

Consider the set Δ of all $x \in X$ such that $|x(s)| \leq \delta(s)$ almost everywhere on Ω, where

$$\delta(s) = \sup\{|z(s)| : z \in B_1(X)\}; \qquad (2.11)$$

here the supremum is taken in the sense of the natural ordering of measurable functions (see Section 1.1). In general, the function (2.11) may be infinite on subsets of Ω of positive measure; in particular, $\delta(s) \equiv \infty$ for all $s \in \Omega_c$ if X is quasi-regular.

Further, we denote by Π the set of all functions $x \in X$ for which there exists a $D \in \mathcal{M}$ such that $\|P_D x\| \le 1$ and $P_{\Omega \backslash D} x \in X^0$. Obviously, the inclusions

$$\{x : x \in X, \pi(x) < 1\} \subseteq \Pi \subseteq \{x : x \in X, \pi(x) \le 1\} \qquad (2.12)$$

hold, where

$$\pi(x) = \overline{\lim_{\lambda(D) \to 0}} \|P_D x\| = \text{dist}\,(x, X^0) \qquad (2.13)$$

is the distance of x to the regular part X^0 of X (see also (2.1)). It is not hard to see that the inclusions

$$\Pi \cap \Delta \subseteq \Sigma \subseteq \Delta \qquad (2.14)$$

hold. If X is regular, both inclusions turn into equalities.

In order to describe the set Σ, it is convenient to introduce the *split functional*

$$H(x) = \inf\Big\{\sum_{i=1}^{n} \|P_{D_i} x\| : \|P_{D_i} x\| \le 1\Big\} \quad (i = 1, \cdots, n), \qquad (2.15)$$

where the infimum is taken over all partitions $\{D_1, \cdots, D_n\}$ of Ω (with n variable). Obviously, this functional takes only nonnegative (maybe, infinite) values on X, and the set Σ consists just of those elements $x \in X$ for which $H(x)$ is finite.

Lemma 2.2 *Let X be an ideal space. Then the split functional H of X has the following properties:*

(a) $H(x) = \|x\|$ for $\|x\| \le 1$, and $H(x) \ge \|x\|$ for $\|x\| > 1$.

(b) H is monotone, i.e. $|x| \le |y|$ $(x, y \in X)$ implies that $H(x) \le H(y)$.

(c) $H((1 - \lambda)x + \lambda y) \le (1 - \lambda)H(x) + \lambda H(y) + 2H(x)H(y)$ for $x, y \in X$ and $0 < \lambda < 1$; in particular, the set Σ is convex.

(d) $H(x + y) \le H(x) + H(y)$ for disjoint $x, y \in X$.

$\Rightarrow$ The statements (a) and (b) follow immediately from the definition of the functional (2.15). To prove (c), we first remark that, given $x \in X$ and $\varepsilon > 0$, among all the partitions $\{D_1, \cdots, D_n\}$ of Ω for which $\|P_{D_i} x\| \le 1$

$(i = 1, \cdots, n)$ and $\sum_{i=1}^{n} \|P_{D_i}x\| \leq H(x) + \varepsilon$, one can always find one such that $n \leq 2H(x) + 1 + 2\varepsilon$. In fact, without loss of generality we can assume that in the partition $\{D_1, \cdots, D_n\}$ we have $\|P_{D_i}x\| \leq \frac{1}{2}$ for only one index i, since, if i' and i'' were two such indices, we could replace the corresponding sets $D_{i'}$ and $D_{i''}$ with their union $D_{i'} \cup D_{i''}$ without changing the properties of the partition claimed in the definition of the functional H. Consequently, we suppose that $\|P_{D_i}x\| > \frac{1}{2}$ holds for all indices i exept one; but this implies that $\frac{n-1}{2} \leq H(x) + \varepsilon$ or, equivalently, $n \leq 2H(x) + 1 + 2\varepsilon$. Now fix $x, y \in X$, $\varepsilon > 0$, and two partitions $\{A_1, \cdots, A_m\}$ and $\{B_1, \cdots, B_n\}$ of Ω such that

$$\|P_{A_i}x\| \leq 1 \ (i = 1, \cdots, m), \ \sum_{i=1}^{m} \|P_{A_i}x\| \leq H(x) + \varepsilon, \ m \leq 2H(x) + 1 + 2\varepsilon,$$

$$\|P_{B_j}x\| \leq 1 \ (j = 1, \cdots, n), \ \sum_{j=1}^{n} \|P_{B_j}y\| \leq H(y) + \varepsilon, \ n \leq 2H(y) + 1 + 2\varepsilon.$$

Then the family of all sets $C_{ij} = A_i \cap B_j$ $(i = 1, \cdots, m; \ j = 1, \cdots, n)$ is also a partition of Ω and has the property that $\|P_{C_{ij}}((1-\lambda)x + \lambda y)\| \leq (1-\lambda)\|P_{C_{ij}}x\| + \lambda\|P_{C_{ij}}y\| \leq (1-\lambda)\|P_{A_i}x\| + \lambda\|P_{B_j}y\|$ $(i = 1, \cdots, m; \ j = 1, \cdots, n)$ and

$$H((1-\lambda)x + \lambda y) \leq \sum_{i=1}^{m}\sum_{j=1}^{n} \|P_{C_{ij}}((1-\lambda)x + \lambda y)\|$$

$$\leq \sum_{i=1}^{m}\sum_{j=1}^{n}[(1-\lambda)\|P_{A_i}x\| + \lambda\|P_{B_j}y\|]$$

$$\leq n(1-\lambda)\sum_{i=1}^{m} \|P_{A_i}x\| + m\lambda\sum_{j=1}^{n} \|P_{B_j}y\|$$

$$\leq (1-\lambda)(H(x) + \varepsilon)(2H(y) + 1 + 2\varepsilon) + \lambda(H(y) + \varepsilon)(2H(x) + 1 + 2\varepsilon)$$

$$= (1-\lambda)H(x) + \lambda H(y) + 2H(x)H(y) + \varepsilon + 2\varepsilon(H(x) + H(y) + \varepsilon).$$

Since $\varepsilon > 0$ is arbitrary, the statement (c) follows; the convexity of the set Σ is an obvious consequence.

Similarly, the statement (d) is proved as follows: given again two partitions $\{A_1, \cdots, A_m\}$ and $\{B_1, \cdots, B_n\}$ of Ω such that

$$\|P_{A_i}x\| \le 1 \quad (i = 1, \cdots, m), \quad \sum_{i=1}^{m} \|P_{A_i}x\| \le H(x) + \varepsilon,$$

$$\|P_{B_j}y\| \le 1 \quad (j = 1, \cdots, n), \quad \sum_{j=1}^{n} \|P_{B_j}y\| \le H(y) + \varepsilon,$$

we can suppose, by the disjointness of x and y, that all the sets A_i ($i = 1, \cdots, m$) and B_j ($j = 1, \cdots, n$) are mutually disjoint. But then the union $\{A_1, \cdots, A_m, B_1, \cdots, B_n\}$ is a partition of Ω, and hence

$$H(x + y) \le \sum_{i=1}^{m} \|P_{A_i}x\| + \sum_{j=1}^{n} \|P_{B_j}y\| \le H(x) + H(y) + 2\varepsilon.$$

Since $\varepsilon > 0$ is arbitrary, the statement (d) follows, and thus the proof is complete. $\Leftarrow$

Unfortunately, the explicit calculation of the split functional H is, in general, even harder than the explicit description of the set Σ. In some special spaces, however, such a calculation is possible. For the moment we restrict ourselves to the remark that

$$H(x) = \begin{cases} \|x\| & \text{if } \sup_{s \in \Omega_d} \|P_{\{s\}}x\| \le 1, \\ \infty & \text{if } \sup_{s \in \Omega_d} \|P_{\{s\}}x\| > 1 \end{cases} \tag{2.16}$$

in the space $L(u_0)$, and

$$H(x) = \begin{cases} \|x\| & \text{if } \|x\| \le 1, \\ \infty & \text{if } \|x\| > 1 \end{cases} \tag{2.17}$$

in the space $M(u_0)$; other examples will be given in subsequent chapters.

It is easy to see that the functional H allows us to describe not only the set $\Sigma = \Sigma(B_1)$, but also the set $\Sigma(x_0 + B_r)$; in fact, $\Sigma(x_0 + B_r)$ consists of all $x \in X$ with $H((x - x_0)/r) < \infty$. In this way, one can get an idea of the set $\Sigma(N)$ even in case N is an F_σ-type or G_δ-type set.

For convenience, a set $N \subset X$ will be called Σ-*stable* if $\Sigma(N) = N$. In this terminology, we may state the following important consequence of Lemma 2.1:

Theorem 2.1   *Let X and Y be two ideal spaces, and let f be a sup-measurable function. Then the domain of definition $\mathcal{D}(F)$ of the superposition operator F, generated by f and considered as an operator between X and Y, is Σ-stable.*

The main part of this chapter is devoted to the basic properties of the superposition operator F, such as boundedness, continuity, etc., between two ideal spaces X and Y, under the hypothesis that $\mathcal{D}(F)$ is "sufficiently large" (for example, has interior points, or even coincides with the whole space X). The problem of finding relations between the function f, on the one hand, and the spaces X and Y, on the other hand, will not yet be considered here; to get significant results in this direction, one must use rather specific properties of concrete spaces, and this will be the purpose of the following chapters.

2.3 Local and global boundedness conditions

Recall that a nonlinear operator F, defined on a subset N of a normed linear space X and taking values in a normed linear space Y, is called *locally bounded* on N if

$$\varlimsup_{x \to x_0} \|Fx\| < \infty \quad (x_0 \in N),$$

i.e. if F is bounded on some neighbourhood of each point $x_0 \in N$. The following theorem gives a condition under which the superposition operator F is locally bounded between ideal spaces:

Theorem 2.2   *Let X and Y be two ideal spaces, with Y being almost perfect. Let f be a sup-measurable function. Suppose that the domain of definition $\mathcal{D}(F)$ of the superposition operator F, generated by f and considered as an operator between X and Y, has interior points. Then F is locally bounded on the interior of $\mathcal{D}(F)$ if and only if, for each $s \in \Omega_d$, the function $f(s, \cdot)$ is bounded on each bounded interval in $\mathbb{R}$.*

$\Rightarrow$ Without loss of generality we may asume that θ is an interior point of $\mathcal{D}(F)$, that F is defined on some ball $B_\delta(X)$, and that $F\theta = \theta$. The last assumption implies, in particular, that F commutes with the operator (1.3) for each $D \in \mathcal{M}$. Moreover, without loss of generality we can consider the two cases $\Omega_d = \emptyset$ and $\Omega_c = \emptyset$ separately.

Suppose first that $\Omega_d = \emptyset$. If F is unbounded at x_0, we can find a sequence x_n in $\mathcal{D}(F)$ such that

$$\|x_n\| \leq 2^{-n}\delta, \ \|Fx_n\| > 2^n n \quad (n = 1, 2, \cdots). \tag{2.18}$$

Since $\Omega_d = \emptyset$, we can find, for each n, a partition of Ω into 2^n subsets of equal λ-measure; for at least one of these subsets (which we denote D_n) we have $\|FP_{D_n}x_n\| > n$. Since $\lambda(D_n) \to 0$ as $n \to \infty$ and Y is almost perfect, we get $\lim_{m\to\infty} \|FP_{D_n\setminus D_m}x_n\| = \|FP_{D_n}x_n\|$ $(n = 1, 2, \cdots)$; consequently, to each n we can associate an $n' \geq n$ such that $\|FP_{D_n\setminus D_{n'}}x_n\| \geq n$. Setting $n_1 = 1$, $n_2 = n_1', \cdots, n_{k+1} = n_k', \cdots$ and denoting $\Omega_k = D_{n_k} \setminus D_{n_{k+1}}$ $(k = 1, 2, \cdots)$ we get from (2.18) that $\|P_{\Omega_k}x_{n_k}\| \leq 2^{-k}\delta$ and $\|FP_{\Omega_k}x_{n_k}\| > k$. Now let x_* $= \sum_{k=1}^{\infty} P_{\Omega_k}x_{n_k}$. On the one hand, we have

$$\|x_*\| \leq \sum_{k=1}^{\infty} \|P_{\Omega_k}x_{n_k}\| \leq \delta;$$

hence, by hypothesis, $x_* \in \mathcal{D}(F)$ and $Fx_* \in Y$. On the other hand,

$$\|Fx_*\| \geq \|FP_{\Omega_k}x_{n_k}\| > k \quad (k = 1, 2, \cdots);$$

hence $Fx_* \notin Y$. This contradiction proves the statement in the case $\Omega_d = \emptyset$.

Now let $\Omega_c = \emptyset$ and suppose first that the function $f(s, \cdot)$ $(s \in \Omega)$ is bounded on each bounded interval in $\mathbb{R}$. Let $\Omega = \{s_1, s_2, \cdots\}$ and $P_n = P_{\{s_{n+1}, s_{n+2}, \cdots\}}$. By the boundedness of $f(s_k, \cdot)$, the numbers

$$c_n = \sum_{k=1}^{n} \sup\{|f(s_k, u)| : |u|\|\chi_{\{s_k\}}\| \leq \delta\}\|\chi_{\{s_k\}}\|$$

are finite, and $\|F(I - P_n)x\| \leq c_n$ for $\|x\| \leq \delta$ and each n. Suppose now that F is unbounded at θ. In this case we can find a sequence x_n in X such that

$$\|x_n\| \leq 2^{-n}\delta, \quad \|Fx_n\| > c_n + n \quad (n = 1, 2, \cdots). \tag{2.19}$$

Since Y is almost perfect, $\lim_{m\to\infty} \|F(I - P_m)x_n\| = \|Fx_n\|$ $(n = 1, 2, \cdots)$; therefore $\|F(I - P_{n'})x_n\| > c_n + n$ for some $n' \geq n$, and hence

$$\|F(P_n - P_{n'})x_n\| \geq \|F(I - P_{n'})x_n\| - \|F(I - P_n)x_n\| \geq n. \tag{2.20}$$

Let the sequence n_k be constructed by induction as above. By (2.19) and (2.20),

$$\|(P_{n_k} - P_{n_{k+1}})x_{n_k}\| \leq 2^{-k}\delta, \quad \|F(P_{n_k} - P_{n_{k+1}})x_{n_k}\| > k.$$

Now let $x_* = \sum_{k=1}^{\infty}(P_{n_k} - P_{n_{k+1}})x_{n_k}$. On the one hand, we have

$$\|x_*\| \leq \sum_{k=1}^{\infty} \|(P_{n_k} - P_{n_{k+1}})x_{n_k}\| \leq \delta,$$

hence $x_* \in \mathcal{D}(F)$ and $Fx_* \in Y$. On the other hand,

$$\|Fx_*\| \geq \|F(P_{n_k} - P_{n_{k+1}})x_{n_k}\| > k \quad (k = 1, 2, \cdots)$$

hence $Fx_* \notin Y$, again a contradiction.

To prove the last part, suppose that $\Omega_c = \emptyset$ and F is locally bounded. Obviously, the function f can be written as composition $f(s,u) = \varepsilon_s \circ F \circ \sigma_s(u)$, where σ_s is defined by $\sigma_s(u) = u\chi_{\{s\}}$, and ε_s is defined by $\varepsilon_s(y) = y(s)$. Since both σ_s and ε_s are bounded, the function $f(s, \cdot)$ is locally bounded for each $s \in \Omega$, and the proof follows from the classical Heine–Borel theorem. $\Leftarrow$

Theorem 2.2 is rather surprising in case $\Omega_d = \emptyset$: the local boundedness of F on an open set $N \subseteq X$ follows already from the acting condition $F(N) \subseteq Y$, provided that Y is almost perfect.

We pass now to the problem of finding conditions for the (global) boundedness of the operator F in ideal spaces. One could expect that F is also bounded on each ball which is entirely contained in the domain of definition $\mathcal{D}(F)$. Simple examples show, however, that this is not so. Consider, for instance, the operator F generated by the function $f(s,u) = u^s$ in the space $X = \ell_1$ of summable sequences. Here we have $\mathcal{D}(F) = X$, and F is locally bounded on $\mathcal{D}(F)$, but F is unbounded on each ball $B_r(X)$ of radius $r > 1$. Analogous counterexamples can be constructed in the case of an atomic-free measure μ. Nevertheless, one can describe large classes of ideal spaces X for which boundedness holds not only on small balls, but on any ball contained in $\mathcal{D}(F)$.

We call an ideal space X a *split space* if one can find a sequence σ_n of natural numbers with the following property: given any sequence $x_n \in B_1(X)$ of disjoint functions, one can construct disjoint functions $x_{n,j}$ ($j = 1, \cdots, \sigma_n$) such that

$$x_n = x_{n,1} + \cdots + x_{n,\sigma_n}, \tag{2.21}$$

and for each choice $\vec{j} = (j_1, j_2, \cdots)$ of natural numbers $1 \leq j_n \leq \sigma_n$, the function $x_{\vec{j}} = \sum_{n=1}^{\infty} x_{n,j_n}$ belongs to $B_1(X)$.

The definition of a split space is, of course, rather technical and therefore difficult to verify in specific ideal spaces. In what follows we shall consider two special classes of split spaces which are much easier to characterize. Before doing so, however, we point out that split spaces are only interesting in the case $\Omega_d = \emptyset$: in fact, if Ω_d consists of a finite number of atoms, every space over Ω_d is a split space. On the other hand, if Ω_d consists of infinitely many atoms $s_1, s_2, \cdots$, the functions $x_n = \|\chi_{\{s_n\}}\|^{-1}\chi_{\{s_n\}}$ are mutually disjoint, and (2.21) implies that all functions $x_{n,j}$ ($j = 1, \cdots, \sigma_n$) are zero except for one function which is equal to x_n. But then the function $x_* = \sum_{n=1}^{\infty} x_n$ belongs to $B_1(X)$ only if $X = M(u_0)$. This shows that $M(u_0)$ is the only split space over a set with infinitely many atoms.

We are now going to characterize an important class of split spaces by means of the split functional (2.15).

Lemma 2.3   *Let X be an ideal space with supp $X = \Omega$, and suppose that the measure μ is atomic-free. Then the following three conditions are equivalent:*

(a) The split functional H is bounded on some ball $B_r(X)$ with radius $r > 1$.

(b) The split functional H is bounded on each ball $B_r(X)$ with radius $r > 1$.

(c) The split functional H is of polynomial growth, i.e.

$$H(x) \leq c(1 + \|x\|^k) \quad (x \in X) \tag{2.22}$$

with some positive constants c and k.

$\Rightarrow$ Since obviously (c) implies (b) and (b) implies (a), it suffices to prove that (a) implies (c). So, let H be bounded on some ball $B_r(X)$ with $r > 1$ by a constant n which we can suppose to be a natural number. This means that, given any $x \in B_r(X)$, one can find a partition $\{D_1, \cdots, D_n\}$ of Ω such that $\|P_{D_i}x\| \leq 1$ for $i = 1, \cdots, n$. Applying this reasoning to the function $rx\|x\|^{-1}$, where $\|x\| = \rho$ for some $\rho > 1$, one sees that the functional H is bounded on $B_{r\rho}$ by mn, whenever H is bounded on B_ρ by m. In particular, H is bounded on each of the balls B_r, B_{r^2}, B_{r^3}, $\cdots$ by n, n^2, n^3, $\cdots$, respectively. Since any ball B_ρ is contained in some ball B_{r^k} (namely for $k = \lceil \log_r \rho \rceil$, where $\lceil \vartheta \rceil$ denotes the smallest integer $k \geq \vartheta$), $H(x)$ is majorized by the constant $c(1 + \rho^k)$, where $c = n$ and $k = \lceil \log_r n \rceil$, and so we are done. $\Leftarrow$

We shall call an ideal space X a *space of power type* or Δ_2*-space*, for short (and write $X \in \Delta_2$) if the split functional H on X has one of the equivalent properties stated in Lemma 2.3.

Lemma 2.4   *Every Δ_2-space is a split space.*

$\Rightarrow$ For $n = 1, 2, \cdots$, let $\sigma_n = \lceil \sup\{H(x) : \|x\| \leq 2^n\} \rceil$. By definition of the functional H, any sequence x_n in $B_1(X)$ of disjoint functions can be written in the form $x_n = x_{n,1} + \cdots + x_{n,\sigma_n}$, where $\|x_{n,j}\| \leq 2^{-n}$ for $j = 1, \cdots, \sigma_n$. For any choice $\overrightarrow{j} = (j_1, j_2, \cdots)$ with $1 \leq j_n \leq \sigma_n$ we therefore have $\|x_{\overrightarrow{j}}\| \leq \sum_{n=1}^{\infty} \|x_{n,j_n}\| \leq 1$, which means that X is a split space. $\Leftarrow$

We remark that there exist split spaces which are not Δ_2-spaces, see Section 4.1.

Next we introduce still another class of split spaces which is larger, but also more technical to describe. We shall say that an ideal space X is a δ_2-*space* if one can define a functional h on the unit ball $B_1(X)$ of X, which takes values in the interval $[0, 1]$ and has the following properties:

(a) For all $x \in B_1(X)$,

$$\lim_{n \to \infty} \sup_{\|x\| \leq 1} \inf_{\{D_1, \cdots, D_n\}} \max_{i=1, \cdots, n} h(P_{D_i} x) = 0\,,$$

where the infimum is taken over all partitions $\{D_1, \cdots, D_n\}$ of Ω (with n fixed).

(b) If x_n is a sequence in $B_1(X)$ of disjoint functions such that

$$\sum_{n=1}^{\infty} h(x_n) \leq 1\,,$$

then the function $x = \sum_{n=1}^{\infty} x_n$ belongs to $B_1(X)$.

Lemma 2.5   *Every δ_2-space is a split space.*

We shall not give the proof of this lemma, since it is analogous to that of Lemma 2.4 (which is a special case, namely $h(x) = \|x\|$). We remark, however, that not every δ_2-space is a Δ_2-space, since there exist functionals h on ideal spaces which are essentially different from the norm and satisfy conditions (a) and (b), see again Section 4.1.

We are now in a position to formulate the main result on the (global) boundedness of the superposition operator between ideal spaces:

Theorem 2.3   *Let X and Y be two ideal spaces, with X being a split space and Y being almost perfect. Let f be a sup-measurable function. Suppose that the domain of definition $\mathcal{D}(F)$ of the superposition operator F, generated by f and considered as an operator between X and Y, has interior points. Then F is bounded on each ball contained in $\mathcal{D}(F)$ if and* only if, for each $s \in \Omega_d$, the function $f(s, \cdot)$ is bounded on each bounded interval in $\mathbb{R}$.

$\Rightarrow$ In the case $\Omega_c = \emptyset$ the proof coincides with that of Theorem 2.2; let us therefore assume that $\Omega_d = \emptyset$. Moreover, we can again suppose that $F\theta = \theta$ and $B_1(X) \subseteq \mathcal{D}(F)$, and consider F on the unit ball $B_1(X)$. If F is unbounded on B_1 we can choose a sequence z_n in B_1 such that $\|Fz_n\| > n2^n\sigma_n$, where σ_n is the numerical sequence occuring in the definition of a split space. Since $\Omega_d = \emptyset$, we can find a partition of Ω into 2^n subsets of equal λ-measure; for at least one of these subsets (which we denote D_n) we have $\|FP_{D_n}z_n\| > n\sigma_n$. As in Theorem 2.2, we construct a sequence n_k inductively such that for $x_k = P_{D_{n_k} \setminus D_{n_{k+1}}} z_{n_k}$ we have $\|x_k\| \leq 1$ and $\|Fx_k\| > k\sigma_k$ Since X is a split space, each x_k can be represented in the form $x_k = x_{k,1} + \cdots + x_{k,\sigma_k}$. The inequality $\|Fx_k\| > k\sigma_k$ implies, in particular, that for each k one can find a j_k such that $\|Fx_{k,j_k}\| > k$. The function $x_* = \sum_{k=1}^{\infty} x_{k,j_k}$ belongs to $B_1(X)$, and hence $Fx_* \in Y$. On the other hand, $\|Fx_*\| \geq \|Fx_{k,j_k}\| > k$ $(k = 1, 2, \cdots)$ which is a contradiction. $\Leftarrow$

In applications it is sometimes necessary to show the boundedness of the superposition operator F on some ball (or other set), where the hypotheses of Theorem 2.3 are not fulfilled. Very little can be said in this situation; sometimes, however, at least some special assertions can be made which build on majorization arguments like the following: if two sup-measurable functions f and g satisfy an estimate $|f(s, u)| \leq g(s, u)$, and if the corresponding superposition operator G is bounded on $N \subseteq \mathcal{D}(G)$, then F is defined and also bounded on N. The "usefulness" of this completely trivial observation consists in the fact that the function g may be essentially simpler than the function f.

To conclude this section, let us introduce the function

$$\mu(F; N) = \sup_{x \in N} \|Fx\| \qquad (2.23)$$

which is finite, of course, only if F is bounded on N. Many problems which are related to the superposition operator lead to the study of the function (2.23). Unfortunately, one can not give much general information about the characteristic (2.23), since it relies heavily on the specific form of the norm under consideration. Nevertheless, some general remarks can be made. For instance, if $\text{ℓu}(N)$ denotes the Luzin hull of a set $N \subset S$ (see Section 1.3), and F takes its values in an almost perfect space Y, we have $\mu(F; \text{ℓu}(N) \cap \mathcal{D}(F)) = \mu(F; N)$ for $N \subseteq \mathcal{D}(F)$. In the most important case,

$N = B_r(X)$, the function (2.23) takes the simpler form

$$\mu_F(r) = \mu(F; B_r) = \sup_{\|x\| \leq r} \|Fx\| \qquad (2.24)$$

and is called the *growth function* of F in the space X; the function μ_F can be regarded as natural analogue to the norm of a linear operator (for F linear we have, of course, $\mu_F(r) = \|F\|r$.). This function gives a precise description of the growth of the superposition operator F on balls in X and is extremely useful in the study of the superposition operator, since many important properties of F, such as continuity, differentiability, analyticity, etc., can be characterized by means of this function.

2.4 Special boundedness properties

Let X and Y be two ideal spaces, and suppose that F, as an operator between X and Y, is bounded on some set $N \subseteq X$. It may happen in this situation that the image $FN \subseteq Y$ is "nicer", in some sense, than the set N itself. For example, for applications it would be very useful, in order to apply classical solvability results for nonlinear equations, if the set FN were to be precompact whenever N is bounded. Unfortunately, this leads to a very strong degeneracy for the superposition operator F, as Theorem 1.8 already suggests. Nevertheless, although not being precompact, the set FN may have some property "between" boundedness and precompactness, if N is only bounded, and thus the operator F may "improve" sets. It is the purpose of this section to give a precise meaning to these vague statements. A set N of an ideal space X is called *U-bounded* if N is bounded with respect to the ordering in X, i.e. if $|x| \leq u_0$ $(x \in N)$ for some $u_0 \in X$. More generally, N is called *W-bounded* if, for some $\varepsilon > 0$, there exists a U-bounded ε-net for N in X; this means, in other words, that for each $\varepsilon > 0$ one can find $u_\varepsilon \in X$ such that $\|x - x_\varepsilon\| \leq \varepsilon$ $(x \in N)$ where

$$x_\varepsilon(s) = \min\{|x(s)|, u_\varepsilon(s)\}\mathrm{sgn}\, x(s). \qquad (2.25)$$

Obviously, every U-bounded set is W-bounded, and every W-bounded set is bounded in norm. The converse is, in general, false; for example, the W-bounded and bounded sets coincide only in the space $X = M(u_0)$.

Although the property of W-boundedness is rather simple, it is somewhat difficult to verify in concrete ideal spaces. An exception is provided by regular spaces; in fact, if X is regular, the concepts of W-boundedness and absolute boundedness are equivalent. Recall that a bounded set $N \subset X$

is called *absolutely bounded* if the elements of N have *uniformly absolutely continuous norms*, i.e.

$$\lim_{\lambda(D)\to 0} \sup_{x\in N} \|P_D x\| = 0 \qquad (2.26)$$

(compare this with (2.2)). In fact, if N is W-bounded and $\varepsilon > 0$ and $x \in N$ are fixed, we have $\|x - x_\varepsilon\| \leq \frac{\varepsilon}{2}$ for some $u_\varepsilon \in X$, x_ε as in (2.25), and hence $\|P_D x\| \leq \|P_D u_\varepsilon\| + \frac{\varepsilon}{2}$. Since X is regular, we have $\|P_D u_\varepsilon\| \leq \frac{\varepsilon}{2}$ if $\lambda(D)$ is sufficiently small, hence $\|P_D x\| \leq \varepsilon$, uniformly in $x \in N$, i.e. N is absolutely bounded. Conversely, let N be absolutely bounded, and let u_0 be any unit in X. Since N is bounded in the space S, we have, by Lemma 1.6,

$$\lim_{h\to\infty} \sup_{x\in N} \lambda(\{s : s \in \Omega, |x(s)| > h u_0(s)\}) = 0. \qquad (2.27)$$

Given $\varepsilon > 0$, we can find $\delta > 0$ such that $\|P_D x\| \leq \varepsilon$ for $x \in N$ and $\lambda(D) \leq \delta$. In particular, by (2.27) the set $\{s : |x(s)| > h_0 u_0(s)\}$ has λ-measure at most δ if h_0 is sufficiently large. This means that the set $N_\varepsilon = \{x : x \in X, |x| \leq h_0 u_0\}$ is a U-bounded ε-net for N in X, and hence N is W-bounded, since $\varepsilon > 0$ is arbitrary.

In general, every precompact set N is W-bounded. If X is regular, this can be strengthened; in this case $N \subset X$ is precompact in X if and only if N is precompact in S and absolutely bounded in X.

Lemma 2.6   *Let X be an ideal space and $N \subseteq X$. Then for the W-boundedness of N it is sufficient (and, in case X is regular, also necessary) that for some unit u_0 in X (for every unit u_0 in X, respectively) the relation*

$$\sup_{x\in N} \|u_0 \Phi(u_0^{-1}|x|)\| < \infty \qquad (2.28)$$

holds, where Φ is some nonnegative increasing function on $[0,\infty)$ such that

$$\lim_{u\to\infty} \frac{\Phi(u)}{u} = \infty. \qquad (2.29)$$

$\Rightarrow$ Suppose that the condition (2.28) holds. To prove the W-boundedness of N, it suffices to show that

$$\lim_{h\to\infty} \sup_{x\in N} \|P_{D(x,h)} x\| = 0, \qquad (2.30)$$

where $D(x,h) = \{s : s \in \Omega, |x(s)| > h u_0(s)\}$. Given $\varepsilon > 0$, denote the left-hand side of (2.28) by a, and choose $h_\varepsilon > 0$ such that $|u| \leq \varepsilon \Phi(u)/a$ for $|u| >$

h_ε. Then for $x \in N$ and $h \geq h_\varepsilon$ we have $\|P_{D(x,h)}x\| \leq \varepsilon \|u_0 \Phi(u_0^{-1}|x|)\|/a \leq \varepsilon$, and hence (2.30) holds.

Suppose now that X is regular, u_0 is some unit in X, and $N \subset X$ is absolutely bounded; hence (2.30) holds. Let ν be some function on $[0, \infty)$ which satisfies $\lim_{h \to \infty} \nu(h) = 0$ and $\|P_{D(x,h)}x\| \leq \nu(h)$ for $x \in N$, and choose a sequence h_n of positive numbers ($h_0 = 0$) such that the series $\sum_{n=1}^{\infty} \nu(h_n)$ converges. Moreover, let ω_n be an increasing numerical sequence, tending to infinity as $n \to \infty$, such that $\sum_{n=1}^{\infty} \omega_n \nu(h_{n-1}) < \infty$. Finally, choose an increasing continuous function ω on $[0, \infty)$ such that $\omega(h_n) = \omega_n$. Then the function $\Phi(u) = \omega(u)u$ satisfies (2.29); moreover, given $x \in N$, we set

$$D_n(x) = \{s : s \in \Omega,\ h_{n-1}u_0(s) < |x(s)| \leq h_n u_0(s)\}$$

and get

$$\left\|\Phi\left[\frac{|x|}{u_0}\right]u_0\right\| \leq \sum_{n=1}^{\infty} \left\|P_{D_n(x)}\left\{\Phi\left[\frac{|x|}{u_0}\right]u_0\right\}\right\| \leq \sum_{n=1}^{\infty} \omega_n \nu(h_{n-1}) < \infty;$$

hence (2.28). $\Leftarrow$

Lemma 2.6 implies immediately the following condition (in general, only sufficient) under which F is W-bounded on a set N:

Theorem 2.4   *Let X and Y be two ideal spaces, and let f be a sup-measurable function. Suppose that the superposition operator F, generated by f and considered as an operator between X and Y, is bounded on some set $N \subseteq X$. Then for the W-boundedness of the set $FN \subset Y$ it is sufficient (and, in case Y is regular, also necessary) that the superposition operator $\widetilde{F}$, generated by the function*

$$\widetilde{f}(s, u) = u_0(s)\Phi[|f(s, u)|u_0(s)^{-1}],$$

is bounded on N, where u_0 is a suitable (an arbitrary, respectively) unit in Y, and Φ is an increasing function on $[0, \infty)$ satisfying (2.29).

Now we pass to a compactness condition for the superposition operator F. As already mentioned above, the compactness of F leads to a strong degeneration of the corresponding function f:

Theorem 2.5   *Let X and Y be two ideal spaces, and let f be a sup-measurable function. Suppose that the superposition operator F generated by f maps every bounded set $N \subset X$ into a precompact set $FN \subset Y$.*

Then the function $f(s, \cdot)$ is constant for each $s \in \Omega_c$, and the operator F is W-bounded.

Conversely, if X and Y are two ideal spaces, with Y regular, and if $f(s, \cdot)$ is constant for each $s \in \Omega_c$ and F is W-bounded on some set $N \subset X$, then $FN \subset Y$ is precompact.

$\Rightarrow$ The proof of the first part repeats literally the proof of Theorem 1.8. The second part is obvious : it follows from the fact that we can restrict ourselves to the case $\Omega_c = \emptyset$, and that precompact and absolutely bounded sets in ideal spaces over a discrete measure coincide. $\Leftarrow$

Theorem 2.5 gives a complete description of compact superposition operators only in case of regular Y (or, more generally, in case of regularity of the subspace Y_d of all functions in Y vanishing on Ω_c). In the general case, one cannot obtain such a complete description, since one does not know appropriate compactness criteria in non-regular spaces.

Let X be a complete metric linear space. We call a nonnegative function ψ on the system of all bounded subsets of X a *Sadovskij functional* if the following two properties are satisfied ($N, N_1, N_2 \subset X$ bounded):

(a) $\psi(\overline{co}\, N) = \psi(N)$.
(b) $\psi(N_1 \cup N_2) = \max\{\psi(N_1), \psi(N_2)\}$.

Moreover, ψ is called a *measure of noncompactness* if $\psi(N) = 0$ holds if and only if N is precompact. If X is a Banach space, one usually requires the additional properties ($\lambda \in \mathbb{R}$):

(c) $\psi(N_1 + N_2) \leq \psi(N_1) + \psi(N_2)$.
(d) $\psi(\lambda N) = |\lambda| \psi(N)$.

Classical examples of measures of noncompactness having all these properties are the *Kuratowski measure of noncompactness,*

$$\gamma(N) = \inf\{d : d > 0, \, N \subseteq \bigcup_{j=1}^{m} N_j, \, \text{diam } N_j \leq d\}, \qquad (2.31)$$

and the *Hausdorff measure of noncompactness,*

$$\alpha(N) = \inf\{r : r > 0, \, N \subseteq B_r(X) + C, \, C \text{ finite}\}, \qquad (2.32)$$

where the infimum may also be taken over all compact subsets C of X. If X is an ideal space it is useful to consider other Sadovskij functionals (which, however, are not measures of noncompactness), like

$$\pi(N) = \varlimsup_{\lambda(D) \to 0} \sup_{x \in N} \|P_D x\|, \qquad (2.33)$$

$$\eta(N) = \inf\{r : r > 0,\ N \subseteq B_r(X) + C,\ C\ U\text{-bounded }\}, \qquad (2.34)$$

$$\omega(N) = \inf\{r : r > 0,\ N \subseteq B_r(X) + C,\ C\ \text{weakly compact}\}, \quad (2.35)$$

and

$$\beta(N) = \sup_{y \in B_1(\widetilde{X})}\ \overline{\lim_{\lambda(D) \to 0}}\ \sup_{x \in N} < x, y >, \qquad\qquad (2.36)$$

where $\widetilde{X}$ denotes the associate space to X and $< \cdot,\cdot >$ is the pairing defined in (2.7). There are some obvious relations between the functionals (2.31) – (2.36): if X is an ideal space we have $\alpha(N) \leq \gamma(N) \leq 2\alpha(N)$, $\beta(N) \leq \pi(N)$ and $\eta(N) \leq \pi(N)$; if X is regular we even have $\beta(N) \leq \eta(N) = \pi(N) \leq \alpha(N) \leq \gamma(N)$. Moreover, as (2.31) and (2.32) measure the "noncompactness" of N, the functionals (2.33) and (2.34) show "to what extent" the set N fails to be absolutely bounded (see (2.13) and (2.26)) and W-bounded, respectively. Finally, (2.35) and (2.36) are certain measures of "weak noncompactness" and "sequential weak noncompactness", respectively.

If X and Y are two ideal spaces, ψ_X and ψ_Y are two Sadovskij functionals on X and Y, respectively, and the superposition operator F is bounded between X and Y, it is useful to consider the function

$$\xi(N;r) = \sup\{\psi_Y(F\widetilde{N}) : \widetilde{N} \subseteq N,\ \psi_X(\widetilde{N}) \leq r\}. \qquad (2.37)$$

In particular, the function

$$\xi_F(r) = \xi(B_r(X);r) = \sup_{N \subseteq B_r(X)} \psi_Y(FN) \qquad\qquad (2.38)$$

plays the same role for "noncompactness" properties of F as the growth function (2.24) for the boundedness of F.

The explicit calculation of the functions (2.37) and (2.38) is very hard, in general, although rather important in applications. The most important case when $\psi = \alpha$ is the Hausdorff measure of noncompactness (2.32) will be dealt with in Section 2.6.

2.5 Continuity conditions

At first glance, it seems very difficult to formulate general continuity conditions for the superposition operator in ideal spaces. However, such conditions exist. To formulate the main result of this section, another auxiliary notion is in order. Let δ be the function defined in (2.18). Given an open subset G of an ideal space X, let

$$\delta(G) = \bigcup_{(x_0, r)} \{(s, u) : s \in \Omega,\ u \in \mathbb{R},\ |u - x_0(s)| \leq r\delta(s)\}, \qquad (2.39)$$

where the union is taken over all pairs $(x_0, r) \in G \times (0, \infty)$ such that the ball $\|x - x_0\|_X < r$ lies in G. It is clear that $\delta(G)$ contains the graphs of all functions $x \in G$ (more precisely, the set $\{s : (s, x(s)) \notin \delta(G)\}$ has measure zero for each $x \in G$). Obviously, $\delta(G) = \Omega \times \mathbb{R}$ if $G = X$; on the other hand, this is true for any nonempty $G \subset X$ if X is quasi-regular.

It is clear from the definition that there is a close "interaction" between the behaviour of the superposition operator F on a set $G \subseteq X$ and that of the generating function f on the set $\delta(G) \subseteq \Omega \times \mathbb{R}$. For instance, if G is the interior of the domain of definition $\mathcal{D}(F)$ of F, the analytical properties of F do not depend on the properties of f outside $\delta(G)$.

We are now in a position to prove our basic result on the continuity of the superposition operator between ideal spaces.

Theorem 2.6   Let f be a sup-measurable function, and suppose that the interior G of the domain of definition $\mathcal{D}(F)$ of the superposition operator F, generated by f and considered as an operator between two ideal spaces X and Y, is non-empty. Then, if the operator F is continuous on G, the function f is sup-equivalent to some Carathéodory function on $\delta(G)$. Conversely, if the function f is sup-equivalent to some Carathéodory function on $\delta(G)$ and the space Y is regular, the operator F is continuous on G.

$\Rightarrow$ The first part follows from Theorem 1.4, since convergence in the norm of an ideal space implies convergence in measure, and the general hypothesis supp $X = \Omega$ guarantees that X is a thick set in the space S.

To prove the second part, we may assume, without loss of generality, that f is a Carathéodory function on $\delta(G)$ and that $B_r(X) \subseteq G$ for some $r > 0$; we show that F is continuous at $x_0 = \theta$.

If this is not so, we can find a sequence x_n such that $\|x_n\|_X \leq 2^{-n}r$ and $\|Fx_n\|_Y \geq \varepsilon$ for some $\varepsilon > 0$. The function $x_* = \sum_{n=1}^{\infty} |x_n|$ belongs then to $B_r(X)$; by the Krasnosel'skij–Ladyzhenskij lemma (see Theorem 6.2), there exists a function $z \in S$ such that $|z| \leq x_*$ (hence $z \in B_r(X)$) and

$$|f(s, z(s))| = \sup \{|f(s, u)| : |u| \leq x_*(s)\},$$

(hence $|Fz| \in Y$), because the set of all (s, u) with $|u| \leq x_*(s)$ is contained in $\delta(G)$ and f is a Carathéodory function on $\delta(G)$. Since x_n converges in X to θ, x_n converges also in measure to θ. But

$$\lim_{\lambda(D)\to 0} \sup_n \|P_D F x_n\|_Y \leq \lim_{\lambda(D)\to 0} \|P_D F z\|_Y = 0 \,,$$

since $Fz \in Y$ and Y is regular; consequently, Fx_n converges in Y to θ, by the remark preceding (2.2), and so we are done. $\Leftarrow$

We point out that the regularity assumption on Y in the second part of Theorem 2.6 is essential, as simple examples show (see Section 4.4). The case when Y is not regular is much more difficult to analyze; in this connection, specific properties of the spaces X and Y are needed. In view of this difficulty, the following general result is somewhat surprising:

Theorem 2.7   *Let X and Y be two ideal spaces, with X being quasi-regular and Y being almost perfect. Let f be a Carathéodory function, and let x_0 be an interior point of the domain of definition $\mathcal{D}(F)$ of the superposition operator F, generated by f and considered as an operator between X and Y. Then F is continuous at x_0 if and only if the superposition operator $\widetilde{F}$ generated by the function (1.11) maps the space X^0 into the space Y^0. Moreover, if F is continuous at x_0, F is continuous at every point of the set $x_0 + X^0$.*

$\Rightarrow$ First, suppose that F is continuous at x_0, and let $x \in X^0$; we show that $\widetilde{F}x \in Y^0$. In fact, if D_n is a sequence with $\lambda(D_n) \to 0$ as $n \to \infty$, we have, by (1.9), that $\|P_{D_n}\widetilde{F}x\| = \|\widetilde{F}P_{D_n}x\| = \|F(x_0 + P_{D_n}x) - Fx_0\|$; since F is continuous at x_0 and $x \in X^0$, we have $\|P_{D_n}\widetilde{F}x\| \to 0$ as $n \to \infty$, hence $\widetilde{F}x \in Y^0$, by the definition (2.1) of the regular part.

Suppose now that $\widetilde{F}(X^0) \subseteq Y^0$. By Theorem 2.6, $\widetilde{F}$ is then continuous at θ, considered between X^0 and Y^0. This means that $\|\widetilde{F}x\| \leq \varepsilon$ if $x \in X^0$ and $\|x\| \leq \delta$. But for any $x \in B_\delta(X)$ one can find a sequence x_n in $B_\delta(X^0)$ which converges to x in S, since X is quasi-regular. By Theorem 1.4, the sequence $\widetilde{F}x_n$ converges to $\widetilde{F}x$ in S, and hence $\|\widetilde{F}x\| \leq \varepsilon$, since Y is almost perfect. This shows that $\widetilde{F}$ is actually continuous at θ, considered between X and Y, and this is in turn equivalent to the continuity of F at x_0. The last part of the theorem follows obviously from the first part. $\Leftarrow$

Theorem 2.7 gives a very convenient continuity condition for F in terms of a simple acting condition for the "translated" operator $\widetilde{F}$. This has some remarkable consequences. We call an ideal space X *completely irregular* if $X^0 = \{\theta\}$, i.e. the regular part of X reduces to the zero function. For example, the space $M = L_\infty$ over some domain in Euclidean space (with the Lebesgue measure) is completely irregular.

The following remarkable result is an immediate consequence of Theorem 2.7:

Theorem 2.8 *Let X and Y be two ideal spaces, with X being quasi-regular and Y being completely irregular. Let f be a Carathéodory function, and let x_0 be an interior point of the domain of definition $\mathcal{D}(F)$ of the superposition operator F, generated by f and considered as an operator between X and Y. Then F is continuous at x_0 if and only if F is constant, i.e. the function f does not depend on u.*

In the first chapter (Theorem 1.7) we showed that every continuous superposition operator F is uniformly continuous on bounded sets in the space S. In ideal spaces, however, this is not true. Consider, for example, the space $X = Y = L$, where the underlying set Ω is either the interval $[0,1]$ (with the Lebesgue measure) or the set $\mathbb{N}$ (with the counting measure). If f is defined by $f(s,u) = u \sin u$ or $f(s,u) = u \sin \pi s u$, respectively, the corresponding operator F is continuous from X into Y, by Theorem 2.6. However, F is not uniformly continuous. To see this, consider in the case $\Omega = [0,1]$ the sequences $x_n = \frac{\pi}{2}(4n+1)\chi_{D_n}$, $y_n = \frac{\pi}{2}(4n-1)\chi_{D_n}$ $(n = 1, 2, \cdots)$, where D_n are subsets of $[0,1]$ with $\lambda(D_n) = \frac{1}{2\pi n}$, and in the case $\Omega = \mathbb{N}$ the sequences $x_n = \frac{2n+1}{2n}\chi_{\{n\}}$, $y_n = \frac{2n-1}{2n}\chi_{\{n\}}$ $(n = 1, 2, \cdots)$. Obviously, these sequences are bounded and satisfy the condition $\lim\limits_{n\to\infty} \|x_n - y_n\| = 0$; on the other hand, the sequence $\|Fx_n - Fy_n\|$ is bounded away from zero, and hence F cannot be uniformly continuous on bounded sets.

The following result is very elementary, but useful for the verification of uniform continuity of concrete operators; we shall not carry out the proof, since it follows immediately from Theorem 1.7.

Theorem 2.9 *Let X and Y be two ideal spaces, and let f be a Carathéodory function. Then the superposition operator F, generated by f and considered as an operator between X and Y, is uniformly continuous on a bounded set $N \subset X$ if and only if*

$$\varlimsup_{\delta \to 0} \sup \{\|P_D Fx - P_D Fy\| : x, y \in N, \ \|x - y\| \leq \delta; \ \lambda(D) \leq \delta\} = 0.$$

This condition implies, in particular, that every superposition operator F which is continuous and absolutely bounded on some set $N \subset X$, is also uniformly continuous on N.

Now we shall discuss conditions for the weak continuity of the superposition operator in ideal spaces. Let X be an ideal space, $\widetilde{X}$ its associate space (see (2.6)), and Γ a *total subspace* of $\widetilde{X}$ (i.e. supp $\Gamma = $ supp $\widetilde{X}$). Recall that a sequence x_n in X is called Γ-*weakly convergent* to $x \in X$ if

$$\lim_{n\to\infty} <x_n, z> = <x, z> \quad (z \in \Gamma).$$

We point out that Γ-weak convergence does not imply convergence in measure, and convergence in measure does not imply Γ-weak convergence. Nevertheless, both types of convergence are compatible in the sense that, if x_n converges Γ-weakly to x and in measure to $\hat{x}$, then $x = \hat{x}$.

In what follows, we shall make use of a special Γ-weakly convergent sequence of measurable functions. Let D be a subset of Ω_c of positive (finite) measure, and let $\{D(\varepsilon_1, \cdots, \varepsilon_n) : \varepsilon_i \in \{0, 1\}\}$ be the sequence of partitions of D constructed in the proof of Theorem 1.8. Further, let D_n^0 and D_n^1 be defined as in (1.32), and let $\theta_n(s) = \chi_{D_n^1}(s) - \chi_{D_n^0}(s)$. Obviously, the functions θ_n satisfy the orthogonality relation $< \theta_j, \theta_k > = \delta_{jk}\mu(D)$. Consequently, we may consider Fourier series expansions with respect to the system $\{\theta_n : n = 1, 2, \cdots\}$: given $x \in L_2(D)$ (which means that x is square-integrable and vanishes outside D), we have

$$x(s) = z(s) + \frac{1}{\mu(D)} \sum_{n=1}^{\infty} \left\{ \int_D \theta_n(s)x(s)ds \right\} \theta_n(s),$$

where z is orthogonal to each θ_n. Since $L_2(D)$ is dense in $L_1(D)$, and every function $x \in L_1$ can be written as sum $x = P_D x + P_{\Omega \setminus D} x$, we get that $\lim_{n \to \infty} < \theta_n, x > = 0$ for all $x \in L_1$.

The following result will be useful in the sequel:

Lemma 2.7 Let X be an ideal space, and suppose that x_1 and x_2 are two functions in X such that the set $D = \{s : x_1(s) \neq x_2(s)\}$ is contained in Ω_c and has finite measure. Then the sequence $z_n(s) = \frac{1+\theta_n(s)}{2}x_1(s) + \frac{1-\theta_n(s)}{2}x_2(s)$ converges $\widetilde{X}$-weakly to the function $z(s) = \frac{x_1(s)+x_2(s)}{2}$.

$\Rightarrow$ The assertion is almost obvious: given $y \in \widetilde{X}$, we have $x_1 y \in L_1$, $x_2 y \in L_1$, and

$$\lim_{n \to \infty} < z_n, y > = < \frac{x_1 + x_2}{2}, y > + \lim_{n \to \infty} < \theta_n, \frac{x_1 - x_2}{2}y > = < z, y > . \Leftarrow$$

Let X and Y be two ideal spaces. The set Y/X consists, by definition, of all $z \in S$ such that $xz \in Y$ for each $x \in X$; equipped with the natural norm

$$\|z\|_{Y/X} = \sup\{\|xz\|_Y : \|x\|_X \leq 1\}, \tag{2.40}$$

Y/X becomes an ideal space, the *multiplicator space* of X with respect to Y. Observe that the adjoint space $\widetilde{X}$ coincides with the multiplicator space L_1/X, as a comparison of (2.8) and (2.40) shows. Many properties of the "numerator space" Y carry over to the multiplicator space Y/X; for example, if Y is (almost) perfect then Y/X is also (almost) perfect.

To give a simple example, consider the spaces $L(u_0)$ and $M(u_0)$ defined in (2.4) and (2.5). If u_1 and u_2 are units over Ω, we have $L(u_1)/L(u_2) = M(u_3)$, $L(u_1)/M(u_2) = L(u_3)$, $M(u_1)/M(u_2) = M(u_3)$ and $M(u_1)/L(u_2) = L(\tilde{u}_3)$, where $u_3(s) = u_2^{-1}(s)u_1(s)$ and $\tilde{u}_3(s) = u_2^{-1}(s)u_1(s)\chi_{\Omega_d}(s)$. Other multiplicator spaces Y/X will be calculated and used in the following two chapters.

We are now ready to formulate a condition (both necessary and sufficient) for the (Γ, Δ)-weak continuity of F between two ideal spaces X and Y; this means that Γ and Δ are total subspaces of $\widetilde{X}$ and $\widetilde{Y}$, respectively, and F maps Γ-weakly convergent sequences into Δ-weakly convergent sequences.

Theorem 2.10 *Let X and Y be two ideal spaces and let Γ and Δ be total subspaces of $\widetilde{X}$ and $\widetilde{Y}$, respectively. Let f be a sup-measurable function, and denote by $\mathcal{D}(F)$ the domain of definition of the superposition operator F, generated by f and considered as an operator between X and Y. Then F is (Γ, Δ)-weakly continuous if and only if the following two conditions hold:*
(a) The restriction of f to $\Omega_c \times \mathbb{R}$ satisfies

$$f(s, u) \simeq a(s) + b(s)u \quad ((s, u) \in \delta(\mathcal{D}(F)) \cap (\Omega_c \times \mathbb{R})), \qquad (2.41)$$

where $a \in Y$ and $b \in Y/X \cap \Gamma/\Delta$.
(b) The restriction of f to $\Omega_d \times \mathbb{R}$ is a Carathéodory function.

$\Rightarrow$ Without loss of generality, we can treat the two cases $\Omega_d = \emptyset$ and $\Omega_c = \emptyset$ separately.

Let first $\Omega_d = \emptyset$. If f has the form (2.41) and x_n converges Γ-weakly to x, then for any $y \in \Delta$ we have $\lim_{n\to\infty} <Fx_n - Fx, y> = \lim_{n\to\infty} <b(x_n - x), y> = \lim_{n\to\infty} <x_n - x, by> = 0$, hence Fx_n converges Δ-weakly to Fx.

The proof of the necessity of (a) is less trivial. First, we claim that the function f is equivalent to an affine function on $\delta(\mathcal{D}(F))$. If this is not so, we can find two functions $x_1, x_2 \in \mathcal{D}(F)$ such that

$$f(s, \tfrac{1}{2}[x_1(s) + x_2(s)]) - \tfrac{1}{2}[f(s, x_1(s)) + f(s, x_2(s))] \neq 0 \quad (s \in D) \quad (2.42)$$

on a set $D \subseteq \Omega_c$ of positive measure. By Lemma 2.7, the sequence $z_n(s) = \frac{1+\theta_n(s)}{2}x_1(s) + \frac{1-\theta_n(s)}{2}x_2(s)$ converges Γ-weakly to the function $z = \frac{1}{2}[x_1 + x_2]$. By hypothesis, F is (Γ, Δ)-weakly continuous, and thus the sequence

Fz_n converges Δ-weakly to the function $Fz = F(\frac{1}{2}[x_1 + x_2])$. On the other hand, again by Lemma 2.7, Fz_n converges Δ-weakly to the function $\frac{1}{2}[Fx_1+Fx_2]$, contradicting (2.42). Consequently, the function f is affine on $\delta(\mathcal{D}(F))$. Moreover, the acting condition $F(\mathcal{D}(F)) \subseteq Y$ and the (Γ, Δ)-weak continuity of F imply that $a \in Y$ and $b \in \Gamma/\Delta \cap Y/X$.

It remains to show that f is a Carathéodory function on $\Omega_d \times \mathbb{R}$. But this is obvious, since Γ-weak convergence implies pointwise convergence. $\Leftarrow$

2.6 Lipschitz and Darbo conditions

In many applications, more than just continuity of nonlinear operators is required in order to make use of the basic principles of nonlinear analysis. For instance, a crucial hypothesis in Banach's contraction mapping principle is the *global Lipschitz condition*

$$\|Fx_1 - Fx_2\|_Y \le k\|x_1 - x_2\|_X \quad (x_1, x_2 \in X), \tag{2.43}$$

or the *local Lipschitz condition*

$$\|Fx_1 - Fx_2\|_Y \le k(r)\|x_1 - x_2\|_X \quad (x_1, x_2 \in B_r(X)). \tag{2.44}$$

Therefore the problem arises of finding conditions (possibly both necessary and sufficient) for (2.43) or (2.44) in terms of the generating function f; this is the purpose of this section. Without loss of generality, we assume that f satisfies a Carathéodory condition.

Theorem 2.11 *Suppose that the superposition operator F generated by f acts between two ideal spaces X and Y, where Y is perfect. Then the following three conditions are equivalent:*
(a) The operator F satisfies the local Lipschitz condition (2.44).
(b) Given any two functions $x_1, x_2 \in B_r(X)$, one can find a function $\xi \in B_{k(r)}(Y/X)$ such that

$$Fx_1(s) - Fx_2(s) = \xi(s)[x_1(s) - x_2(s)]. \tag{2.45}$$

(c) The function f satisfies a local Lipschitz condition

$$|f(s,u) - f(s,v)| \le g(s,w)|u - v| \quad (|u|,|v| \le w), \tag{2.46}$$

where the function g generates a superposition operator G which maps the ball $B_r(X)$ into the ball $B_{k(r)}(Y/X)$.

$\Rightarrow$ Let us show that (a) implies (b). To this end, we suppose first that the derivative $\widehat{f}(s, u) = \frac{\partial}{\partial u} f(s, u)$ of f with respect to the second argument exists. In this case

$$f(s, u) - f(s, v) = (u - v) \int_0^1 \widehat{f}(s, (1 - t)u + tv) dt;$$

to prove (b) it is therefore sufficient to show that the function $\xi(s) = \int_0^1 \widehat{f}[s, (1 - t)x_1(s) + tx_2(s)] dt$ belongs to the ball $B_{k(r)}(Y/X)$. Let $z \in B_r(X)$ and $h \in X$ such that $z + th$ belongs to $B_r(X)$ for small t. Then the equality $\widehat{f}(s, z(s))h(s) = \text{S-}\lim_{t \to 0} \frac{1}{t}[f(s, z(s) + th(s)) - f(s, z(s))]$ holds, where S-lim denotes the limit taken in measure (i.e. with respect to the metric (1.1)). By assumption (a), we have $\frac{1}{t}\|F(z + th) - Fz\|_Y \le k(r)\|h\|_X$; if we denote by $\widehat{F}$ the superposition operator generated by $\widehat{f}$, we get, since Y is perfect, that

$$\|(\widehat{F}z)h\|_Y \le \lim_{t \to 0} \frac{1}{t}\|F(z + th) - Fz\|_Y \le k(r)\|h\|_X.$$

This means, by definition of the norm (2.40), that $\widehat{F}z$ belongs to the ball $B_{k(r)}(Y/X)$. From the definition of ξ and the perfectness of Y/X it follows that

$$\|\xi\|_{Y/X} \le \int_0^1 \|\widehat{F}((1 - t)x_1 + tx_2)\|_{Y/X} dt \le k(r),$$

as claimed.

To prove (b) in the general case, we associate to f a sequence of functions

$$f_n(s, u) = \int_0^1 f\left[s, u + \frac{t}{n} u_0(s)\right] dt, \tag{2.47}$$

where u_0 is an arbitrary unit in X. The relation

$$f_n(s, u) = n u_0(s)^{-1} \int_u^{u + u_0(s)/n} f(s, v) dv$$

shows that f_n is differentiable with respect to the second argument, and its derivative $\widehat{f}_n(s, u) = \frac{\partial}{\partial u} f_n(s, u)$ satisfies $\widehat{f}_n(s, u) = n u_0(s)^{-1}[f(s, u + u_0(s)/n) - f(s, u)]$. Moreover, the superposition operators F_n generated by

f_n also satisfy the Lipschitz condition (a) since

$$\|F_n x_1 - F_n x_2\|_Y \le \int_0^1 \|F(x_1 + \frac{t}{n} u_0) - F(x_2 + \frac{t}{n} u_0)\|_Y \, dt$$

$$\le k(r) \int_0^1 \|x_1 - x_2\|_X \, dt = k(r)\|x_1 - x_2\|_X \,.$$

In virtue of the first part we know that, for any $x_1, x_2 \in B_r(X)$, the functions $\xi_n(s) = [x_1(x) - x_2(s)]^{-1}[F_n x_1(s) - F_n x_2(s)]$ belong to $B_{k(r)}(Y/X)$. But (2.47) shows that $\xi(s) = [x_1(s) - x_2(s)]^{-1}[Fx_1(s) - Fx_2(s)] = \underset{n\to\infty}{\text{S-lim}}\,\xi_n(s)$; since Y/X is perfect, the function ξ belongs to the ball $B_{k(r)}(Y/X)$, and thus (b) holds also in the general case.

Now we show that (b) implies (c). Let

$$g(s, w) = \sup_{\substack{|u|,|v| \le w \\ u \neq v}} (u - v)^{-1}(f(s, u) - f(s, v));$$

it is not difficult to see that the function g generates a superposition operator G from X into S. By Sainte-Beuve's selection theorem, for each $w \in B_r(X)$ we can find functions $u, v \in S$ such that $|u| \le w$, $|v| \le w$, and $g(s, w(s)) = [u(s) - v(s)]^{-1}[Fu(s) - Fv(s)]$. But this means that u and v belong to the ball $B_r(X)$, and hence Gw belongs to the ball $B_{k(r)}(Y/X)$.

The fact that (c) implies (a) is obvious (even without the perfectness assumption on Y). $\Leftarrow$

Observe that there seems to be a slight technical flaw in the proof of Theorem 2.11: in the definition of the operators F_n, we considered the functions $x + \frac{t}{n} u_0$ which do not belong any more to the ball $B_r(X)$ if $\|x\| = r$. Nevertheless, we have $x + \frac{t}{n} u_0 \in B_r(X)$ whenever we choose x in the interior of $B_r(X)$, and this can always be done without loss of generality, since the Lipschitz condition (2.44) obviously does not change when passing from an open set to its closure.

Theorem 2.11 has some important consequences: for example, the "size" of the class of Lipschitz-continuous superposition operators F between two ideal spaces X and Y heavily depends on the relation of X and Y, rather than on the structure of the generating function f. To state this more precisely, we introduce an auxiliary notion. Let us say that two ideal spaces (X, Y) form a V-pair (a V_0-pair, respectively) if $\Omega_d = \emptyset$, X is quasi-regular, and there exist two positive functions $u_0 \in X^0$ and $v_0 \in \widetilde{Y}$ such that $\|P_D u_0\|_X \|P_D v_0\|_{\widetilde{Y}} = O(\mu(D))$ ($\|P_D u_0\|_X \|P_D v_0\|_{\widetilde{Y}} = o(\mu(D))$, respectively) as $\mu(D) \to 0$. The following lemma exhibits a close relation between the fact that (X, Y) is a V-pair (or V_0-pair) and the "degeneracy" of the multiplicator space Y/X:

Lemma 2.8   If (X, Y) is a V-pair, then $(Y/X)^0 = \{\theta\}$. If (X, Y) is a V_0-pair, then $Y/X = \{\theta\}$.

$\Rightarrow$ If (X, Y) is a V-pair and $z \in (Y/X)^0$ is a non-zero function, we can find a $\gamma > 0$ such that $z(s)u_0(s)v_0(s) \geq \gamma$ on a subset $D_0 \subseteq \Omega$ of positive measure. Then, for any $D \subseteq D_0$, we have $\gamma\mu(D) \leq\ < P_D z P_D u_0, P_D v_0 > \leq \|P_D z P_D u_0\|_Y \|P_D v_0\|_{\widetilde{Y}} \leq \|P_D z\|_{Y/X} \|P_D u_0\|_X \|P_D v_0\|_{\widetilde{Y}} = o(\mu(D))$ as $\mu(D) \to 0$, a contradiction.

Similarly, if (X, Y) is a V_0-pair and $z \in Y/X$ is a non-zero function, we have with γ, D_0 and D as above, that $\gamma\mu(D) \leq\ < z P_D u_0, P_D v_0 > \leq \|z P_D u_0\|_Y \|P_D v_0\|_{\widetilde{Y}} \leq \|z\|_{Y/X} \|P_D u_0\|_X \|P_D v_0\|_{\widetilde{Y}} = o(\mu(D))$ as $\mu(D) \to 0$, the same contradiction. $\Leftarrow$

Combining Lemma 2.8 and Theorem 2.11, we get the following:

Theorem 2.12   Let X and Y be two ideal spaces, with Y being perfect, and suppose that (X, Y) is a V_0-pair. Let f be a sup-measurable function, and let F be the superposition operator generated by f between X and Y. Then F satisfies the Lipschitz condition (2.43) if and only if

$$f(s, u) \simeq a(s) \qquad (2.48)$$

for some $a \in Y$, i.e. F is constant.

Theorems 2.11 and 2.12 give a complete characterization of Lipschitz continuous superposition operators between ideal spaces. Apart from the contraction mapping principle, there are other fixed-point theorems which require special properties of the nonlinear operators involved. A classical example is Schauder's fixed-point principle. Unfortunately, Theorem 2.5 shows that Schauder's theorem does not apply, in general, to nonlinear problems involving superposition operators in ideal spaces. A more general fixed-point principle which has found many applications, is the Darbo–Sadovskij principle, which essentially uses the *global Darbo condition*

$$\alpha(FN) \leq k\alpha(N) \quad (N \subset X \text{ bounded}), \qquad (2.49)$$

or the *local Darbo condition*

$$\alpha(FN) \leq k(r)\alpha(N) \quad (N \subseteq B_r(X)), \qquad (2.50)$$

where α is the Hausdorff measure of noncompactness (2.32). It is clear that the Lipschitz condition (2.44) is sufficient for (2.50) if the operator F

is defined on the whole space; in fact, if $\{z_1, \cdots, z_m\}$ is a finite ε-net for $N \subseteq B_r(X)$, then $\{Fz_1, \cdots, Fz_m\}$ is a finite $k(r)\varepsilon$-net for $FN \subseteq FB_r(X)$. The question arises whether (2.44) is also necessary for (2.50). It is evident that, in general, (2.50) does not imply (2.44) if the measure μ has atoms in Ω; therefore we shall assume in the sequel that μ is atomic-free on Ω, i.e. $\Omega_d = \emptyset$.

Surprisingly enough, it turns out that (2.44) and (2.50) are in fact equivalent in many ideal spaces. The proof of this relies on the construction of special sets which provide a simple connection between the norm and the measure of noncompactness. Given an ideal space X and two fixed functions $x_1, x_2 \in X$, consider the *random interval*

$$R[x_1, x_2] = \{P_D x_1 + P_{\Omega \setminus D} x_2 : D \in \mathcal{M}\},\qquad (2.51)$$

where P_D is, as usual, the multiplication operator (1.3). Obviously, $\alpha(R[x_1, x_2]) \leq \frac{1}{2}\|x_1 - x_2\|_X$, since the singleton $\{\frac{1}{2}(x_1 + x_2)\}$ is a finite $\frac{1}{2}\|x_1 - x_2\|$-net for $R[x_1, x_2]$ in X. We shall call an ideal space X α-*nondegenerate* if for any $x_1, x_2 \in X$ the equality $\alpha(R[x_1, x_2]) = \frac{1}{2}\|x_1 - x_2\|_X$ holds.

Theorem 2.13   *Suppose that the superposition operator F generated by f acts between two ideal spaces X and Y, where Y is α-nondegenerate. Then the following two conditions are equivalent:*
(a) The operator F satisfies the local Lipschitz condition (2.44).
(b) The operator F satisfies the local Darbo condition (2.50).

$\Rightarrow$ The proof is almost obvious: if F does not satisfy (2.44), then $\|Fx_1 - Fx_2\|_Y > k(r)\|x_1 - x_2\|_X$ for some $x_1, x_2 \in B_r(X)$. Let $N = R[x_1, x_2]$. By the disjoint additivity of the superposition operator (see Lemma 1.1) we have $FN = FR[x_1, x_2] = R[Fx_1, Fx_2]$. Since Y is α-nondegenerate, we get in addition that $\alpha(R[Fx_1, Fx_2]) = \frac{1}{2}\|Fx_1 - Fx_2\|_Y > \frac{1}{2}k(r)\|x_1 - x_2\|_X \geq k(r)\alpha(R[x_1, x_2]) = k(r)\alpha(N)$ contradicting (2.50). $\Leftarrow$

We remark that one can prove a slight generalization of Theorem 2.13, which states that $\alpha(FN) \leq \phi(\alpha(N))$ for some convex function ϕ if and only if $\|Fx_1 - Fx_2\|_Y \leq 2\phi(\frac{1}{2}\|x_1 - x_2\|_X)$.

Obviously, the problem now arises to give a precise characterisation of all ideal spaces which are α-nondegenerate. A complete answer to this problem seems impossible. Nevertheless, we can describe a large class of α-nondegenerate spaces which contains most ideal spaces arising in applications.

We call an almost perfect ideal space X *average-stable* if the *averaging operator*

$$P_\omega x(s) = \sum_{n=1}^\infty \left\{ \frac{1}{\mu(D_n)} \int_{D_n} x(s) ds \right\} \chi_{D_n}(s) \qquad (2.52)$$

associated to some finite or countable partition $\omega = \{D_1, D_2, \cdots\}$ of Ω, acts from X into the second associate space $\overset{\approx}{X}$ and has norm 1. Several examples of average-stable spaces will be given in the following chapters (see e.g. Section 5.1).

Lemma 2.9 *Every average-stable ideal space whose underlying measure is atomic-free is α-nondegenerate.*

$\Rightarrow$ Obviously, it suffices to consider the case $x_2 = -x_1 = u_0$, since one can always pass from $R[x_1, x_2]$ to $R[-u_0, u_0]$ by putting $u_0 = \frac{1}{2}(\max\{x_1, x_2\} - \min\{x_1, x_2\})$. Thus, let $\{z_1, \cdots, z_m\}$ be some finite ε-net for $R[-u_0, u_0]$ in X; we have to show that $\varepsilon \geq \|u_0\|$. Without loss of generality, we may suppose that all functions $u_0, z_1, \cdots, z_m$ are of the form

$$u_0 = \sum_{n=1}^\infty c_n \chi_{D_n}, \quad z_k = \sum_{n=1}^\infty d_{k,n} \chi_{D_n} \quad (k = 1, \cdots, m), \qquad (2.53)$$

where $\omega = \{D_1, D_2, \cdots\}$ is some partition of Ω, since the set of all such functions is dense in X. By hypothesis, the measure μ is atomic-free, hence we can divide each D_n into two subsets A_n and B_n of equal measure. Then $x_* = \sum_{n=1}^\infty c_n(\chi_{A_n} - \chi_{B_n})$ belongs to $R[-u_0, u_0]$, hence $\|z_k - x_*\| \leq \varepsilon$ for some k. On the one hand,

$$|z_k - x_*| = \sum_{n=1}^\infty \left[|d_{k,n} - c_n| \chi_{A_n} + |d_{k,n} + c_n| \chi_{B_n} \right],$$

and, on the other,

$$P_\omega(|z_k - x_*|) = \frac{1}{2} \sum_{n=1}^\infty \left[|d_{k,n} - c_n| + |d_{k,n} + c_n| \right] \chi_{D_n},$$

where P_ω is the operator (2.52). But both functions $|z_k - x_*|$ and $P_\omega(|z_k - x_*|)$ belong to X (since $u_0 \in X$ and $z_k \in X$), and hence

$$\varepsilon \geq \|z_k - x_*\| \geq \|P_\omega(z_k - x_*)\|$$

$$= \frac{1}{2} \left\| \sum_{n=1}^\infty \left[|d_{k,n} - c_n| + |d_{k,n} + c_n| \right] \chi_{D_n} \right\| \geq \left\| \sum_{n=1}^\infty c_n \chi_{D_n} \right\| = \|u_0\|$$

as claimed. $\Leftarrow$

2.7 Differentiability conditions

Recall that on operator F between two normed spaces X and Y is called
(Fréchet-) differentiable at an interior point x of $\mathcal{D}(F)$ if the increment
$F(x+h) - Fx$ can be represented in the form

$$F(x+h) - Fx = Ah + \omega(h)$$

where A is a continuous linear operator from X into Y, and ω satisfies

$$\lim_{\|h\| \to 0} \frac{\|\omega(h)\|}{\|h\|} = 0 .$$

The linear operator A is called the *(Fréchet) derivative* $F'(x)$ of F at x, and
the value $F'(x)h$ of $F'(x)$ at h the *(Fréchet) differential* of F at x along h;
the formula

$$F'(x)h = \lim_{t \to 0} \frac{1}{t}[F(x+th) - Fx] \quad (h \in X) \tag{2.54}$$

holds.

Now, let X and Y be two ideal spaces, and F the superposition operator
generated by some sup-measurable function f. Then the local determination
of F (see Lemma 1.1) and (2.54) imply the local determination of $F'(x)$;
this simple observation allows us to give the explicit form of the derivative
$F'(x)$ of F (if it exists!).

Let u be some positive function in X, and let $a_u(s) = u(s)^{-1}Au(s)$. Since
the operator $A = F'(x)$ is locally determined, we have $Ah(s) = a_u(s)h(s)$
for any function h of the form $h = zu$ with $z \in S_0$ (the space of simple
functions over Ω, see (1.4)). Since A is continuous from X into Y, and the
multiplication operator is continuous in the space S, the equality $Ah = a_u h$
holds also on the closure X_u of the set of all $h = zu$, with $z \in S_0$.

Now let u_1 be another positive function in X with $u_1 \leq u$. Then $u_1 \in X_u$
and thus $Au_1(s) = a_u(s)u_1(s)$, hence $a_{u_1}(s) = a_u(s)$. Further, if u_1 and
u_2 are two arbitrary positive functions in X, for $u = u_1 + u_2$ we have
$a_{u_1}(s) = a_u(s) = a_{u_2}(s)$. In other words, the function a_u does not depend
on u. Since the whole space X can be represented as union over all sets X_u,
the derivative $A = F'(x)$ (if it exists, of course) has necessarily the form

$$F'(x)h(s) = a(s)h(s), \tag{2.55}$$

i.e. $F'(x)$ is always a *multiplication operator* by some measurable function. More precisely, (2.54) shows also that

$$a(s) = \text{S-}\lim_{u \to 0} \frac{1}{u}[f(s, x(s) + u) - f(s, x(s))]. \qquad (2.56)$$

By Theorem 1.3, the S-limit in (2.56) can be replaced by the usual limit for almost all $s \in \Omega$, if f is a Shragin function (a fortiori, if f is a Carathéodory function).

Consider the function

$$g(s, u) = \begin{cases} \frac{1}{u}[f(s, x(s) - u) + f(s, x(s))] & \text{if } u \neq 0, \\ a(s) & \text{if } u = 0; \end{cases} \qquad (2.57)$$

this is obviously a sup-measurable function (respectively, a Shragin function) if f is a sup-measurable function (respectively, a Shragin function). The corresponding superposition operator $Gh(s) = g(s, h(s))$ allows us to formulate a sufficient differentiability condition for F. In fact, since $F(x + h) - Fx - (G\theta)h = (Gh - G\theta)h$, the operator F is differentiable at x if the operator G is continuous between X and Y/X at θ. For further reference, we summarize with the following:

Theorem 2.14 *Let X and Y be two ideal spaces, and let f be a sup-measurable function. Suppose that the superposition operator F generated by f is differentiable at $x \in X$, considered as an operator between X and Y. Then the derivative $F'(x)$ has the form (2.55), where the function a given by (2.56) belongs to the multiplicator space Y/X. Moreover, the function g given by (2.57) is sup-measurable (and even a Shragin function if f is so).*

Conversely, if $x \in X$, $Fx \in Y$, and the superposition operator G generated by the function g acts between X and Y/X and is continuous at θ, then F is differentiable at x, and formula (2.55) holds.

Observe that the first part of Theorem 2.14 gives only a necessary differentiability condition, while the second part gives only a sufficient condition (an example which shows that this latter condition is not necessary will be given in Section 3.7). Conditions which are both necessary and sufficient can be obtained only for special ideal spaces (see e.g. Theorem 3.13 and Theorem 4.12). However, the general statement of Theorem 2.14 suffices for most applications.

Now we give a "degeneracy" result for superposition operators which plays the same role for differentiability as Theorem 2.12 plays for Lipschitz continuity.

Theorem 2.15   *Let X and Y be two ideal spaces, and suppose that (X, Y) is a V-pair. Let f be a sup-measurable function, and assume that the superposition operator F generated by f is differentiable between X and Y at some point $x \in X$. Then the function f satisfies*

$$f(s, u) \simeq a(s) + b(s)u \tag{2.58}$$

(i.e. is sup-equivalent to an affine function in u), where $a \in Y$ and $b \in Y/X$. Moreover, if (X, Y) is a V_0-pair, then f even satisfies (2.48), i.e. F is constant. Finally, if f is a Carathéodory function, (2.58) may be replaced by

$$f(s, u) = a(s) + b(s)u \,. \tag{2.59}$$

$\Rightarrow$ We may assume that f is a Carathéodory function and that F is defined on some ball $B_\delta(X)$ for small $\delta > 0$. Moreover, by (2.55) and (2.56), we may suppose without loss of generality that $x = \theta$, $F\theta = \theta$, and $F'(\theta) = \theta$, hence

$$\lim_{\|h\| \to 0} \frac{\|Fh\|}{\|h\|} = 0 \,. \tag{2.60}$$

To prove the assertion we must show that F is identically zero. Choose $u_0 \in X^0$ and $v_0 \in \tilde{Y}$ as in the definition of a V-pair. It suffices to show that F vanishes on the set N of all functions $h \in X$ for which $|h| \leq \lambda u_0$ for some $\lambda > 0$, since any $h \in X$ differs from such a function only on a set of arbitrarily small λ-measure. Given $h \in N$, we have $\|P_D h\|_X \leq \delta$ for small $\lambda(D)$ (recall that X is quasi-regular), hence $F P_D h$ is defined. Moreover, (2.60) implies that

$$
\begin{aligned}
< FP_D h, P_D v_0 > &\leq \|FP_D h\|_Y \|P_D v_0\|_{\tilde{Y}} \\
&= o(\|P_D h\|_X) \|P_D v_0\|_{\tilde{Y}} = o(\|P_D u_0\|_X) \|P_D v_0\|_{\tilde{Y}} \\
&= o(\mu(D)) \quad (\mu(D) \to 0) \,.
\end{aligned}
$$

Letting $\mu(D)$ tend to zero, we get $Fh = \theta$ as claimed. The last statement follows trivially from Lemma 2.8. $\Leftarrow$

    Theorem 2.15 enables us to study the differentiability of the superposition operator F at single points. Now we shall be concerned with the case when F is continuously differentiable on some open subset O of X.

Theorem 2.16   *Let X and Y be two ideal spaces, and suppose that both the functions f and the function $\hat{f}(s, u) = \frac{\partial}{\partial u} f(s, u)$ are Carathéodory functions. Assume that the superposition operator F generated by f maps* some open set $O \subset X$ into Y. Then F, considered as an operator between X and Y, is continuously differentiable on O if and only if the superposition operator $\widehat{F}$ generated by $\widehat{f}$ and considered as an operator between X and Y/X, is continuous on O. In this case, the equality

$$F'(x)h = \widehat{F}(x)h \quad (x \in O, \ h \in X) \tag{2.61}$$

holds.

$\Rightarrow$ Suppose first that $\widehat{F}$ is continuous on O. For any $h \in X$ we can find a measurable function ϑ such that $0 \le \vartheta(s) \le 1$ almost everywhere on Ω and $f[s, x(s) + h(s)] - f(s, x(s)) - \widehat{f}(s, x(s))h(s) = [\widehat{f}(s, x(s) + \vartheta(s)h(s)) - \widehat{f}(s, x(s))]h(s)$.
This implies that $\|F(x + h) - Fx - \widehat{F}(x)]h\| \le \|\widehat{F}(x + \vartheta h) - \widehat{F}x\| \|h\|$, and hence, by the continuity of $\widehat{F}$ at x, $\|F(x + h) - Fx - \widehat{F}(x)h\| = o(\|h\|)$ ($\|h\| \to 0$). This shows that F is differentiable at x with (2.61).

Conversely, if F is differentiable at each point $x \in O$, its derivative $F'(x)$ has the form $F'(x)h(s) = a(s)h(s)$, where, by Theorem 2.14, the function a is given by (2.56).
The existence of the partial derivative $\widehat{f}$ implies that $a(s) = \widehat{f}(s, x(s))$. Thus, from Theorem 2.14 it follows that the corresponding superposition operator $\widehat{F}$ maps O into the multiplicator space Y/X. The continuity of $\widehat{F}$ follows from the fact that any multiplication operator by some function a, considered as operator from X into Y, has operator norm $\|a\|_{Y/X}$. $\Leftarrow$

We conclude with another result which also follows quite easily from Theorem 2.14. This result enables us to establish the differentiability of the superposition operator on a dense subset of X.

Theorem 2.17 *Let X and Y be two ideal spaces, and suppose that both f and the partial derivative $\widehat{f}(s, u) = \frac{\partial}{\partial u} f(s, u)$ are Carathéodory functions. Assume that the operator F generated by f maps some set $N \subseteq X$ into Y. Moreover, suppose that the superposition operator $G(x, h)(s) = g(s, x(s), h(s))$ defined by the function*

$$g(s, u, v) = \begin{cases} \frac{1}{v}[f(s, u + v) - f(s, u)] & \text{if } v \ne 0, \\ \widehat{f}(s, u) & \text{if } v = 0, \end{cases} \tag{2.62}$$

has the property that, for $x \in N$, the operator $G(x, \cdot)$ is continuous between X and Y/X at θ. Then the operator F is defined on some neighbourhood O of N, is differentiable at every point $x \in N$, and satisfies (2.61) for $x \in N$ and $h \in X$.

A notion which is related to differentiability of an operator is that of asymptotic linearity. Recall that an operator F between two normed spaces X and Y is called *asymptotically linear* (or *differentiable at infinity*) if F can be represented, for sufficiently large x, in the form

$$Fx = Ax + \omega(x),$$

where A is a continuous linear operator from X into Y, and ω satisfies

$$\lim_{\|x\| \to \infty} \frac{\|\omega(x)\|}{\|x\|} = 0.$$

The linear operator A is called the *asymptotic derivative* $F'(\infty)$ of F; the formula

$$F'(\infty)h = \lim_{t \to \infty} \frac{1}{t} F(th) \quad (h \in X) \tag{2.63}$$

holds.

If X and Y are two ideal spaces and F is the superposition operator generated by some sup-measurable function f, the asymptotic linearity of F can be analyzed by an analogous reasoning as above for the differentiability at "finite points". In particular, an analogue of Theorem 2.14 is the following:

Theorem 2.18   *Let X and Y be two ideal spaces, and let f be a sup-measurable function. Suppose that the superposition operator F generated by f is asymptotically linear, considered as an operator between X and Y. Then the asymptotic derivative $F'(\infty)$ has the form*

$$F'(\infty)h(s) = a_\infty(s)h(s), \tag{2.64}$$

where the function

$$a_\infty(s) = \operatorname*{S-lim}_{u \to \infty} \frac{1}{u} f(s, u) \tag{2.65}$$

belongs to the multiplicator space Y/X. Conversely, if the function (2.65) belongs to Y/X, and if there exists a function $b \in Y/X$ with the property that for each $\varepsilon > 0$ one can find a function $a_\varepsilon \in Y$ such that

$$|f(s, u) - a_\infty(s)u| \leq a_\varepsilon(s) + \varepsilon b(s)|u|, \tag{2.66}$$

then F is asymptotically linear between X and Y, and its asymptotic derivative is given by (2.64).

$\Rightarrow$ The proof of the first part is almost literally the same as that of Theorem 2.14, and therefore we do not present it. For proving the second part, we remark that (2.66) implies that $\|Fx - Ax\|_Y \leq \|a_\varepsilon\|_Y + \varepsilon\|b\|_{Y/X}\|x\|_X$, where $Ah(s) = a_\infty(s)h(s)$, hence $\varlimsup\limits_{\|x\| \to \infty} \|x\|^{-1}\|Fx - Ax\| \leq \varepsilon\|b\|$. This shows that F is asymptotically linear with $F'(\infty) = A$. $\Leftarrow$

2.8 Higher derivatives and analyticity

There are several ways to define higher derivatives for nonlinear operators; one of the most useful definitions is usually referred to as Fréchet–Taylor derivative. An operator F between two Banach spaces X and Y is called *n-times (Fréchet–Taylor) differentiable* at an interior point x of $\mathcal{D}(F)$ if the increment $F(x + h) - Fx$ can be represented in the form

$$F(x + h) - Fx = \sum_{k=1}^{n} A_k h^k + \omega(h),$$

where A_k $(k = 1, \cdots, n)$ is a continuous k-form (i.e. $A_k h^k = \widetilde{A}_k(h, \cdots, h)$, where $\widetilde{A}_k$ is multilinear from $X \times \cdots \times X$ into Y), and ω satisfies

$$\lim_{\|h\| \to 0} \frac{\|\omega(h)\|}{\|h\|^n} = 0.$$

The operators $k!A_k$ $(k = 1, \cdots, n)$ are called the *k-th (Fréchet–Taylor) derivatives* $F^{(k)}(x)$ of F at x.

Existence conditions for higher derivatives of the superposition operator F between two ideal spaces X and Y can be found by similar methods as those developed in the previous section.

We say that two ideal spaces (X, Y) form a V^n-*pair* (a V_0^n-*pair*, respectively) if $\Omega_d = \emptyset$, X is quasi-regular, and there exist two positive functions $u_0 \in X^0$ and $v_0 \in \widetilde{Y}$ such that $\|P_D u_0\|_X^n \|P_D v_0\|_{\widetilde{Y}} = O(\mu(D))$ ($\|P_D u_0\|_X^n \|P_D v_0\|_{\widetilde{Y}} = o(\mu(D))$, respectively) as $\mu(D) \to 0$.

Given two ideal spaces X and Y, we denote by Y/X^n the set of all functions $z \in S$ with the property that, for any $x \in X$, the product $x^n z$ belongs to the space Y. Equipped with the natural norm

$$\|z\|_{Y/X^n} = \sup\{\|x^n z\|_Y : \|x\|_X \leq 1\},\tag{2.67}$$

Y/X^n becomes an ideal space, the *n-th multiplicator space* of X with respect to Y; for $n = 1$ we get the usual multiplicator space Y/X (see (2.40)), for $n = 0$ the space Y itself.

The higher multiplicator spaces Y/X^n have an important *interpolation property* which we state as follows:

Lemma 2.10 *Let X and Y be two ideal spaces and $0 \le m < k < n$. Then for $x_0 \in Y/X^m$ and $x_1 \in Y/X^n$ the inequality*

$$\| |x_0|^{\frac{n-k}{n-m}} |x_1|^{\frac{k-m}{n-m}} \| \le \|x_0\|^{\frac{n-k}{n-m}} \|x_1\|^{\frac{k-m}{n-m}} \qquad (2.68)$$

holds.

$\Rightarrow$ Fix $h \in X$ with $\|h\| \le 1$. By the classical Hölder inequality $uv \le \frac{1}{p}u^p + \frac{1}{q}u^q$ ($\frac{1}{p} + \frac{1}{q} = 1$) applied to $u = t|x_0(s)|^{\frac{n-k}{n-m}} |h^m(s)|^{\frac{n-k}{n-m}}$, $v = t^{-1}|x_1(s)|^{\frac{k-m}{n-m}} |h^n(s)|^{\frac{k-m}{n-m}}$, and $p = \frac{n-m}{n-k}$, $q = \frac{n-m}{k-m}$, we get

$$|x_0^{\frac{n-k}{n-m}} x_1^{\frac{k-m}{n-m}} h^k| \le \frac{n-k}{n-m} t^{\frac{n-m}{n-k}} |x_0| |h|^m + \frac{k-m}{n-m} t^{-\frac{n-m}{k-m}} |x_1| |h|^n .$$

Since Y is an ideal space, this implies that

$$\| |x_0|^{\frac{n-k}{n-m}} |x_1|^{\frac{k-m}{n-m}} h^k \| \le \frac{n-k}{n-m} t^{\frac{n-m}{n-k}} \|x_0 h^m\| + \frac{k-m}{n-m} t^{-\frac{n-m}{k-m}} \|x_1 h^n\| .$$

Passing in this inequality to the infimum with respect to $t > 0$, we obtain (2.68), since $h \in B_1(X)$ is arbitrary. $\Leftarrow$

Lemma 2.10 implies, in particular, the following: if $Y/X^k = \{\theta\}$ for some k, then also $Y/X^n = \{\theta\}$ for all $n > k$. In fact, if x_0 is a unit in Y and x_1 is an arbitrary function in Y/X^n we get, applying (2.68) with $m = 0$, that $|x_0|^{1-k/n}|x_1|^{k/n} \in Y/X^k$, and hence $x_1 = \theta$, since x_0 is different from zero almost everywhere.

We remark that a similar relation holds for higher multiplicator spaces Y/X^n and V^n-pairs to that established in Lemma 2.8. Now we state some results on higher derivatives of the superposition operator. Since the proofs are completely parallel to those of the corresponding theorems for the first derivative, we shall drop them.

Theorem 2.19 *Let X and Y be two ideal spaces, and let f be a supmeasurable function. Suppose that the superposition operator F generated by f is n-times differentiable at $x \in X$, considered as operator between X and Y. Then the derivative $F^{(k)}(x)$ ($k = 1, \cdots, n$) has the form*

$$F^{(k)}(x)h^k(s) = a_k(s)h(s)^k \quad (h \in X), \qquad (2.69)$$

where the function

$$a_k(s) = \text{S-}\lim_{u \to 0} \frac{1}{u^k}\Big[f(s, x(s) + u) - f(s, x(s)) - \sum_{j=1}^{k-1} a_j(s)\frac{u^j}{j!}\Big] \qquad (2.70)$$

belongs to the multiplicator space Y/X^k. Moreover, the function

$$g_n(s, u) = \begin{cases} \frac{1}{u^n}\Big[f(s, x(s) + u) - f(s, x(s)) - \sum_{k=1}^{n} a_k(s)\frac{u^k}{k!}\Big] & \text{if } u \neq 0, \\ 0 & \text{if } u = 0 \end{cases}$$

is sup-measurable (and even a Shragin function if f is so).

Conversely, if $x \in X$, $Fx \in Y$, and the superposition operator G_n generated by the function g_n acts between X and Y/X^n and is continuous at θ, then F is n times differentiable at x, and formula (2.69) holds.

Theorem 2.20   Let X and Y be two ideal spaces, and suppose that (X, Y) is a V^n-pair. Let f be a sup-measurable function, and assume that the superposition operator F generated by f is n times differentiable between X and Y at some point $x \in X$. Then the function f satisfies

$$f(s, u) \simeq a_0(s) + a_1(s)u + \cdots + \frac{a_n(s)}{n!}u^n \qquad (2.71)$$

(i.e. is sup-equivalent to a polynomial in u), where $a_k \in Y/X^k$ ($k = 0, 1, \cdots, n$).

Theorem 2.21   Let X and Y be two ideal spaces, and suppose that the function f and all functions $\widehat{f}_k(s, u) = \frac{\partial^k}{\partial u^k}f(s, u)$ ($k = 1, \cdots, n$) are Carathéodory functions. Assume that the superposition operator F generated by f maps some open set $O \subset X$ into Y. Then F, considered as an operator between X and Y, is n times continuously differentiable on O if and only if all superposition operators $\widehat{F}_k$ generated by $\widehat{f}_k$ and considered as operators between X and Y/X^k ($k = 1, \cdots, n$) are continuous on O. In this case, the equality $F^{(k)}(x)h^k(s) = \widehat{F}_k(x)h(s)^k$ ($x \in O$, $h \in X$) holds.

In view of several applications in nonlinear analysis, it is desirable to have analyticity conditions for the superposition operator at hand.

Recall that an operator F between two Banach spaces X and Y is called *analytic* at an interior point x of $\mathcal{D}(F)$ if the increment $F(x + h) - Fx$ can be represented as convergent series

$$F(x + h) - Fx = \sum_{k=1}^{\infty} A_k h^k \qquad (2.72)$$

for sufficiently small $\|h\|$, where A_k $(k = 1, 2, \cdots)$ is a continuous k-form. The series (2.72) converges uniformly on some ball $B_\rho(X)$; the supremum ρ_u of all such ρ is called the radius of uniform convergence of (2.72) and can be calculated by the formula

$$\rho_u = \left[\varlimsup_{k \to \infty} \sup_{\|h\| \le 1} \|A_k h^k\|^{1/k} \right]^{-1} \qquad (2.73)$$

Moreover, the series (2.72) converges absolutely on the star-shaped domain $S_a = \{h : \rho_a(h) \le 1\}$, where

$$\rho_a(h) = \left[\varlimsup_{k \to \infty} \|A_k h^k\|^{1/k} \right]^{-1} ;$$

in particular, absolute convergence holds inside the ball $B_{\rho_a}(X)$, where

$$\rho_a = \inf\{\rho_a(h) : \|h\| \le 1\}. \qquad (2.74)$$

We remark that these radii are related by the estimates $0 < \rho_u \le \rho_a \le \infty$, where strict inequality may occur.

A very striking result in the theory of analytic superposition operators in ideal spaces is the fact that an analytic superposition operator which is defined on some Δ_2-space (see Section 2.3) necessarily reduces to a polynomial:

Theorem 2.22   *Let X and Y be two ideal spaces, and suppose that $\Omega_d = \emptyset$ and $X \in \Delta_2$. Let f be a sup-measurable function, and suppose that the superposition operator F, generated by f and considered as an operator between X and Y, is analytic in a neighbourhood of some point $x \in X$. Then F is a polynomial operator on X.*

$\Rightarrow$ Without loss of generality, we may assume that $F\theta = \theta$ and that F is bounded and analytic on some ball $B_r(X)$. By Lemma 2.1, F is then also defined on the set $\Sigma(B_r(X))$ (see (2.10)), and this set coincides with the whole space since $X \in \Delta_2$. Moreover, we have, by the definition (2.15) of the split functional H, that $\|Fx\| \le m(1 + H(x))$ for all $x \in X$. By Lemma 2.3, the operator F has at most polynomial growth. On the other hand, again by the definition of $\Sigma(B_r(X))$, F is analytic on the whole space X. By Liouville's theorem, any analytic operator of polynomial growth at infinity is a polynomial, and so we are done. $\Leftarrow$

The hypothesis $X \in \Delta_2$ is sufficient to produce the "degeneracy" of analytic superposition operators on X, but not necessary. More generally, a similar degeneration occurs if the multiplicator space Y/X^n is trivial for large n; in this case, the "global" relation $\Sigma(B_r(X)) = X$ does not hold, but there is always at least a "local degeneration": if F is analytic at x, then F reduces to a polynomial in some neighbourhood of x.

We give now a necessary analyticity condition which follows from Theorem 2.19:

Theorem 2.23   *Let X and Y be two ideal spaces, and let f be a sup-measurable function. Suppose that the superposition operator F, generated by f and considered as an operator between X and Y, is analytic at $x \in X$. Then f is sup-equivalent to some function $\tilde{f}$ which is, for almost all $s \in \Omega$, analytic at $x(s)$, i.e.*

$$\tilde{f}(s, x(s) + u) - \tilde{f}(s, x(s)) = \sum_{k=1}^{\infty} a_k(s) u^k \,, \tag{2.75}$$

where the series (2.75) is convergent for $|u| < \delta(s)\rho_a$ (δ as in (2.11) and ρ_a as in (2.74)), and the function a_k belongs to the space Y/X^k ($k = 1, 2, \cdots$).

$\Rightarrow$ For the sake of simplicity, assume that $F\theta = \theta$ and F is analytic at $x = \theta$. Since F is infinitely often differentiable, Theorem 2.10 ensures that $a_k \in Y/X^k$ with a_k given by (2.70) ($k = 1, 2, \cdots$). By the analyticity of F at θ, the series $Fh = \sum_{k=1}^{\infty} a_k h^k$ converges absolutely for $\|h\| < \rho_a$; therefore the series

$$\xi = \sum_{k=1}^{\infty} a_k(s) h^k(s) \tag{2.76}$$

which is obtained from the numerical series

$$w(s, u) = \sum_{k=1}^{\infty} a_k(s) u^k \tag{2.77}$$

by the formal substitution $u = h(s)$, converges absolutely in Y. Consequently, the series $\xi^* = \sum_{k=1}^{\infty} |a_k(s)| \, |h(s)|^k$ of nonnegative terms converges in Y, therefore almost everywhere; hence (2.76) converges almost everywhere as well. For fixed $h \in B_{\rho_a}(X)$ we have $\rho_a(s) \geq |h(s)|$, where $\rho_a(s)$ is

the radius of absolute convergence of the series (2.77) for fixed $s \in \Omega$. By
definition of δ, this implies that $\rho_a(s) \geq \rho_a \delta(s)$. It remains to observe that
the function

$$\widetilde{f}(s, u) = \begin{cases} w(s, u) & \text{if } |u| < \rho_a \delta(s), \\ f(s, u) & \text{if } |u| \geq \rho_a \delta(s) \end{cases}$$

is sup-equivalent to the function f and has the required properties. $\Leftarrow$

Theorem 2.23 allows us to formulate quite easily an analyticity condition
for F which is both necessary and sufficient:

Theorem 2.24   *Let X and Y be two ideal spaces, and let f be a Carathéodory function which can be represented in the form*

$$f(s, x(s) + u) - f(s, x(s)) = \sum_{k=1}^{\infty} a_k(s) u^k, \qquad (2.78)$$

*where $x \in X$, $Fx \in Y$, $a_k \in Y/X^k$, and (2.78) converges for almost all
$s \in \Omega$ and $|u| < a\delta(s)$ ($a > 0$ fixed). Then the corresponding superposition
operator F is analytic at x if and only if*

$$0 < \left[\varlimsup_{k \to \infty} \|a_k\|_{Y/X^k}^{1/k} \right]^{-1} \leq a. \qquad (2.79)$$

$\Rightarrow$ The "if" part is obvious, while the "only if" part follows from Theorem
2.23. $\Leftarrow$

In order to verify condition (2.79) it is necessary to calculate the norms
of the functions a_k in the space Y/X^k. If the functions a_k are constant, for
example, this amounts to studying the numerical sequence $\gamma_n = \|1\|_{Y/X^n}$.
Precise information about this sequence is contained in the following:

Lemma 2.11   *The sequence γ_n is logarithmically convex. If the space X
is not imbedded into the space M, and the multiplicator spaces Y/X^n are
nontrivial for each n, then the limit $\gamma_* = \lim\limits_{n \to \infty} \gamma_n^{1/n}$ exists and is infinite.*

$\Rightarrow$ The logarithmic convexity of γ_n is equivalent to the inequality $\gamma_k^{n-m} \leq
\gamma_m^{n-k} \gamma_n^{k-m}$ which was established in Lemma 2.10. The logarithmic convexity
implies also the existence of the limit γ_*. Now, if γ_* were finite, we would
have

$$\lim_{n \to \infty} \|h^n\|_Y^{1/n} \leq \gamma_* \|h\|_X \quad (h \in X). \qquad (2.80)$$

But the left-hand side of (2.80) is just the norm of h in the space M, and
thus X is continuously imbedded into M, in contradiction to our hypothesis.
$\Leftarrow$

Theorem 2.24 provides a necessary and sufficient condition for the analyticity of the superposition operator between two given ideal spaces. Now we
shall invert the point of view: given a nonlinearity f, we provide a "recipe"
for constructing a pair of ideal spaces X and Y such that the superposition operator F generated by f will be analytic between these spaces. To
this end, consider again the function w given in (2.77). If w is an entire
analytic function with measurable coefficients, we may choose appropriate
ideal spaces X and Y such that the superposition operator W generated by
w becomes analytic at the zero function θ of the space X. In fact, consider
the growth function

$$m(s, u) = \max_{|v|=u} |w(s, v)| \qquad (2.81)$$

of w, and the superposition operator M generated by m. Given an ideal
space Y, the linear set of all $x \in S$ for which the norm

$$\|x\| = \inf\{\lambda : \lambda > 0, \ \|M(x/\lambda)\|_Y \leq 1\} \qquad (2.82)$$

is finite, is an ideal space $X = M^{-1}[Y]$ which we call the *inverse M-
transform* of the space Y. The obvious estimate $|a_n(s)u^n| \leq m(s, |u|)$ $(n =
1, 2, \cdots)$ implies that $\|a_n\|_{Y/X^n} \leq 1$ $(n = 1, 2, \cdots)$; hence the radius ρ_u of
uniform convergence of the operator series

$$Wh = \sum_{n=1}^{\infty} A_n h^n, \quad A_n h^n(s) = a_n(s)h(s)^n$$

is equal to 1. By Theorem 2.24, the superposition operator M generated by
the function (2.81) is analytic at θ, considered as an operator from X into Y.
Simple examples show that this operator is, in general, not entire analytic,
even if the generating function is so (see Section 4.6). We summarize our
results in the following:

Theorem 2.25   *Suppose that, for almost all $s \in \Omega$, the function $f(s, \cdot)$
is entire analytic, and let Y be an ideal space. Let $x_0 \in S$ be fixed, let
$X = M^{-1}[Y]$ be the inverse M-transform of Y, with m given by (2.81), and
assume that $x_0 \in X$ and $Fx_0 \in Y$. Then the operator F generated by f is
analytic at x_0, considered as an operator from X into Y.*

The growth function (2.81) is of course rather difficult to calculate explicitly. However, one may consider, instead of m, the functions

$$m_-(s, u) = \sup_n |a_n(s)|u^n, \quad m_+(s, u) = \sum_{n=1}^{\infty} |a_n(s)|u^n,$$

which are easier to compute and equivalent to m in the sense that

$$m_-(s,u) \le m(s,u) \le m_+(s,u) \le m_-(s,2u).$$

The inverse transforms $X = M^{-1}[Y]$, $X_- = M_-^{-1}[Y]$ and $X_+ = M_+^{-1}[Y]$ coincide as linear spaces, and their norms are equivalent:

$$\|x\|_{X_-} \le \|x\|_X \le \|x\|_{X_+} \le 2\|x\|_{X_-}.$$

2.9 Notes, remarks and references

1. The notion of an ideal space arose in the early fifties (under the name "ideal structure", "Banach structure", "Banach lattice" etc.) and was studied by I. Amemiya, J. Dieudonné, H.W. Ellis, I. Halperin, Ju.I. Gribanov, G. Lorentz, W.A.J. Luxemburg, D. Wertheim, A.C. Zaanen, P.P. Zabrejko, and others. The most complete and detailed presentation of the theory can be found in the books [202], [369] and [370], see also [283]. We followed the presentation of the monograph [375] which is most appropriate for the application of ideal spaces to linear and nonlinear integral equations ([373], see also [371]). The proofs of all facts mentioned in Section 2.1 may be found in [375]. [384] is one of the first papers on ideal spaces of vector-valued functions; for new results parallel to [375] see e.g. [379], [383]. We remark that the theory of ideal spaces exhibits in the "vector case" a lot of interesting and difficult new features.

2. The sets $\Sigma(N)$ were introduced and studied by P.P. Zabrejko [371] in connection with the domain of definition of disjointly additive nonlinear operators; the usefulness of the sets Δ and Π (see (2.11) and (2.13)) was also shown by P.P. Zabrejko [375]. In the meantime, there is already a considerable amount of work on disjointly additive operators on ideal spaces and, in particular, so called *modular spaces* (see e.g. [100–103], [155], [163], [203–205]). The *split functional H* (see (2.15)) was introduced in [32] in connection with analyticity conditions for the superposition operator; explicit formulas and estimates for this functional will be given in subsequent chapters.

3. Theorem 2.2 was first obtained (in the case $\Omega = \Omega_c$) by P.P. Zabrejko ([373], see also [23]); the proof in the case $\Omega = \Omega_d$ is new. The class of split spaces was introduced in [35], the class of Δ_2-spaces in [32] and [33] (where also Lemma 2.3 is proved), and the class of δ_2-spaces in [35]. The basic boundedness result Theorem 2.3 is taken from [35]; a weaker form (namely for Δ_2-spaces X) was proved in [23], a still weaker form (namely for regular Δ_2-spaces) in [161], see also [367] and [393].

Important examples of split spaces are Orlicz spaces (with Luxemburg norm; see Section 4.1), and, more generally, any inverse F-transform $F^{-1}[Y]$ of a Δ_2-space Y, provided the corresponding function $f(s, \cdot)$ is even and convex for almost all $s \in \Omega$.

The growth function (2.23) appears first in its special form (2.24) in the papers [14], [30] and [31] for Lebesgue spaces; this function will be of fundamental importance in the following chapters.

4. The notions of U-boundedness and W-boundedness were studied in detail in connection with nonlinear operator equations in [371] and [373]. An important application of superposition operators which map bounded sets into absolutely bounded sets is given in [392], where such operators are called "improving". These applications are based on the observation that large classes of linear *integral operators*

$$Ky(s) = \int_{\Omega} k(s,t)y(t)dt, \qquad (2.83)$$

although not being compact, map absolutely bounded sets into precompact sets, and thus one may apply the classical Schauder fixed-point principle to obtain solutions of the nonlinear Hammerstein integral equation $x = KFx$ (see e.g. [268], [374], [385-387]).

Absolutely bounded and U-bounded sets were also studied in detail in the book [182], W-bounded operators in [371] and [373]. Theorem 2.4, which is the main result on W-bounded superposition operators, is new.

The axiomatic definition of a *measure of noncompactness* in its general form presented above is due to B.N. Sadovskij [280]. The special examples (2.31) and (2.32), however, go back to K. Kuratowski and Hausdorff [186]. The functional (2.33) was introduced by N. Jerzakova [152], [153] and, independently, in [22]. The functionals (2.34) and (2.36) were introduced and studied in [21]; some applications may be found in [15], [18], [29], and [42]; the functional (2.35) was introduced in [92]. Detailed information on the functional (2.38) in various function spaces may be found in [20].

5. The first general continuity theorem for the superposition operator with values in a regular ideal space is [179]. Our Theorem 2.6 goes much farther, since we do not assume that f is a Carathéodory function, and since we also get "almost" a necessary condition.

Continuity conditions for the superposition operator at single points, as those given in Theorem 2.7, may be found in [206] and [249]. The fact that a continuous superposition operator with values in a completely irregular space is necessarily constant was observed by V.A. Bondarenko [54]; in the case $Y = L_\infty$ this is well-known [380˘].

Various examples of continuous superposition operators which are not uniformly continuous on bounded sets were obtained by M.M. Vajnberg [347], M.A. Krasnosel'skij [180], F. Dedagić [93] and P.P. Zabrejko [94]. The general Theorem 2.9 is new.

The first result on the *weak continuity* and sequentially weak continuity of the superposition operator between two perfect ideal spaces is contained in [293], for Lebesgue spaces in [289]. Our Theorem 2.10 covers also non-Carathéodory functions and refers to weak topologies with respect to some total subspace of the associate space $\widetilde{X}$ (see Section 2.1 and [375]); we followed the presentation in [36].

6. The *multiplicator space* Y/X was studied, for example, in [371] and [373], where also some important examples are given; we shall discuss such examples in the following chapters, mainly in connection with Lipschitz conditions and differentiability results.

Some important properties of special superposition operators may be expressed by means of multiplicator spaces. For instance, the (linear) *multiplication operator* $Fx(s) = a(s)x(s)$ is bounded between X and Y if and only if $\|a\|_{Y/X} < \infty$; in particular, X is imbedded in Y if and only if $L_\infty \subseteq Y/X$.

The results given in Section 2.6 are presented, together with examples, in [27]. The notions of a V-pair (V_0-pair, V^n-pair, V_0^n-pair) of ideal spaces were introduced in [34] in connection with differentiability properties of the superposition operator. The Darbo condition (2.49) appears first (for the Kuratowski measure of noncompactness (2.31)) in Darbo's pioneering paper [89]. In the meantime, the literature on the theory and applications of nonlinear operators satisfying (2.50) is vast; we just mention the monographs [7], [8], [280].

7. Various pathological phenomena in connection with Fréchet differentiability of the superposition operator in concrete ideal spaces (above all, Lebesgue and Orlicz spaces) were observed by M.M. Vajnberg, M.A. Krasnosel'skij and Ja.V. Rutitskij [166], [175], [182], [347]. Earlier differentiability conditions were either only sufficient, or only necessary. Both Theorems 2.14 and 2.16 were obtained for Lebesgue spaces (and later for Orlicz spaces) by W. S. Wang [363]. The degeneracy result given in Theorem 2.15 was obtained for Lebesgue spaces in case $\Omega_d = \emptyset$ by M.M. Vajnberg [347], in its general form in [34]. A variant of Theorem 2.17 may be found in [371] and [373]. General results on uniform continuity and differentiability of nonlinear operators emphasize the necessity of considering superposition operators "of several arguments" like $G(x,y)(s) = g(s,x(s),y(s))$ (see e.g. (2.62)) between ideal spaces $X \times Y$ and Z, for details see [376–378]. The first systematic account on continuous, uniformly continuous, bounded, W-bounded, or asymptotically linear superposition operators, as well as many applications to nonlinear Hammerstein integral equations, may be found in [387–390].

Asymptotically linear operators were first considered by M.A. Krasno-sel'skij [166]; the knowledge of both the derivative of a nonlinear operator at a single point and the asymptotic derivative is useful in the theory of general operator equations. Suppose, for instance, that a trivial solution ($x = \theta$, say) of a nonlinear operator equation is known in advance, and one is interested in a second (i.e. nontrivial) solution. If the derivative of the nonlinear operator involved at zero and infinity exists, and the sums of the multiplicities of their eigenvalues bounded away from zero have different parity, it follows from standard degree-theoretic arguments that there is a second solution (see e.g. [181]). Thus, not only the mere differentiability of a nonlinear operator is important, but also the knowledge of the explicit form of the derivative, as given in Theorems 2.14 and 2.18.

General conditions for the differentiability or asymptotic linearity of the superposition operator F may be equivalently formulated as conditions on the asymptotic behaviour of the growth function (2.24) of some auxiliary superposition operator $\widetilde{F}$. Thus, the superposition operator F generated by f is differentiable at some point x (is asymptotically linear, respectively) if and only if the growth function $\mu_{\widetilde{F}}$ of the superposition operator $\widetilde{F}$ generated by $\widetilde{f}(s, u) = f(s, x(s) + u) - f(s, x(s)) - a(s)u$ ($\widetilde{f}(s, u) = f(s, u) - a_\infty(s)u$, respectively), where the functions a and a_∞ are given by (2.56) and (2.65), respectively, satisfies the relation $\mu_{\widetilde{F}}(r) = o(r)$ as $r \to 0$ (as $r \to \infty$, respectively). This observation reduces the problem of finding necessary and sufficient differentiability conditions to that of calculating or estimating the growth function (2.24). Unfortunately, explicit formulas for this function are available only in very exceptional cases (e.g. in Lebesgue and Orlicz spaces). We shall return to this in Chapters 3 and 4; see Theorems 3.13 and 4.12.

Apart from Fréchet differentiablity, other derivatives are of great importance in nonlinear analysis. Similar results to those given in Section 2.7, but for Gâteaux or weak Δ-derivatives, may be found in [34].

8. Concerning higher derivatives, we could repeat many remarks made above for the first (Fréchet) derivative. We just confine ourselves to a short comparison of the Taylor-type n-th derivative with the more common (at least in finite-dimensional spaces) Newton–Leibniz definition of the n-th derivative $F^{(n)}(x_0)$ as the first derivative of $F^{(n-1)}(x_0)$. It is well known that the n-th Newton–Leibniz derivative (if it exists) is also the n-th Taylor derivative; the converse is false. Nevertheless, if the n-th Taylor derivative exists in some neighbourhood of x_0 and is locally bounded there, the n-th

Newton–Leibniz derivative exists as well, and both definitions are nearly equivalent. However, for the Newton–Leibniz derivative one must consider powers $X^k = X + \cdots + X$ (k-times) of the space X, while for the Taylor derivative it suffices to work in the space X.

The main results on the analyticity of the superposition operator between ideal spaces are contained in the papers [32] and [33]. In particular, we emphasize again that Theorem 2.22 shows that the class of analytic superposition operators between two ideal spaces X and Y, with X being a Δ_2 space, is extremely narrow. This may be the reason that analytic superposition operators were not considered between, say, Lebesgue spaces.

In this connection, we mention the classical example of a superposition operator F which is analytic at θ in the space ℓ_1 (i.e. the space L_1 with $\Omega = \mathbb{N}$ and μ being the counting measure), generated by the function $f(s, u) = u^s$. The generating function f is entire analytic, and the radii of uniform and absolute convergence of the expansion of F at zero

$$Fh = \sum_{n=1}^{\infty} A_n h^n, \quad A_n h^n(s) = \chi_{\{n\}}(s) h(s)^n \quad (n = 1, 2, \cdots)$$

are $\rho_u = 1$ and $\rho_a = \infty$, respectively (see (2.73) and (2.74)).

9. The construction of ideal spaces X and Y with the property that the superposition operators F generated by a given nonlinearity f becomes analytic, is closely related to the notion of the so-called $\mathcal{L}$-*characteristic* of the superposition operator. Given some property $\mathcal{P}$ (acting, boundedness, continuity, differentiability etc.), we denote by $\mathcal{L}(F; \mathcal{P})$ the set of all pairs (X, Y) of ideal spaces such that F has the property $\mathcal{P}$, considered as an operator between X and Y. To illustrate this concept, let us assume, for simplicity, that f is a Carathéodory function and that $f(s, 0) = 0$.

Let X be an ideal space. We are interested in finding another ideal space Y (if it exists!) such that F, as an operator between X and Y, is defined on some subset N of X and has the property $\mathcal{P}$. Here we confine ourselves to the case that N is the unit ball $B_1(X)$, and that F is bounded on $B_1(X)$; in view of Theorems 2.2 and 2.3, this is a natural assumption. Thus, the image $FB_1(X)$ of $B_1(X)$ should be contained in some ball $B_r(Y)$. This in turn means that the convex hull $coFB_1(X)$ is bounded in the space S, i.e. $FB_1(X)$ is co-bounded in the terminology of Section 1.5. But this is not always the case: just the simple function $f(s, u) = |u|^k$ ($k > 1$) generates an operator F in the space $X = L_1$ with the property that the convex hull of $FB_1(L_1)$ is unbounded in S if $\Omega_c \neq \emptyset$! This shows that the class $\mathcal{Y}(X, F)$ of all ideal spaces Y for which F maps $B_1(X)$ into Y and is bounded, may be in fact empty.

Observe that, if such spaces Y exist in $\mathcal{Y}(X, F)$, the class $\mathcal{Y}(X, F)$ also contains spaces of type $L(u_0)$; this follows from the maximality of such spaces (see Section 2.1).

Suppose that the class $\mathcal{Y}(X, F)$ is nonempty. In this case, one can choose even a minimal (with respect to inclusion) space Y in $\mathcal{Y}(X, F)$. In fact, one may take as Y the space of all measurable functions y for which the norm

$$\|y\| = \inf\{\lambda : \lambda > 0, \, \lambda^{-1}y \in id\,coFB_1(X)\}$$

makes sense and is finite; here by $id\,N$ we denote the *ideal hull* of the set $N \subseteq S$, i.e. the set

$$id\,N = \bigcup_{u \in N} \{x : x \in S, |x| \leq u\}. \tag{2.84}$$

The ideal space Y obtained in this way is usually called the *direct F-transform* [373] of the ideal space X and denoted by $Y = F[X]$. For example, if the function $f(s, \cdot)$ is nonnegative, even, and concave on $[0, \infty)$, then the direct F-transform $F[X]$ of X consists precisely of all functions $y \in S$ for which the norm

$$\|y\| = \inf\{\lambda : \lambda > 0, \, \lambda^{-1}y \in FB_1(X)\}$$

is finite. Moreover, the space $F[X]$ is complete (perfect; almost perfect; regular) if the space X is complete (perfect; almost perfect; regular, respectively).

Let Y be an ideal space. Analogously to the above reasoning, one could try to find another ideal space X (if it exists!) such that F, as an operator between X and Y, is defined on $B_1(X)$ and has some property $\mathcal{P}$. If $\mathcal{P}$ means the boundedness of F on $B_1(X)$, the class $\mathcal{X}(F, Y)$ of such spaces X is always nonempty: in fact, one may choose $X = M(u_0)$, where u_0 is some measurable function such that $F_* u_0 \in Y$; here F_* denotes the superposition operator generated by the Carathéodory function

$$f_*(s, u) = \max_{|v| \leq u} |f(s, v)|. \tag{2.85}$$

In this case, however, one can not guarantee the existence of a maximal space with this property, as in the class $\mathcal{Y}(X, F)$. In fact, the unit ball $B_1(X)$ of X should be contained in the pre-image $F^{-1}(B_r(Y))$ of some ball $B_r(Y)$, but the sets $F^{-1}(B_r(Y))$ are, in general, not convex, and contain many "maximal" convex sets which may be chosen as the unit ball of some ideal space. Nevertheless, if the set $F^{-1}(B_r(Y))$ is convex and bounded in S for some $r > 0$, one may find an ideal space X such that $B_1(X) = F^{-1}(B_r(Y))$ for this r. The ideal space X obtained in this way is usually called the *inverse F-transform* [373] of the ideal space Y and denoted by $X = F^{-1}[Y]$. For example, if the function $f(s, \cdot)$ is nonnegative, even, and convex on $[0, \infty)$, then the inverse F-transform $F^{-1}[Y]$ of Y consists precisely of all functions $x \in S$ for which the norm

$$\|x\| = \inf\{\lambda : \lambda > 0, \|f(\cdot, \lambda^{-1}x(\cdot))\|_Y \leq 1\} \qquad (2.86)$$

is finite. Moreover, the space $F^{-1}[Y]$ is complete (perfect; almost perfect; regular) if the space Y is complete (perfect; almost perfect; regular, respectively). Observe that the appropriate spaces in Theorem 2.23 were constructed in this way, as a comparison of (2.82) and (2.86) shows. We remark that special direct and inverse transforms (mostly of polynomial type) have been considered, for instance, in [76], [183], [184], or [197–199] in connection with interpolation theory.

The above reasoning makes it possible to describe the $\mathcal{L}$-characteristic $\mathcal{L}(F; \mathcal{P})$ in case $\mathcal{P}$ means boundedness on the unit ball. For instance, in the special case $Y = \tilde{X}$ (see (2.8)) one may show that $(X, \tilde{X}) \in \mathcal{L}(F; \mathcal{P})$ if and only if $X \in \mathcal{X}(H, L_1)$, where H is the superposition operator generated by the function $h(s, u) = uf_*(s, u)$ with f_* given by (2.85). In fact, if F is bounded between $B_1(X)$ and $\tilde{X}$, then F_*, the superposition operator generated by (2.85), is also bounded between $B_1(X)$ and $\tilde{X}$, and hence H is bounded from $B_1(X)$ into L_1. Conversely, if H maps X into L_1, then $(X, \tilde{X}) \in \mathcal{L}(F; \mathcal{P})$ since $|f_*(s, u)v| \leq |h(s, u)| + |h(s, v)|$; for more information, see [381].

Chapter 3

The superposition operator
in Lebesgue spaces

By reformulating the general results of Chapter 2, one gets many results on the superposition operator in Lebesgue spaces. On the other hand, the theory in Lebesgue spaces is much richer than in general ideal spaces. The most interesting (and pleasant) fact is that one can give an acting condition for F, in terms of the generating function f, which is both necessary and sufficient. It follows, in particular, that F is always bounded and continuous, whenever F acts from some L_p into L_q (for $1 \leq p \leq \infty$, $1 \leq q < \infty$ and $\Omega_d = \emptyset$); the corresponding problems in the case $\Omega_c = \emptyset$ are more delicate.

Apart from continuity and boundedness conditions, we provide a concrete "recipe" to calculate the growth function of the superposition operator on balls in L_p. Moreover, criteria for absolute boundedness and uniform continuity are given, as well as two-sided estimates for the modulus of continuity of F.

As immediate consequences of some results of the preceding chapter, we get that F is weakly continuous from L_p into L_q if and only if f is affine in u. Further, it turns out that the Darbo or Lipschitz condition for F is equivalent to a Lipschitz condition for the function f with respect to u. Hölder continuity of F is also briefly discussed.

Another pleasant fact concerns differentiability: in Lebesgue spaces one can give conditions for differentiability, asymptotic linearity, and higher differentiability which are both necessary and sufficient. In particular, every differentiable superposition operator from L_p into L_q is affine in case $p = q$, and even constant in case $p < q$. A similar "degeneracy" occurs for analyticity: every analytic superposition operator from L_p into L_q reduces to a polynomial, whatever p and q are.

Most results refer to the case $1 \leq p, q < \infty$, but carry over to the case $p = \infty$ or $q = \infty$ as well; a special section is devoted to this latter case. Finally, in the last section we prove a result which enlightens the "geometric" structure of the $\mathcal{L}$–characteristic of the superposition operator in Lebesgue spaces.

3.1 Lebesgue spaces

As before, let Ω be an arbitrary set, $\mathcal{M}$ some σ-algebra of subsets of Ω, and μ a σ-finite and countably additive measure on $\mathcal{M}$; together with μ we shall sometimes consider an equivalent normalized countably additive measure λ (see Section 1.1).

By $L_p = L_p(\Omega, \mathcal{M}, \mu)$ we denote the set of all (equivalence classes of) μ-measurable functions x on Ω for which

$$\int_\Omega |x(s)|^p ds < \infty \tag{3.1}$$

for $1 \leq p < \infty$ (where, as usual, we write ds instead of $d\mu$), and

$$\operatorname*{ess\,sup}_{s \in \Omega} |x(s)| < \infty \tag{3.2}$$

for $p = \infty$, respectively. Equipped with the natural algebraic operations and the norm

$$\|x\|_p = \begin{cases} \left(\int_\Omega |x(s)|^p ds \right)^{1/p} & \text{if } 1 \leq p < \infty, \\ \operatorname*{ess\,sup}_{s \in \Omega} |x(s)| & \text{if } p = \infty, \end{cases} \tag{3.3}$$

the set L_p becomes an ideal space which is completely regular for $1 \leq p < \infty$, and perfect for $p = \infty$.

Let us recall some simple properties of the spaces L_p. First of all, we mention the important *Hölder inequality*

$$|<x,y>| \leq \|x\|_p \|x\|_{\tilde{p}} \tag{3.4}$$

where the pairing $< \cdot, \cdot >$ is given in (2.7), $1 \leq p \leq \infty$, and $\tilde{p} = p/(1-p)$. This inequality turns into an equality if and only if $a|x(s)|^p + b|y(s)|^{\tilde{p}} = 0$, where a and b are two numbers which are not both zero. The inequality

$$\|xy\|_{pq/(p+q)} \leq \|x\|_p \|y\|_q \quad (1 \leq p, q \leq \infty) \tag{3.5}$$

which is equivalent to (3.4) is also sometimes called Hölder inequality; in (3.5) we have equality if and only if for some a and b, not both zero,

$a|x(s)|^p + b|y(s)|^q = 0$. The estimate (3.5) implies, in particular, that, for any measurable function x on Ω, the function $\varphi_x(p) = \|x\|_p$ is logarithmically convex, i.e. $\|x\|_{p_\lambda} \leq \|x\|_{p_0}^{1-\lambda}\|x\|_{p_1}^{\lambda}$ where $p_\lambda^{-1} = (1-\lambda)p_0^{-1} + \lambda p_1^{-1}$. This in turn shows that the function $\varphi_x(p) = \|x\|_p$ is finite on some subinterval of $[1, \infty]$, is continuous inside this subinterval, and is infinite outside the closure of this subinterveal. Finally, the Hölder inequality (3.5) may also be written as an *interpolation inequality*

$$\| \, |x_0|^{1-\lambda}|x_1|^\lambda \, \|_{p_\lambda} \leq \|x_0\|_{p_0}^{1-\lambda}\|x_1\|_{p_1}^{\lambda}. \qquad (3.6)$$

From the classical Riesz theorem it follows that $\tilde{L}_p = L_{\tilde{p}}$ $(1 \leq p \leq \infty)$, i.e. $L_{\tilde{p}}$ is the associate space to L_p (see (2.8)). Since L_p is completely regular for $1 \leq p < \infty$, we have in addition that $L_p^* = L_{\tilde{p}}$ for $1 \leq p < \infty$; moreover, L_p is reflexive for $1 < p < \infty$. On the other hand, $L_1 = \tilde{L}_\infty$ is a proper subspace of L_∞^*, since L_∞ is not regular. We denote the regular part L_∞^0 of L_∞ (see (2.1)) by $c_0 = c_0(\Omega, \mathcal{M}, \mu)$; this space consists of all functions $x \in S$ which vanish on Ω_c and satisfy $\lim_{\Omega_d} x(s) = 0$; the notation $\lim_{\Omega_d}$ means limit with respect to the Cauchy filter of all subsets of Ω_d with finite complement. In particular, the space L_∞ is quasi-regular in the case $\Omega_c = \emptyset$, and completely irregular in the case $\Omega_d = \emptyset$ (see Section 2.5). The dual space L_∞^* of L_∞ is not a space of measurable functions, but the dual space $(L_\infty^0)^*$ of L_∞^0 consists of all functions $x \in L_1$ which vanish outside Ω_d.

Now we describe the multiplicator spaces L_q/L_p. To this end, we denote by $Z_p = Z_p(\Omega, \mathcal{M}, \mu)$ the set of all functions $x \in S$ which vanish on Ω_c and satisfy the relation $|x(s)| = O(\mu(s)^{1/p})$ for $s \in \Omega_d$; here and in what follows we write $\mu(s)$ instead of $\mu(\{s\})$. With the natural algebraic operations and the norm

$$\|x\|_{(p)} = \sup_{s \in \Omega_d} |x(s)|\mu(s)^{-1/p} \quad (0 < p < 1), \qquad (3.7)$$

the set Z_p becomes an ideal space. Further, for $1 \leq p < q < \infty$ we write $Z_{p,q}$ for the set of all functions $x \in S$ such that $|x|^q \in Z_{p/q}$, equipped with the norm $\|x\|_{p,q} = \| \, |x|^q \, \|_{(p/q)}^{1/q}$. Finally, let $Z_{p,\infty} = Z_p$. In this terminology, we have the following formula for the multiplicator space L_q/L_p (see (2.40)):

$$L_q/L_p = \begin{cases} L_{pq/(p-q)} & \text{if } p \geq q, \\ Z_{p,q} & \text{if } p < q. \end{cases} \qquad (3.8)$$

More generally, for the n-th multiplicator space L_q/L_p^n (see (2.67)) we have

$$
L_q/L_p^n = \begin{cases} L_{pq/(p-nq)} & \text{if } p \geq nq, \\ Z_{p/n,q} & \text{if } p < nq. \end{cases} \tag{3.9}
$$

Since the imbedding $L_p \subseteq L_q$ is equivalent to the fact that the constant function $x(s) \equiv 1$ belongs to L_q/L_p, formula (3.8) gives the following imbedding criterion: in the case $p \geq q$, the imbedding $L_p \subseteq L_q$ holds if and only if $\mu(\Omega) < \infty$; here

$$
\|x\|_q \leq [\mu(\Omega)]^{\frac{1}{q}-\frac{1}{p}} \|x\|_p . \tag{3.10}
$$

In the case $p \leq q$, the imbedding $L_p \subseteq L_q$ holds if and only if $\Omega_c = \emptyset$ and $\mu(s) \geq \mu_0 > 0$ for $s \in \Omega_d$; here

$$
\|x\|_q \leq \left[\inf_{s \in \Omega_d} \mu(s) \right]^{-1} \|x\|_p . \tag{3.11}
$$

We now pass to the description of the sets Σ, Π, Δ introduced in Section 2.2; to be specific, we denote the corresponding sets and functions in L_p by a subscript p.

First of all, for the functions (2.11) and (2.13) we have

$$
\delta(s) = \delta_p(s) = \begin{cases} \infty & \text{if } 1 \leq p < \infty, \ s \in \Omega_c, \\ \mu(s)^{-1/p} & \text{if } 1 \leq p < \infty, \ s \in \Omega_d, \\ 1 & \text{if } p = \infty, \end{cases} \tag{3.12}
$$

and

$$
\pi(x) = \pi_p(x) = \begin{cases} 0 & \text{if } 1 \leq p < \infty, \\ \overline{\lim_{\Omega_d}} |x(s)| & \text{if } p = \infty, \end{cases} \tag{3.13}
$$

respectively. From (3.12) it follows that

$$
\Delta = \Delta_p = \begin{cases} L_p & \text{if } \Omega_d = \emptyset, \\ \{x : x \in L_p, \ \mu(s)|x(s)|^p \leq 1\} & \text{if } \Omega_c = \emptyset. \end{cases}
$$

On the other hand, from (3.13) it follows that

$$
\Pi = \Pi_p = \begin{cases} L_p & \text{if } 1 \leq p < \infty, \\ \{x : x \in L_\infty, \ |x(s)| \leq 1\} & \text{if } p = \infty; \end{cases}
$$

in particular, $\Pi_\infty = B_1(L_\infty)$ if the underlying measure is atomic-free. Since in all cases we have $\Pi_p \cap \Delta_p = \Delta_p$, formula (2.14) implies that $\Sigma_p = \Delta_p$ for $1 \le p \le \infty$.

We give now a description of the split functional (2.15) in Lebesgue spaces. If μ is atomic-free, a simple calculation shows that

$$H(x) = H_p(x) = \|x\|_p \lceil \|x\|_p \rceil^{p-1} \quad (1 \le p < \infty), \tag{3.14}$$

where $\lceil \vartheta \rceil$ denotes the smallest natural number $n \ge \vartheta$. For practical reasons, instead of (3.14) it is convenient to use the two-sided estimate

$$\|x\|_p^p \le H_p(x) \le 1 + \|x\|_p^p \quad (x \in L_p). \tag{3.15}$$

The case $p = \infty$ is covered, of course, by formula (2.17). If μ is purely atomic, it is not possible to give explicit formulas for the functional H. We just remark that the estimate

$$\|x\|_p^p \le H_p(x) \le 1 + 2\|x\|_p^p \quad (x \in \Delta_p) \tag{3.16}$$

holds. Outside the set Δ_p, the functional H_p is of course infinite.

The estimates (3.15) and (3.16) show that L_p is a Δ_2 space (and hence a δ_2 space and split space, by Lemma 2.5) if $1 \le p < \infty$. Finally, we mention the following obvious fact: if $\Omega_d = \emptyset$, the pair (L_p, L_q) is a V-pair (V_0-pair, V^n-pair, V_0^n-pair, respectively) if and only if $p \le q$ ($p < q$, $p \le nq$, $p < nq$, respectively). Moreover, the space L_p is average-stable and hence, by Lemma 2.9, α-nondegenerate.

3.2 Acting conditions

In Lebesgue spaces it is fairly simple to give acting conditions for the superposition operator which are both necessary and sufficient. First of all, it is important to note that the space L_p is thick (see Section 1.3) in S; consequently, if the superposition operator F acts between two Lebesgue spaces, the generating function f is sup-measurable, and this will always be assumed in this chapter. Moreover, we suppose throughout that the indices p and q belong to the interval $[1, \infty)$; the case $p = \infty$ or $q = \infty$ will be treated in Section 3.7.

Recall (see Section 1.3) that we write $f(s, u) \preceq g(s, u)$ if $f(s, x(s)) \le g(s, x(s))$ for all $x \in S$. The following can be regarded as the Fundamental Theorem on the superposition operator in Lebesgue spaces:

Theorem 3.1   *The superposition operator F generated by f maps L_p into L_q if only if there exist a function $a \in L_q$ and constants $b \geq 0$, $\delta > 0$ such that*

$$|f(s, u)| \preceq a(s) + b|u|^{p/q} \tag{3.17}$$

for all $(s, u) \in \Omega \times \mathbb{R}$ such that

$$\lambda(s) \leq \delta, \quad \mu(s)|u|^p \leq \delta^p. \tag{3.18}$$

$\Rightarrow$ As usual, it suffices to consider the two cases $\Omega_d = \emptyset$ and $\Omega_c = \emptyset$ separately.

Suppose first that $\Omega_d = \emptyset$. In this case (3.18) holds for all $(s, u) \in \Omega \times \mathbb{R}$, since $\lambda(s) = \mu(s) = 0$. The sufficiency of (3.17) follows from the obvious relation

$$\|Fx\|_q \leq \|a\|_q + b\|x\|_p^{p/q}.$$

To prove the necessity of (3.17), suppose that F maps L_p into L_q. By Theorem 2.2, F is locally bounded at zero, hence we can find $b \geq 0$ such that $\|Fx\|_q \leq b$ for $\|x\|_p \leq 1$. Let $a(s, u) = \max\{0, |f(s, u)| - b|u|^{p/q}\}$, which is obviously a nonnegative sup-measurable function. Given $x \in L_p$, denote by $D(x)$ the set of all $s \in \Omega$ for which $Ax(s) > 0$, where A is the superposition operator generated by a, and let $m = \lceil \|P_{D(x)}x\|_p^p \rceil$. By assumption, the measure μ is atomic-free on Ω, hence we can find a partition $\{D_1, \cdots, D_m\}$ of $D(x)$ such that $\|P_{D_j}x\|_p \leq 1$ for $j = 1, \cdots, m$; consequently,

$$\int_\Omega Ax(s)^q ds = \int_{D(x)} Ax(s)^q ds = \sum_{j=1}^m \int_{D_j} a(s, x(s))^q ds$$

$$\leq \sum_{j=1}^m \int_{D_j} |f(s, x(s))|^q ds - b^q \sum_{j=1}^m \int_{D_j} |x(s)|^p ds \leq mb^q - (m-1)b^q = b^q.$$

Applying Lemma 1.3 to the function a, we find a function $\bar{a} \in L_1$ such that $\|\bar{a}\|_1 \leq b^q$ and $a(s, u)^q \preceq \bar{a}(s)$. It remains to choose $a(s) = \bar{a}(s)^{1/q}$, and the estimate (3.17) holds by definition.

Suppose now that $\Omega_c = \emptyset$. If (3.17) holds for $(s, u) \in \Omega \times \mathbb{R}$ with (3.18), and if $x \in L_p$ is fixed, there are only finitely many atoms $s \in \Omega$ for which either $\lambda(s) > \delta$ or $\mu(s)^{1/p}|x(s)| > \delta$; denote the set of all other $s \in \Omega$ by $D(x)$. We have $|f(s, x(s))| \leq a(s) + b|x(s)|^{p/q}$ for $s \in D(x)$; hence $Fx \in L_q$, since $\Omega \setminus D(x)$ is finite.

The proof of the necessity of (3.17) is again somewhat harder than sufficiency. Observe first that, if F maps L_p into L_q, we can find two numbers
$\varepsilon, \delta > 0$ and a finite set $D \subseteq \Omega$ such that $x(s) = 0$ on D, and $\|Fx\|_q \leq \varepsilon$ for
$\|x\|_p \leq \delta$.

Let

$$a(s, u) = \begin{cases} \max\{0, |f(s,u)| - 2^{1/q}\delta^{-p/q}\varepsilon|u|^{p/q}\} & (s \notin D), \\ 0 & (s \in D), \end{cases}$$

which is again a nonnegative sup-measurable function. Given $x \in L_p$, define
$D(x)$ as before, and let $m = \lceil \delta^{-p}\|x\|_p^p \rceil$. Divide the set $D(x)$ into $2m + 1$
subsets $D_0, \cdots, D_{2m}$ such that $\|P_{D_j}x\|_p \leq \delta$ for $j = 0, \cdots, 2m$. Then the
superposition operator $Ax(s) = a(s, x(s))$ satisfies

$$\int_\Omega Ax(s)^q ds = \sum_{s \in \Omega} Ax(s)^q \mu(s) = \sum_{s \in D(x)} Ax(s)^q \mu(s)$$

$$\leq \sum_{s \in D(x)} |f(s, x(s))|^q \mu(s) - 2\delta^{-p}\varepsilon^q \sum_{s \in D(x)} |x(s)|^p$$

$$\leq \sum_{j=0}^{2m} \sum_{s \in D_j} |f(s, x(s))|^q \mu(s) - 2\delta^{-p}\varepsilon^q \|x\|_p^p$$

$$\leq (2m + 1)\varepsilon^q - 2\varepsilon^q \delta^{-p} m \delta^p = \varepsilon^q .$$

The rest of the proof follows from Lemma 1.3 as before. $\Leftarrow$

As already observed in the proof, in the case of an atomic-free measure μ
the restriction (3.18) is superfluous, and (3.17) holds for all $(s, u) \in \Omega \times \mathbb{R}$.
Roughly speaking, this means that the class of functions f which generate
a superposition operator F between L_p and L_q is in case $\Omega_d = \emptyset$ rather narrow: it contains only functions of polynomial growth in u (with "maximal"
exponent p/q). On the other hand, if μ is a counting measure (i.e. $\mu(s) = 1$
for $s \in \Omega = \Omega_d$), condition (3.17) must hold for $s \in \Omega \setminus D$ and $|u| \leq \delta$, with
$\delta > 0$ and D finite.

We point out that in the case when f is a Carathéodory function or, more
generally, a Shragin function, (3.17) becomes the usual growth condition

$$|f(s, u)| \leq a(s) + b|u|^{p/q} , \tag{3.19}$$

by Lemma 1.5.

3.3 The growth function

In this section we shall be concerned with several boundedness properties of the superposition operator in Lebesgue spaces. First of all, Theorem 2.2 implies immediately the following:

Theorem 3.2   If F acts from L_p into L_q, then F is locally bounded if and only if the function $f(s, \cdot)$ is bounded for each $s \in \Omega_d$.

If the underlying measure μ is atomic-free, Theorem 3.2 is of course not interesting, since F is in this case even bounded on each ball in L_p, by Theorem 2.3. In the case $\Omega_c = \emptyset$, however, F may fail to be bounded even if F is locally bounded at each point of its domain of definition. Consider, for example, the set $\Omega = \mathbb{N}$ with the counting measure μ, and the function $f(s, u) = u^s$ ($s \in \mathbb{N}$, $u \in \mathbb{R}$). This function generates a superposition operator from any L_p into any L_q which is bounded on each ball of radius $r < 1$, but unbounded on each ball of radius $r > 1$.

The next theorem gives a necessary and sufficient boundedness condition for the superposition operator. Recall that by μ_F we denote the growth function (2.24) of F; moreover, for $r > 0$ let

$$\nu_f(r) = \inf\{\|a_r\|_1 + b_r r^p\}, \qquad (3.20)$$

where the infimum is taken over all pairs $(a_r, b_r) \in L_1 \times [0, \infty)$ such that

$$|f(s, u)|^q \preceq a_r(s) + b_r |u|^p \qquad (3.21)$$

for $\mu(s)|u|^p \leq r^p$.

Theorem 3.3   If F acts from L_p into L_q, then F is bounded on bounded sets if and only if for each $r > 0$ there exist a function $a_r \in L_1$ and a constant $b_r \geq 0$ such that (3.21) holds for all $(s, u) \in \Omega \times \mathbb{R}$ such that $\mu(s)|u|^p \leq r^p$. Moreover, the two-sided estimate

$$\mu_F(r) \leq \nu_f(r) \leq \|F\theta\|_q + 3\mu_F(r) \qquad (3.22)$$

holds, where $\mu_F(r)$ is defined by (2.24) and $\nu_f(r)$ by (3.20).

$\Rightarrow$ The sufficiency of (3.21) for the boundedness of F and the estimate $\mu_F(r) \leq \nu_f(r)$ are obvious. To prove the necessity, we define for $r > 0$ the auxiliary function

$$a_r(s, u) = \max\left\{0, |f(s, u)|^q - |f(s, 0)|^q - 2r^{-p}\mu_F(r)^q |u|^p\right\}.$$

Given $x \in L_p$ with $\mu(s)|x(s)|^p \leq r^p$, denote by $D(x)$ the set of all $s \in \Omega$ for which $A_r x(s) > 0$, where A_r is the superposition operator generated by a_r. By (3.15) and (3.16), we can write $P_{D(x)} x$ as sum of at most m disjoint functions $x_1, \cdots, x_m$ with $\|x_j\|_p \leq r$, where $m = \lceil 2r^{-p}\|P_{D(x)}x\|_p^p \rceil$. Since $a_r(s, 0) = 0$, we have

$$\int_\Omega A_r x(s)ds = \int_{D(x)} A_r x(s)ds$$

$$= \sum_{j=1}^m \int_\Omega |f(s, x_j(s))|^q ds - \sum_{j=1}^m |f(s, 0)|^q ds - 2r^{-p}\mu_F(r)^q \sum_{j=1}^m \int_\Omega |x_j(s)|^p ds$$

$$\leq (2r^{-p}\|P_{D(x)}x\|_p^p + 1)\mu_F(r)^q - 2r^{-p}\mu_F(r)^q\|P_{D(x)}x\|_p^p \leq \mu_F(r)^q \, ;$$

hence

$$\int_\Omega A_r x(s)ds \leq \mu_F(r)^q \quad (x \in L_p, \, \mu(s)|x(s)|^p \leq r^p).$$

Lemma 1.3 implies that $a_r(s, u) \preceq \bar{a}_r(s)$ $(\mu(s)|u|^p \leq r^p)$ for some $\bar{a}_r \in L_1$ with $\|\bar{a}_r\|_1 \leq \mu_F(r)^q$. But this means that $|f(s, u)|^q \preceq |f(s, 0)|^q + \bar{a}_r(s) + 2r^{-p}\mu_F(r)^q|u|^p$ for $\mu(s)|u|^p \leq r^p$; consequently, we may take $a_r(s) = |f(s, 0)|^q + \bar{a}_r(s)$ and $b_r = 2r^{-p}\mu_F(r)^q$. Moreover, by definition (3.20) we get

$$\nu_f(r)^q \leq \|F\theta\|_q^q + \mu_F(r)^q + 2r^{-p}\mu_F(r)^q r^p = \|F\theta\|_q^q + 3\mu_F(r)^q \, ,$$

which proves the second estimate in (3.22). $\Leftarrow$

We point out that Theorem 3.3 simplifies in case $\Omega_d = \emptyset$: by Theorem 2.3, F is always bounded from L_p into L_q; in fact, since $\mu(s) = 0$ in this case, condition (3.21) reads

$$|f(s, u)|^q \preceq a(s) + b|u|^p \, , \tag{3.23}$$

which is just the acting condition (3.17) of Theorem 3.1.

Evidently, for many applications it is desirable to find explicit formulas for the growth function (2.24), rather than just estimates. Surprisingly, it is in fact possible to find such formulas. We shall now describe this in detail in the case $p = q = 1$; afterwards we shall show that this is not really a restriction.

For brevity, we shall call a Carathéodory function f admissible if $f(s, u) \geq 0$ and $f(s, 0) \equiv 0$ (nonnegativity), $f(s, -u) = f(s, u)$ (evenness), and $f(s, (1 - \lambda)u_0 + \lambda u_1) \geq (1 - \lambda)f(s, u_0) + \lambda f(s, u_1)$ for $0 \leq \lambda \leq 1$(concavity).

For such functions, the limits $\alpha(s) = \lim_{u \to \infty} \frac{1}{u} f(s, u)$, and $\beta(s) = \lim_{u \to 0} \frac{1}{u} f(s, u)$ exist, and the left and right derivatives of f with respect to u satisfy $\alpha(s) \leq f_u^{(r)}(s, u) \leq f_u^{(\ell)}(s, u) \leq \beta(s)$; it is convenient to assume here that $f_u^{(\ell)}(s, 0) = \infty$. By $f_u'(s, u)$ we denote the interval $[f_u^{(r)}(s, u), f_u^{(\ell)}(s, u)]$, i.e. we identify the number $f_u'(s, u) = \frac{\partial}{\partial u} f(s, u)$ and the singleton $\{f_u'(s, u)\}$ if the derivative exists. With this notation, the relation $\lambda \in f_u'(s, u)$ certainly has a solution for every $\lambda > \alpha(s)$, and sometimes also for $\lambda = \alpha(s)$. Let $H(\lambda)$ be the set of all function $w \in S$ for which $\lambda \in f_u'(s, w(s))$, and let Λ be the set of all $\lambda \geq 0$ such that $H(\lambda) \not\subseteq \emptyset$. Obviously, Λ is either the interval (α_*, ∞), or the interval $[\alpha_*, \infty)$, where

$$\alpha_* = \|\alpha\|_\infty = \operatorname*{ess\,sup}_{s \in \Omega} \alpha(s) .$$

It is not hard to see that the set $H(\lambda)$ is closed and convex, and in fact coincides with the set of all $w \in S$ such that $w_-(\lambda, s) \leq w(s) \leq w_+(\lambda, s)$, where $w_-(\lambda, s)$ and $w_+(\lambda, s)$ are the minimal and maximal solutions, respectively, of the relation $\lambda \in f_u'(s, u)$.

We still introduce the multi-valued function $\omega(\lambda) = [\omega_-(\lambda), \omega_+(\lambda)]$ $(\lambda \in \Lambda)$, where $\omega_-(\lambda) = \inf\{\|w\|_1 : w \in H(\lambda)\}$, and $\omega_+(\lambda) = \sup\{\|w\|_1 : w \in H(\lambda)\}$. By construction, for each $r \in \omega(\lambda)$ one can find a function $w \in H(\lambda)$ such that $\|w\|_1 = r$.

Observe that $\lim_{\lambda \to \infty} \omega_-(\lambda) = 0$; unfortunately, the analogous relation $\lim_{\lambda \to \alpha_*} \omega_+(\lambda) = \infty$ is not always true. Let $r_* = \lim_{\lambda \to \alpha_*} \omega_+(\lambda)$; if r_* is finite, we must still introduce the set $H(\alpha_*, \varepsilon)$ of all $w \in S$ for which

$$\alpha_* - \varepsilon \leq f_u^{(r)}(s, w(s)) \leq f_u^{(\ell)}(s, w(s)) \leq \alpha_* + \varepsilon .$$

These sets are also nonempty for any $\varepsilon > 0$; moreover, for each $r \geq r_*$ one can find a function $w_\varepsilon \in H(\alpha_*, \varepsilon)$ such that $\|w_\varepsilon\|_1 = r$.

The following theorem gives an explicit formula for calculating the growth function (2.24) of F in the space L_1. To this end, let

$$W_r = \bigcup_{r \in \omega(\lambda)} H(\lambda) \cap B_r(L_1) \quad (r < r_*)$$

and

$$W_{r,\varepsilon} = H(\alpha_*, \varepsilon) \cap B_r(L_1) \quad (r \geq r_*) .$$

Theorem 3.4   Let f be an admissible function, and suppose that the superposition operator F generated by f acts in the space L_1. Then the *growth function (2.24) of F is given by*

$$\mu_F(r) = \begin{cases} \|Fw_r\|_1 & \text{if } r < r_*, \\ \lim_{\varepsilon \to 0} \|Fw_{r,\varepsilon}\|_1 & \text{if } r \geq r_*, \end{cases}$$

where $w_r \in W_r$ and $w_{r,\varepsilon} \in W_{r,\varepsilon}$ are arbitrary. Moreover, the equality

$$\mu_F(r) = \nu_f(r) \tag{3.24}$$

holds, where

$$\nu_f(r) = \inf\{\|a\|_1 + br : |f(s,u)| \leq a(s) + b|u|\}. \tag{3.25}$$

$\Rightarrow$ Suppose first that $r < r_*$ and $\|x\|_1 \leq r$. Then for any $w \in W_r$ we have $f(s,u) \leq f(s,w(s)) + \lambda(u - w(s))$, and hence, for $u = x(s)$, $f(s,x(s)) \leq f(s,w(s)) + \lambda(|x(s)| - w(s))$, i. e. $\|Fx\|_1 \leq \|Fw\|_1 + \lambda(\|x\|_1 - r) \leq \|Fw\|_1$, which implies that $\mu_F(r) \leq \|Fw\|_1$. The converse inequality follows trivially from the fact that $\|w\|_1 = r$.

Now let $r \geq r_*$. Then for $w \in W_{r,\varepsilon}$ we have $f(s,u) \leq f(s,w(s)) + \alpha_*(u - w(s)) + \varepsilon(u + w(s))$, and hence, for $u = x(s)$, $f(s,x(s)) \leq f(s,w(s)) + \alpha_*(|x(s)| - w(s)) + \varepsilon(|x(s)| + w(s))$, i.e. $\|Fx\|_1 \leq \|Fw\|_1 + \alpha_*(\|x\|_1 - r) + 2\varepsilon r \leq \|Fw\|_1 + 2\varepsilon r$, which implies that $\mu_F(r) \leq \lim_{\varepsilon \to 0} \|Fw\|_1$. The inequality $\mu_F(r) \geq \overline{\lim_{\varepsilon \to 0}} \|Fw\|_1$ is obvious.

To show (3.24), it suffices to choose $a(s) = f(s,w(s)) - \lambda w(s)$, $b = \lambda$ in the case $r < r_*$, and $a(s) = f(s,w(s)) - (\alpha_* - \varepsilon)w(s)$, $b = \alpha_* + \varepsilon$ in the case $r \geq r_*$. $\Leftarrow$

Theorem 3.4 gives explicit formulas for the calculation of $\mu_F(r)$ if the function f is admissible. The general case can always be reduced to this special one, provided the measure μ is atomic-free. To see this, let f be a sup-measurable function which generates a superposition operator F in L_1. By Theorem 3.1, we then have

$$|f(s,u)| \preceq a(s) + b|u| \tag{3.26}$$

for some $a \in L_1$ and $b \geq 0$. Denote by $a_b(s)$ the infimum (in the space S) of all functions $a \in L_1$ for which (3.26) holds for fixed $b \geq 0$. Then (3.26) is true for $a = a_b$, the function a_b depends monotonically on b, and the function

$$\tilde{f}(s,u) = \inf_{b \geq 0}\{a_b(s) + b|u|\}$$

is sup-measurable, satisfies the estimate $|f(s, u)| \leq \widetilde{f}(s, u)$, and defines a superposition operator $\widetilde{F}$ in L_1. Moreover, the function $\widetilde{f}(s, u) - \widetilde{f}(s, 0)$ is admissible; therefore we call $\widetilde{f}$ the *admissible majorant* of f.

The following theorem, which is proved by standard methods of convex analysis, essentially completes Theorem 3.4.

Theorem 3.5 *Let f be a sup-measurable function, and suppose that the superposition operator F generated by f acts in the space L_1, where the underlying measure μ is atomic-free. Let $\widetilde{f}$ be the admissible majorant of f, and $\widetilde{F}$ the superposition operator generated by $\widetilde{f}$ in L_1. Then $\mu_{\widetilde{F}}(r) = \mu_F(r)$ and $\nu_{\widetilde{f}}(r) = \nu_f(r)$; consequently, (3.24) holds.*

So far we have only dealt with the case $p = q = 1$. It is not hard to see, however, that Theorem 3.4 and 3.5 carry over to the general case $1 \leq p < \infty$, $1 \leq q < \infty$. This follows from the following elementary lemma:

Lemma 3.1 *If a sup-measurable function f generates a superposition operator F from L_p into L_q, the function $g(s, u) = |f(s, |u|^{1/p}\mathrm{sgn}\, u)|^q$ generates a superposition operator G in L_1. Conversely, if a sup-measurable function g generates a superposition operator G in L_1, the function $f(s, u) = |g(s, |u|^p \mathrm{sgn}\, u)|^{1/q}$ generates a superposition operator F from L_p into L_q.*

3.4 Absolute boundedness and uniform continuity

Recall that the operator F is called absolutely bounded from L_p into L_q if, whenever N is a bounded subset of L_p, FN is an absolutely bounded subset of L_q (see (2.26)). If F is compact, F is also absolutely bounded; the converse is only true in the case $\Omega_c = \emptyset$ (see Theorem 2.5). Roughly speaking, compactness of F may be considered as a combination of two properties, namely absolute boundedness and "constancy on Ω_c", i.e. $f(s, u) \simeq c(s)$ ($s \in \Omega_c$).

Theorem 3.6 *Let f be a sup-measurable function, and suppose that the superposition operator F generated by f acts from L_p into L_q. Then the following three conditions are equivalent:*

(a) F is absolutely bounded.

(b) For each $r > 0$ there exists a monotonically increasing function Φ_r on $[0, \infty)$ such that (2.29) holds, and the superposition operator $\widetilde{F}_r$ generated by the function $\widetilde{f}_r(s, u) = u_0(s)\Phi_r[|f(s, u)|u_0(s)^{-1}]$ (with u_0 an arbitrary unit in L_q) is bounded on the ball $B_r(L_p)$.

(c) For each $r > 0$ and $\varepsilon > 0$ one can find a function $a_\varepsilon \in L_1$ such that

$$|f(s, u)|^q \preceq a_\varepsilon(s) + \varepsilon|u|^p \quad (\mu(s)|u|^p \leq r^p). \tag{3.27}$$

$\Rightarrow$ The fact that (a) implies (b) follows from Theorem 2.4. Suppose that (b) holds. By Theorem 3.3 we have $[u_0(s)\Phi_r(u_0(s)^{-1}|f(s,u)|)]^q \preceq a_r(s) + b_r|u|^p$ for $\mu(s)|u|^p \leq r^p$, where $a_r \in L_1$ and $b_r \geq 0$. Given $\varepsilon > 0$, choose $\omega = \omega(\varepsilon) > 0$ such that $\Phi_r(t) \geq b_r^{1/q}\varepsilon^{-1/q}t$ for $t \geq \omega$, which is possible by (2.29). For $u_0(s)^{-1}|f(s,u)| \geq \omega$ we get then $|f(s,u)|^q \leq b_r^{-1}\varepsilon[u_0(s)\Phi_r(u_0(s)^{-1}|f(s,u)|)]^q \preceq b_r^{-1}\varepsilon a_r(s) + \varepsilon|u|^p$, while for $u_0(s)^{-1}$ $|f(s,u)| < \omega$ we get $|f(s,u)|^q \leq \omega^q u_0(s)^q$. Thus choosing $a_\varepsilon(s) = b_r^{-1}\varepsilon a_r(s)$ $+\omega^q u_0(s)^q$ yields (3.27), i.e. (c) holds.

Finally, suppose that (c) is true, and fix $x \in B_r(L_p)$. For any $D \in \mathcal{M}$ we have, by (3.27), that $\|P_D Fx\|_q^q \leq \|P_D a_\varepsilon\|_1 + \varepsilon r^p$. Since L_1 is regular, we have

$$\lim_{\lambda(D)\to 0} \sup_{x\in B_r(L_p)} \|P_D Fx\|_q^q \leq \varepsilon r^p,$$

and thus $FB_r(L_p)$ is absolutely bounded in L_q, since $\varepsilon > 0$ is arbitrary. $\Leftarrow$

We pass now to continuity properties of the superposition operator in Lebesgue spaces. The general Theorems 1.4 and 2.6 imply immediately the following:

Theorem 3.7 *Let f be a sup-measurable function, and suppose that the superposition operator F generated by f acts from L_p into L_q. Then F is continuous if and only if f is sup-equivalent to some Carathéodory function.*

As the example after Theorem 2.8 shows, continuity of F does not imply uniform continuity (on bounded sets). In Lebesgue spaces, however, one can give a necessary and sufficient condition for uniform continuity which is more explicit than Theorem 2.9. To state this condition, we recall that the *modulus of continuity* of an operator F is defined by

$$\omega_F(r,\delta) = \sup\{\|Fx_1 - Fx_2\| : \|x_1\| \leq r, \|x_2\| \leq r, \|x_1 - x_2\| \leq \delta\}. \quad (3.28)$$

Moreover, for $r > 0$ and $\delta > 0$ let

$$\nu_f(r,\delta) = \inf\{\|a_\varepsilon\|_q + 2\varepsilon r^{p/q} + b_\varepsilon \delta^{p/q}\}, \quad (3.29)$$

where the infimum is taken over all triples $(a_\varepsilon, \varepsilon, b_\varepsilon) \in L_q \times (0,\infty) \times [0,\infty)$ such that

$$|f(s,u) - f(s,v)| \preceq a_\varepsilon(s) + \varepsilon(|u|^{p/q} + |v|^{p/q}) + b_\varepsilon|u - v|^{p/q} \quad (3.30)$$

for $\mu(s)|u|^p \leq r^p$, $\mu(s)|v|^p \leq r^p$, and $\mu(s)|u - v|^p \leq \delta^p$.

Theorem 3.8   *Let f be a sup-measurable function, and suppose that the superposition operator F generated by f acts from L_p into L_q. Then F is uniformly continuous on bounded sets if and only if for each $\varepsilon > 0$ and $r > 0$ one can find $b_\varepsilon \geq 0$, $\delta > 0$ and $a_\varepsilon \in L_q$ with $\|a_\varepsilon\|_q \leq \varepsilon$ such that (3.30) holds for $\mu(s)|u|^p \leq r^p$, $\mu(s)|v|^p \leq r^p$, and $\mu(s)|u-v|^p \leq \delta^p$. Moreover, the two-sided estimate*

$$\omega_F(r,\delta) \leq \nu_f(r,\delta) \leq (1 + 2^{1+1/q})\omega_F(r,\delta) \tag{3.31}$$

holds, where $\omega_F(r,\delta)$ is defined by (3.28) and $\nu_f(r,\delta)$ by (3.29).

⇒ We shall not prove the sufficiency of (3.30) and the first inequality in (3.31), since they are completely analogous to the corresponding statements in Theorem 3.3.

   To prove the necessity of (3.30), we may assume that f is a Carathéodory function; we claim that for every $x \in L_p$

$$|f(s, x(s) + u) - f(s, x(s))| \leq a(s) + 2^{1/q}\delta^{-p/q}\omega_F(r,\delta)|u|^{p/q}$$

for $\mu(s)|u|^p \leq \delta^p$ and $\mu(s)|x(s)+u|^p \leq r^p$, where $a \in L_q$ (possibly depending on x) has norm $\|a\|_q \leq \omega_F(r,\delta)$. In fact, let

$$g_x(s, u) = \max\left\{0, |f(s, x(s) + u) - f(s, x(s))| - 2^{1/q}\delta^{-p/q}\omega_F(r,\delta)|u|^{p/q}\right\},$$

and denote by G_x the superposition operator generated by g_x. Given $h \in L_p$ such that $\mu(s)|h(s)|^p \leq \delta^p$ and $\mu(s)|x(s) + h(s)|^p \leq r^p$, let $D(h)$ be the set of all $s \in \Omega$ for which $G_x h(s) > 0$. As in Theorem 3.3, we can write $P_{D(h)}h$ as sum of at most m disjoint functions $h_1, \cdots, h_m$, where $m = \lceil 2\delta^{-p}\|P_{D(h)}h\|_p^p \rceil$, such that $\|h_j\|_p \leq \delta$ and $|x(s) + h_j(s)| \leq r$. But then $\|F(x + h_j) - Fx\|_q \leq \omega_F(r,\delta)$, and hence

$$\int_\Omega G_x h(s)^q ds = \sum_{j=1}^m \int_\Omega G_x h_j(s)^q ds$$

$$= \sum_{j=1}^m \int_\Omega [|f(s, x(s) + h_j(s)) - f(s, x(s))| - 2^{1/q}\delta^{-p/q}\omega_F(r,\delta)|h_j(s)|^{p/q}]^q ds$$

$$\leq \sum_{j=1}^m \int_\Omega |f(s, x(s) + h_j(s)) - f(s, x(s))|^q ds - 2\delta^{-p}\omega_F(r,\delta)^q \int_{D(h)} |h(s)|^p ds$$

$$\leq (2\delta^{-p}\|P_{D(h)}h\|_p^p + 1)\omega_F(r,\delta)^q - 2\delta^{-p}\omega_F(r,\delta)^q\|P_{D(h)}h\|_p^p \leq \omega_F(r,\delta)^q.$$

This shows that $\|G_x h\|_q \le \omega_F(r, \delta)$, and thus (3.30) holds.

Further, the function

$$g(s, u) = \sup_t \max\{0, |f(s, u+t) - f(s, t)| - 2^{1/q}\delta^{-p/q}\omega_F(r, \delta)|t|^{p/q}\}, \quad (3.32)$$

where the supremum is taken over all t such that $\mu(s)|t|^p \le r^p$ and $\mu(s)|x(s) + t|^p \le r^p$, generates a superposition operator G which maps the ball $B_r(L_p)$ into the ball $B_\omega(L_q)$ with $\omega = \omega_F(r, \delta)$. By the above reasoning, we see that the function (3.32) satisfies the estimate $|g(s, u)| \le a(s) + 2^{1/q}r^{-p/q}\omega_F(r, \delta)$ $|u|^{p/q}$ for $\mu(s)|u|^p \le r^p$, where $a \in L_q$ has norm $\|a\|_q \le \omega_F(r, \delta)$. Now, this implies that

$$\begin{aligned}|f(s, u) - f(s, v)| \le &a(s) + 2^{1/q}r^{-p/q}\omega_F(r, \delta)(|u|^{p/q} + |v|^{p/q}) \\ &+ 2^{1/q}\delta^{-p/q}\omega_F(r, \delta)|u - v|^{p/q}.\end{aligned} \quad (3.33)$$

Note that (3.33) is just (3.30), where only ε is replaced by $2^{1/q}r^{-p/q}\omega_F(r, \delta)$ and b_ε by $2^{1/q}\delta^{-p/q}\omega_F(r, \delta)$. Since $\lim_{\delta \to 0} \omega_F(r, \delta) = 0$, by assumption, (3.30) holds for any given $\varepsilon > 0$, provided $\delta > 0$ is sufficiently small. Moreover, by (3.29), we conclude that $\nu_f(r, \delta) \le \|a\|_q + 2^{1/q}r^{-p/q}\omega_F(r, \delta)r^{p/q} + 2^{1/q}\delta^{-p/q}\omega_F(r, \delta)\delta^{p/q} \le (1 + 2^{1+1/q})\omega_F(r, \delta)$, which completes the proof. $\Leftarrow$

We point out that the conditions of Theorem 3.8 are again much easier formulated in the case $\Omega_d = \emptyset$.

We conclude this section with a criterion for weak continuity of the superposition operator in Lebesgue spaces which is just a reformulation of Theorem 2.10.

Theorem 3.9 *Let f be a sup-measurable function, and suppose that the superposition operator F generated by f acts from L_p into L_q. Then F is weakly continuous if and only if the restriction of f to $\Omega_c \times \mathbb{R}$ satisfies*

$$f(s, u) \simeq a(s) + b(s)u \quad (s \in \Omega_c, u \in \mathbb{R}), \quad (3.34)$$

with $a \in L_q$ and $b \in L_q/L_p$, and the restriction of f to $\Omega_d \times \mathbb{R}$ is a Carathéodory function.

3.5 Lipschitz and Darbo conditions

In this section we shall be concerned with superposition operators which satisfy a Lipschitz or Darbo condition in Lebesgue spaces. By Theorem 3.7, we may assume without loss of generality that the generating function f satisfies the Carathéodory conditions.

The following result is an immediate consequence of Theorems 2.11 and 2.12:

Theorem 3.10   *Let f be a Carathéodory function, and suppose that the superposition operator F generated by f acts from L_p into L_q. Then the following three conditions are equivalent:*
(a) The operator F satisfies a Lipschitz condition

$$\|Fx_1 - Fx_2\|_q \leq k(r)\|x_1 - x_2\|_p \quad (x_1, x_2 \in B_r(L_p)). \qquad (3.35)$$

(b) Given two functions $x_1, x_2 \in B_r(L_p)$, one can find a function $\xi \in B_{k(r)}(L_q/L_p)$ such that (2.45) holds.
(c) The function f satisfies a Lipschitz condition

$$|f(s,u) - f(s,v)| \leq g(s,w)|u - v| \quad (|u|, |v| \leq w), \qquad (3.36)$$

where the function g generates a superposition operator G which maps the ball $B_r(L_p)$ into the ball $B_{k(r)}(L_q/L_p)$.

Moreover, in the case $p = q$ ($p < q$, respectively) the function g in (3.36) is essentially bounded (is identically zero, respectively). Finally, in the case $p > q$, each of the above conditions is equivalent to the following one:
(d) For each $\lambda > 0$, there exists a function $a_\lambda \in L_1$ such that $c(\rho) = \sup_{\lambda \geq \rho} \|a_\lambda\|_1 < \infty$ ($\rho > 0$) and

$$|f(s,u) - f(s,v)|^q \leq \lambda^{-q} a_\lambda(s) + \lambda^{p-q}|u - v|^p. \qquad (3.37)$$

$\Rightarrow$ Only the equivalence of (a) and (d) is not contained in Theorems 2.11 and 2.12 and therefore has to be proved. Now, (3.37) implies that for $x_1, x_2 \in B_r(L_p)$ and $\lambda = \|x_1 - x_2\|_p^{-1}$ we have $\|Fx_1 - Fx_2\|_q^q \leq \|x_1 - x_2\|_p^q c(1/2r) + \|x_1 - x_2\|_p^q$, i.e. (a) holds with $k(r) = (c(1/2r) + 1)^{1/q}$. Conversely, if (3.35) holds for $x_1, x_2 \in B_r(L_p)$, then for $\lambda > 0$ let

$$g_\lambda(s, u, v) = \max\left\{0, |f(s,u) - f(s,v)|^q - \lambda^{p-q}|u - v|^p\right\}.$$

Given two functions $x_1, x_2 \in B_r(L_p)$, let $D(x_1, x_2)$ denote the set of all $s \in \Omega$ such that $G_\lambda(x_1, x_2)(s) > 0$, where G_λ is the superposition operator (of two variables) generated by g_λ. For $m = \lceil k(r)^{pq/(p-q)}\lambda^{-p}\rceil$ choose a partition $\{D_1, \cdots, D_m\}$ of Ω such that

$$\int_{D_j} |x_1(s) - x_2(s)|^p ds \leq k(r)^{pq/(p-q)} \lambda^{-p} \quad (j = 1, \cdots, m).$$

Then (3.35) implies that

$$\int_{D_j} |Fx_1(s) - Fx_2(s)|^q ds \leq k(r)^{pq/(p-q)} \lambda^{-q},$$

hence $\|G_\lambda(x_1, x_2)\|_1 \leq mk(r)^{pq/(p-q)} \lambda^{-q} - (m-1)k(r)^{pq/(p-q)} \lambda^{-q} = k(r)^{pq/(p-q)} \lambda^{-q}$. Consequently, the function

$$a_\lambda(s) = \lambda^q \sup_{u,v>0} g_\lambda(s, u, v)$$

belongs to L_1 and $c(\rho) = \sup_{\lambda \geq \rho} \|a_\lambda\|_1 \leq k(1/2\rho)^{pq/(p-q)}$. $\Leftarrow$

We remark that results analogous to those in Theorem 3.10 may be formulated for a *Hölder condition*

$$\|Fx_1 - Fx_2\|_q \leq k(r)\|x_1 - x_2\|_p^\alpha \quad (x_1, x_2 \in B_r(L_p)) \tag{3.38}$$

with Hölder exponent $\alpha \in (0, 1]$. For instance, if $p < \alpha q$, (3.38) leads to the following condition: $f(s, \cdot)$ is constant for $s \in \Omega_c$, and $|f(s, u) - f(s, v)| \leq k(r)|u - v|^\alpha \mu(s)^{\alpha/p}$ for $s \in \Omega_d$ with $\mu(s)|u|^p \leq r^p$ and $\mu(s)|v|^p \leq r^p$. On the other hand, if $p > \alpha q$, (3.38) is equivalent to the condition

$$|f(s, u) - f(s, v)| \leq \frac{p - \alpha q}{p} \varepsilon^{-p/(p-\alpha q)} g_\varepsilon(s, w) + \frac{\alpha q}{p} \varepsilon^{p/\alpha q} |u - v|^{p/q}$$

for $|u|, |v| \leq w$, $\mu(s)|u|^p \leq r^p$, $\mu(s)|v|^p \leq r^p$, and $\mu(s)|u-v|^p \leq r^p$, where the functions g_ε generate a family of uniformly bounded superposition operators G_ε $(0 < \varepsilon < 1)$ from L_p into L_q. The condition (3.37) of Theorem 3.10 becomes

$$|f(s, u) - f(s, v)|^q \leq \lambda^{-\alpha q} a_\lambda(s) + \lambda^{p-\alpha q}|u - v|^p.$$

Such results may be obtained by either adapting the proof for the Lipschitz condition (3.35) (i.e. $\alpha = 1$), or observing that (3.38) means that $\omega_F(r, \delta) = O(\delta^\alpha)$, which by (3.31) is in turn equivalent to $\nu_f(r, \delta) = O(\delta^\alpha)$, with $\nu_f(r, \delta)$ as given in (3.29).

Finally, we want to point out a special aspect of condition (3.35). If $f(s, 0) = 0$, (3.35) implies a *sublinear growth estimate* for F, i.e.

$$\|Fx\|_q \le k(r)\|x\|_p \quad (x \in B_r(L_p)). \tag{3.39}$$

Suppose now that the function f satisfies a sublinear growth estimate with respect to u, i.e.

$$|f(s, u)| \le \ell(s)|u| \tag{3.40}$$

for some $\ell \in L_{pq/(p-q)}$ $(p \ge q)$. Obviously, it follows from the Hölder inequality (3.5) that (3.40) is sufficient for (3.39) (with $k(r) = \|\ell\|_{pq/(p-q)}$). Surprisingly, the converse is not true, at least in the case $p > q$, as the following example shows: let $\Omega = [0, 1]$, with the Lebesgue measure μ, and let

$$f(s, u) = \begin{cases} |u|^{p/q}s^{-(p-q)/q^2}[1 - \log|u|^{p/q}s^{-(p-q)/q^2}] & \text{if } |u|^{p/q} > s^{(p-q)/q^2}, \\ 1 & \text{if } |u|^{p/q} \le s^{(p-q)/q^2}. \end{cases}$$

An easy computation shows that

$$f(s, u) = \inf_{\eta > 0}\{\ell_\eta(s) + \eta u^{p/q}\},$$

where $\ell_\eta(s) = \exp[-\eta s^{(p-q)/q^2}]$; thus $\|Fx\|_q \le \|\ell_\eta\|_q + \eta\|x\|_p^{p/q} = O(\eta^{-q/(p-q)}) + \eta\|x\|_p^{p/q}$. Taking the minimum of the right side (for η) yields $\eta_{\min} \sim \|x\|_p^{-(p-q)/q}$; hence $\|Fx\|_q = O(\|x\|_p)$, i.e. (3.39) holds. On the other hand, for large $|u|$ we have

$$\frac{1}{|u|}|f(s, u)| \sim \frac{1}{|u|} \sim s^{-(p-q)/pq},$$

which does not belong to $L_{pq/(p-q)}$ over $\Omega = [0, 1]$; thus, (3.40) fails.

To conclude this section, we want to compare the Lipschitz condition (3.35) with the Darbo condition

$$\alpha_q(FN) \le k(r)\alpha_p(N) \quad (N \subseteq B_r(L_p)), \tag{3.41}$$

where α_p denotes the Hausdorff measure of noncompactness (2.32) in the space L_p. Since the Lebesgue space L_q is α-nondegenerate for $1 \le q < \infty$, Theorem 2.13 implies immediately the following:

Theorem 3.11   *Let f be a Carathéodory function, and suppose that the superposition operator F generated by f acts from L_p into L_q. Then the Lipschitz condition (3.35) and the Darbo condition (3.41) are equivalent.*

Theorem 3.11 implies again that every *compact* superposition operator in L_p is necessarily constant, in accordance with Theorem 2.5.

3.6 Differentiability conditions and analyticity

In this section we shall formulate the differentiability results of Sections 2.7 and 2.8 for the superposition operator between Lebesgue spaces. It turns out that more precise statements are possible than those given in Chapter 2 for general ideal spaces. For example, one can give differentiability conditions for the operator F in terms of the function f which are both necessary and sufficient. Without loss of generality, we assume again that f is a Carathéodory function.

We begin by rephrasing Theorems 2.14 and 2.15, which now read as follows:

Theorem 3.12   *Let f be a Carathéodory function, and suppose that the superposition operator F generated by f acts from L_p into L_q. If F is differentiable at $x \in L_p$, the derivative has the form (2.55), where the function a given by (2.56) belongs to the multiplicator space L_q/L_p. In particular, in the case $p = q$ the function f has the form (2.59) for $s \in \Omega_c$ with $a \in L_q$ and $b \in L_\infty$; in the case $p < q$ the function $f(s, \cdot)$ is constant for $s \in \Omega_c$.*

Conversely, if the superposition operator G generated by the function

$$g(s, u) = \begin{cases} \frac{1}{u}[f(s, x(s) + u) - f(s, x(s))] & \text{if } u \neq 0, \\ a(s) & \text{if } u = 0, \end{cases} \tag{3.42}$$

acts from L_p into L_q/L_p and is continuous at θ, then F is differentiable at x and (2.55) holds.

Observe that the preceding theorem has two parts: the first one gives a necessary condition, the second one a sufficient condition for differentiability. In particular, the following example shows that the acting condition $G(L_p) \subseteq L_q/L_p$ is sufficient, but not necessary for the differentiability of the operator F at x.

Let $\Omega = (0,1]$, with the Lebesgue measure μ, and

$$f(s,u) = \begin{cases} 0 & \text{if } |u| < 1, \\ \left|s \log \frac{s}{2}\right|^{-1/2}(|u| - 1) & \text{if } 1 \leq |u| \leq 2, \\ \left|s \log \frac{s}{2}\right|^{-1/2} & \text{if } |u| > 2. \end{cases}$$

It is not hard to see that the corresponding superposition operator F acts from L_2 into L_1 and is differentiable at $x = \theta$ with $F'(\theta) \equiv \theta$. Moreover, the function (3.42) reads

$$g(s,u) = \begin{cases} \frac{1}{u} f(s,u) & \text{if } u \neq 0, \\ 0 & \text{if } u = 0. \end{cases}$$

Consequently, any constant function $x(s) = c$ with $c \geq 2$ is mapped by G into the function $g(s,c) = \left|s \log \frac{s}{2}\right|^{-1/2}$, which does not belong to the space $L_1/L_2 = L_2$.

Now we give a necessary and sufficient differentiability condition for F from L_p into L_q. We shall restrict ourselves to the case $p > q$; the "degeneracy" statements in Theorem 3.12 show that this is the only interesting case.

Theorem 3.13 *Let f be a Carathéodory function, and suppose that the superposition operator F generated by f acts from L_p into L_q $(p > q)$. Then F is differentiable at $x \in L_p$ if and only if the limit (2.56) exists, belongs to $L_{pq/(p-q)}$, and satisfies the following condition: for each $\lambda > 0$, there exists a function $a_\lambda \in L_1$ such that $\|a_\lambda\|_1 \to 0$ $(\lambda \to \infty)$ and*

$$|f(s,x(s) + u) - f(s,x(s)) - a(s)u|^q \leq \lambda^{-q}a_\lambda(s) + \lambda^{p-q}|u|^p. \qquad (3.43)$$

$\Rightarrow$ Define a function $\tilde{f}$ by $\tilde{f}(s,u) = f(s,x(s) + u) - f(s,x(s)) - a(s)u$, and let $\tilde{F}$ be the superposition operator generated by $\tilde{f}$ from L_p into L_q. The differentiability of F at x is then equivalent to the condition $\|\tilde{F}x\|_q = o(\|x\|_p)$ $(\|x\|_p \to 0)$, i.e. $\mu_{\tilde{F}}(r) = o(r)$ $(r \to 0)$; by the two-sided estimate (3.22) (observe that $\tilde{F}\theta = \theta$), this is in turn equivalent to $\nu_{\tilde{f}}(r) = o(r)$ $(r \to 0)$, and from this the assertion follows easily. $\Leftarrow$

Observe that condition (3.43) is rather similar to (3.37), except for the fact that here $\|a_\lambda\|_1 = o(1)$ instead of $O(1)$.

An analogous result holds for asymptotic differentiability. First of all, the asymptotic derivative (if it exists!) has always the form (2.64), where the function a_∞ is given by (2.65).

Theorem 3.14   *Let f be a Carathéodory function, and suppose that the superposition operator F generated by f acts from L_p into L_q ($p > q$). Then F is asymptotically linear if and only if the limit (2.65) exists, belongs to $L_{pq/(p-q)}$, and satisfies the following condition: for each $\lambda > 0$, there exists a function $a_\lambda \in L_1$ such that $\|a_\lambda\|_1 \to 0$ ($\lambda \to 0$) and*

$$|f(s,u) - a_\infty(s)u|^q \le \lambda^{-q} a_\lambda(s) + \lambda^{p-q}|u|^p . \qquad (3.44)$$

$\Rightarrow$ The proof follows again from the fact that the asymptotic differentiability of F is equivalent to $\nu_{\tilde f}(r) = o(r)$ ($r \to \infty$) where $\tilde f(s,u) = f(s,u) - a_\infty(s)u$ and $\nu_{\tilde f}(r)$ is given in (3.20). $\Leftarrow$

Now we pass to existence results for higher derivatives. First of all, observe that, by formula (3.9) and Theorem 2.20, the existence of the n-th derivative $F^{(n)}(x)$ of F in the case $p = nq$ implies that

$$f(s,u) = a_0(s) + a_1(s)u + \cdots + \frac{a_n(s)}{n!}u^n$$

and in the case $p < nq$ implies that

$$f(s,u) = a_0(s) + a_1(s)u + \cdots + \frac{a_{n-1}(s)}{(n-1)!}u^{n-1},$$

where a_k is given by (2.70). For $p > nq$, we get the following "higher order analogue" of Theorem 3.13 which is proved in the same way:

Theorem 3.15   *Let f be a Carathéodory function, and suppose that the superposition operator F generated by f acts from L_p into L_q ($p > nq$). Then F is differentiable at $x \in L_p$ if and only if the limits (2.70) exist for $k = 0, \cdots, n$, belong to $L_{pq/(p-kq)}$, and satisfy the following condition: for each $\lambda > 0$, there exists a function $a_\lambda \in L_1$ such that $\|a_\lambda\|_1 \to 0$ ($\lambda \to \infty$) and*

$$|f(s,x(s)+u) - f(s,x(s)) - a_1(s)u - \cdots - a_n(s)u^n|^q$$
$$\le \lambda^{-nq} a_\lambda(s) + \lambda^{p-nq}|u|^p .$$

To conclude this section, we recall that L_p is a Δ_2 space for each $p \ge 1$, and hence there are no nontrivial (i.e. non-polynomial) analytic superposition operators in Lebesgue spaces (see Theorem 2.20); we state this more precisely as follows:

Theorem 3.16   *Let f be a sup-measurable function, and suppose that the superposition operator F generated by f acts from L_p into L_q. If F is analytic in a neighbourhood of some point $x \in L_p$, the function $f(s, \cdot)$, for $s \in \Omega_c$, is a polynomial of degree at most $\lceil p/q \rceil$.*

3.7 The case $p = \infty$ or $q = \infty$

So far, we have restricted ourselves to the case when both p and q are real numbers in the interval $[1, \infty)$; now we shall briefly summarize some results in the case when at least one of these numbers is ∞. As pointed out in Section 3.1, the properties of L_∞ are rather different from those of L_p for $p < \infty$; recall, for instance, that L_∞ is completely irregular for $\Omega_d = \emptyset$. In the sequel we shall always assume that f is a Carathéodory function. Since the proofs are in part obvious, in part similar to those of the corresponding theorems for $p, q < \infty$, we shall not present them. To make the statements more transparent, we treat the cases $\Omega_d = \emptyset$ and $\Omega_c = \emptyset$ separately.

Theorem 3.17   *Let $\Omega_d = \emptyset$ and $1 \le p, q < \infty$. The superposition operator F generated by f maps L_p into L_∞ if and only if*

$$|f(s, u)| \le a(s) \quad (u \in \mathbb{R}) \tag{3.45}$$

for some $a \in L_\infty$; in this case, F is always bounded; F is continuous if and only if F is constant (i.e. f does not depend on u). The operator F maps L_∞ into L_q if and only if for all $r > 0$ there exists $a_r \in L_q$ such that

$$|f(s, u)| \le a_r(s) \quad (|u| \le r); \tag{3.46}$$

in this case, F is always bounded and continuous. The operator F maps L_∞ into L_∞ if and only if (3.46) holds with $a_r \in L_\infty$; in this case, F is always bounded; F is continuous if and only if for all $r > 0$ there exists a continuous function b_r with $b_r(0) = 0$ such that

$$|f(s, u) - f(s, v)| \le b_r(|u - v|) \quad (|u|, |v| \le r). \tag{3.47}$$

Theorem 3.18   *Let $\Omega_c = \emptyset$ and $1 \le p, q < \infty$. The superposition operator F generated by f maps L_p into L_∞^0 (L_∞, respectively) if and only if*

$$\lim_{\substack{s \to \infty \\ u \to 0}} f(s, u) = 0 \quad (\overline{\lim_{\substack{s \to \infty \\ u \to 0}}} |f(s, u)| < \infty, \text{ respectively}); \tag{3.48}$$

in this case, F is always bounded; F is continuous if and only if $f(s, \cdot)$ is continuous for all $s \in \Omega_d$ (uniformly with respect to $s \in \Omega_d$, respectively). The operator F maps L^0_∞ into L_q (L^0_∞; L_∞, respectively) if and only if there exist $n \in \mathbb{N}$ and $\delta > 0$ such that

$$|f(s, u)| \leq a(s) \quad (s \geq n, |u| \leq \delta), \tag{3.49}$$

for some $a \in L_q$ ($a \in L^0_\infty$; $a \in L_\infty$, respectively); in this case, F is always bounded; F is continuous if and only if $f(s, \cdot)$ is continuous for all $s \in \Omega_d$ (for all $s \in \Omega_d$; uniformly with respect to $s \in \Omega_d$, respectively). The operator F maps L_∞ into L_q (L^0_∞; L_∞, respectively) if and only if for all $r > 0$ there exists $a_r \in L_q$ ($a_r \in L^0_\infty$; $a_r \in L_\infty$, respectively) such that (3.46) holds; in this case, F is always bounded; F is continuous if and only if $f(s, \cdot)$ is continuous for all $s \in \Omega_d$ (for all $s \in \Omega_d$; uniformly with respect to $s \in \Omega_d$, respectively).

Let us make some comments on Theorems 3.17 and 3.18. First, they easily provide elementary formulas for the growth function (2.24). For example, if F maps $X = L_\infty$ into $Y = L_q$ or $Y = L_\infty$ (Ω arbitrary), then

$$\mu_F(r) = \|a_r\|_Y, \tag{3.50}$$

where

$$a_r(s) = \sup_{|u| \leq r} |f(s, u)|. \tag{3.51}$$

The fact that any continuous superposition operator F from L_p into L_∞ ($1 \leq p < \infty$, $\Omega_d = \emptyset$) degenerates is a simple consequence of Theorem 2.8.

Finally, we point out that continuity of the function $f(s, \cdot)$ for $s \in \Omega$ (i.e. not uniformly with respect to $s \in \Omega$) does not suffice to ensure the continuity of the corresponding superposition operator F from $X = L_p$ or $X = L_\infty$ into $Y = L_\infty$ (Ω arbitrary). A simple example is the operator generated by the function $f(s, u) = \sin(u/s)$ (just take $x_n(s) = 1/n$).

3.8 The $\mathcal{L}$-characteristic

In this final section, we describe some properties of the $\mathcal{L}$-characteristics of the superposition operator introduced in Section 2.9. To this end, we first point out that many of the preceding results carry over to the case when F acts from L_p into L_q for $0 < p, q < \infty$. The linear space L_p is not normable for $0 < p < 1$, but the functional

$$[x]_p = \int_\Omega |x(s)|^p ds \tag{3.52}$$

defines a p-norm on L_p which makes L_p a completely regular ideal p-normed space. It turns out that Theorem 3.1 holds also in this more general framework: the operator F maps L_p into L_q if and only if (3.17) holds for some $a \in L_q$ and some $b \geq 0$.

Let f be a sup-measurable function, and let F be the superposition operator generated by f. By $\mathcal{L}(F; \text{act.})$ we denote the set of all points $(\alpha, \beta) \in [0, \infty) \times [0, \infty)$ such that F acts from the space $L_{1/\alpha}$ into $L_{1/\beta}$. Given $(\alpha_0, \beta_0), (\alpha_1, \beta_1) \in [0, \infty) \times [0, \infty)$, where without loss of generality $\alpha_1 \leq \alpha_0$, let

$$\Sigma(\alpha_0, \beta_0; \alpha_1, \beta_1) = \begin{cases} \{(\alpha, \beta) : (\alpha, \beta) \in [0, \infty) \times [0, \infty), \\ \quad \beta_0 \leq \beta \leq \beta_1, \frac{\beta_0}{\alpha_0} \leq \frac{\beta}{\alpha} \leq \frac{\beta_1}{\alpha_1}\} & \text{if } \beta_0 \leq \beta_1, \\[2mm] \{(\alpha_0, \beta_0), (\alpha_1, \beta_1)\} & \text{if } \beta_0 > \beta_1. \end{cases}$$
$$(3.53)$$

Thus, the set $\Sigma(\alpha_0, \beta_0; \alpha_1, \beta_1)$ is a quadrangular domain in the plane if $\beta_0 < \beta_1$, a horizontal segment joining (α_0, β_0) with (α_1, β_1) if $\beta_0 = \beta_1$, and a set of two points if $\beta_0 > \beta_1$.

Lemma 3.2   If $(\alpha_0, \beta_0), (\alpha_1, \beta_1) \in \mathcal{L}(F; \text{act.})$, then $\Sigma(\alpha_0, \beta_0; \alpha_1, \beta_1) \subseteq \mathcal{L}(F; \text{act.})$.

$\Rightarrow$ For $(\alpha_0, \beta_0), (\alpha_1, \beta_1) \in \mathcal{L}(F; \text{act.})$ we have, by (3.17),

$$|f(s, u)| \preceq a_j(s) + b_j |u|^{\beta_j/\alpha_j} \quad (j = 0, 1)$$

for some $a_j \in L_{1/\beta_j}$ and $b_j \geq 0$. Consequently,

$$|f(s, u)| \preceq \min\{a_0(s) + b_0 |u|^{\beta_0/\alpha_0}, a_1(s) + b_1 |u|^{\beta_1/\alpha_1}\}$$
$$\leq \min\{a_0(s), a_1(s)\} + \min\{a_0(s), b_1 |u|^{\beta_1/\alpha_1}\}$$
$$+ \min\{a_1(s), b_0 |u|^{\beta_0/\alpha_0}\} + \min\{b_0 |u|^{\beta_0/\alpha_0}, b_1 |u|^{\beta_1/\alpha_1}\}.$$

Therefore we can write the function f as a sum $f = f_1 + f_2 + f_3 + f_4$, where

$$|f_1(s, u)| \preceq \min\{a_0(s), a_1(s)\}, \quad |f_2(s, u)| \preceq \min\{a_0(s), b_1 |u|^{\beta_1/\alpha_1}\},$$

$$|f_3(s, u)| \preceq \min\{a_1(s), b_0 |u|^{\beta_0/\alpha_0}\}, \quad |f_4(s, u)| \preceq \min\{b_0 |u|^{\beta_0/\alpha_0}, b_1 |u|^{\beta_1/\alpha_1}\},$$

hence $\mathcal{L}(F; \text{act.}) \supseteq \mathcal{L}(F_1; \text{act.}) \cap \mathcal{L}(F_2; \text{act.}) \cap \mathcal{L}(F_3; \text{act.}) \cap \mathcal{L}(F_4; \text{act.})$. Now, the $\mathcal{L}$-characteristic of F_1 includes the set of all (α, β) such that $\beta_0 \leq \beta \leq \beta_1$ or $\beta_1 \leq \beta \leq \beta_0$, that of F_4 the set of all (α, β) such that $\beta_0 \alpha_0^{-1} \leq \beta \alpha^{-1} \leq \beta_1 \alpha_1^{-1}$. Similarly, the $\mathcal{L}$-characteristic of F_2 (of F_3, respectively) includes the set of all (α, β) such that $0 \leq \alpha \leq \beta_0 \alpha_1 \beta_1^{-1}$ and $\alpha \beta_1 \alpha_1^{-1} \leq \beta \leq \beta_0$, or $\beta_0 \alpha_1 \beta_1^{-1} \leq \alpha < \infty$ and $\beta_0 \leq \beta \leq \alpha \beta_1 \alpha_1^{-1}$ (the set of all (α, β) such that $0 \leq \alpha \leq \beta_1 \alpha_0 \beta_0^{-1}$ and $\alpha \beta_0 \alpha_0^{-1} \leq \beta \leq \beta_1$, or $\beta_1 \alpha_0 \beta_0^{-1} \leq \alpha < \infty$ and $\beta_1 \leq \beta \leq \alpha \beta_0 \alpha_0^{-1}$, respectively). The intersection of these four sets is just the set $\Sigma(\alpha_0, \beta_0; \alpha_1, \beta_1)$. $\Leftarrow$

We remark that the sets $\mathcal{L}(F; \mathrm{act.})$ and $\Sigma(\alpha_0, \beta_0; \alpha_1, \beta_1)$ may coincide, and thus a more precise statement than Lemma 3.2 is, in general, not possible. For example, let $\Omega = \mathbb{R}$, with the Lebesgue measure μ, and let c_* be a L_1-function which does not belong to any L_p for $p \neq 1$. If for some fixed $(\alpha_0, \beta_0), (\alpha_1, \beta_1) \in [0, \infty) \times [0, \infty)$ we set

$$f(s, u) = \min\{c_*(s)^{\alpha_0} + |u|^{\beta_0/\alpha_0}, c_*(s)^{\alpha_1} + |u|^{\beta_1/\alpha_1}\},$$

then the $\mathcal{L}$-characteristic of the corresponding superposition operator F coincides precisely with the set $\Sigma(\alpha_0, \beta_0; \alpha_1, \beta_1)$.

Under additional assumptions, however, more can be said about the $\mathcal{L}$-characteristic of the superposition operator. Suppose, for instance, that $\mu(\Omega) < \infty$. In this case L_p is continuously imbedded in L_q for $p \geq q$ (see (3.10)). From this and (3.17) it follows that $\mathcal{L}(F; \mathrm{act.})$ includes the set

$$\begin{aligned} \Sigma_c(\alpha_0, \beta_0) = &\{(\alpha, \beta) : 0 \leq \alpha \leq \alpha_0, \ \beta \geq \beta_0\} \\ &\cup \{(\alpha, \beta) : \alpha_0 \leq \alpha < \infty, \ \beta \geq \beta_0 \alpha_0^{-1} \alpha\} \end{aligned} \qquad (3.54)$$

whenever $(\alpha_0, \beta_0) \in \mathcal{L}(F; \mathrm{act.})$. On the other hand, if $\Omega_c = \emptyset$ and $\mu(s) \geq \mu_0 > 0$, L_p is continuously imbedded in L_q for $p \leq q$ (see (3.11)). Here $\mathcal{L}(F; \mathrm{act.})$ includes the set

$$\begin{aligned} \Sigma_d(\alpha_0, \beta_0) = &\{(\alpha, \beta) : 0 \leq \alpha \leq \alpha_0, \ \beta \geq \beta_0 \alpha_0^{-1} \alpha\} \\ &\cup \{(\alpha, \beta) : \alpha_0 \leq \alpha < \infty, \ \beta \leq \beta_0\} \end{aligned} \qquad (3.55)$$

whenever $(\alpha_0, \beta_0) \in \mathcal{L}(F; \mathrm{act.})$.

We summarize our results in the following:

Theorem 3.19   *If* $(\alpha_0, \beta_0), (\alpha_1, \beta_1) \in \mathcal{L}(F; \mathrm{act.})$, *the inclusion* $\Sigma(\alpha_0, \beta_0; \alpha_1, \beta_1) \subseteq \mathcal{L}(F; \mathrm{act.})$ *holds, where* $\Sigma(\alpha_0, \beta_0; \alpha_1, \beta_1)$ *is defined by (3.53).* *Moreover, if* $\mu(\Omega) < \infty$, *the inclusion* $\Sigma_c(\alpha_0, \beta_0) \subseteq \mathcal{L}(F; \mathrm{act.})$ *holds, where* $\Sigma_c(\alpha_0, \beta_0)$ *is defined by (3.54). Finally, if* $\Omega_c = \emptyset$ *and* $\mu(s) \geq \mu_0 > 0$ *for* $s \in \Omega_d$, *the inclusion* $\Sigma_d(\alpha_0, \beta_0) \subseteq \mathcal{L}(F; \mathrm{act.})$ *holds, where* $\Sigma_d(\alpha_0, \beta_0)$ *is defined by (3.55).*

3.9 Notes, remarks and references

1. Most facts about Lebesgue spaces mentioned in Section 3.1 may be found in [104]. Unfortunately, the space L_p is not normable for $0 < p < 1$; for this reason, we restricted ourselves to the case $1 \le p \le \infty$ in order to apply the general results of Chapter 2. Nevertheless, many results carry over to the case when F acts in the p-normed space L_p for $0 < p < 1$, for instance, the basic Theorem 3.1. Moreover, it turns out that the associate space $L_{\tilde{p}}$ in case $0 < p < 1$ is the Banach space Z_p defined by the norm (3.7).

2. Lebesgue spaces have been the first function spaces in which the superposition operator was studied systematically. The first results on continuity and boundedness of this operator in Lebesgue spaces over bounded domains Ω in Euclidean space are due to M.A. Krasnosel'skij [164–167] and, independently, M.M. Vajnberg [341–347]. For instance, the fact that the growth condition (3.19) is sufficient for F to act from L_p into L_q is mentioned in [341–343]; the necessity of (3.19) (which is of course the nontrivial part) was first proved by M.A. Krasnosel'skij [166]. For functions f without Carathéodory condition, Theorem 3.1 is mentioned in [171] (see also the end of §23 in [174]).

As an immediate consequence of Theorem 3.1, one can give a necessary and sufficient condition under which F maps the *weighted Lebesgue space* $L_p(\sigma)$ of all functions $x \in S$ with

$$\|x\|_{p,\sigma}^p = \int_\Omega |x(s)|^p \sigma(s) ds < \infty$$

into another weighted Lebesgue space $L_q(\tau)$, namely [44]

$$|f(s,u)| \le a(s) + b\sigma(s)^{1/q} \tau(s)^{-1/q} |u|^{p/q} .$$

As already mentioned, the literature on the superposition operator in Lebesgue spaces over bounded domains Ω in Euclidean space, equipped with the Lebesgue measure μ, is vast (see e.g. [166] or [182]), while the case of discrete measures μ (i.e. sequence spaces ℓ_p) has been treated only recently. For instance, all results given for the discrete case in this chapter can be found in [94], detailed proofs are given in [93]; for similar results in general sequence spaces see also [273].

3. The boundedness Theorems 3.2 and 3.3 follow from the corresponding theorems in ideal spaces (Theorem 2.2 and 2.3). A great advantage, however, consists in the fact one can calculate *explicitly* the growth function

(2.24) in Lebesgue spaces. All results of the second part of Section 3.3, as well as many examples and comments, can be found in [30]. We remark that the problem of calculating the function $\mu_F(r)$ generalizes the classical one-dimensional allocation problem of dynamic programming (see e.g. [43]) which is obtained for $p = q = 1$ and Ω finite.

4. Theorem 3.6 may be considered as a consequence of the classical criterion of De la Vallée–Poussin for the absolute boundedness of subsets of L_1 (see e.g. [235]). The importance of absolutely bounded superposition operators in the theory of nonlinear integral equations of Hammerstein type was mentioned in Section 2.9; for Lebesgue spaces, this idea is due to P.P. Zabrejko and Je.I. Pustyl'nik [392].

The first (sufficient) condition for the *uniform continuity* of the superposition operator in Lebesgue spaces seems to be due to M.M. Vajnberg [346]. The general condition given in Theorem 3.8 is new, as well as the two-sided estimate (3.51) for the modulus of continuity of F; for the case $\Omega_c = \emptyset$ see [94].

The fact that a weakly continuous superposition operator from L_p into L_q is linear was proved in the case $\Omega_d = \emptyset$ in [289]; in its general form it follows, of course, from Theorem 2.10.

5. Lipschitz continuous superposition operators in Lebesgue spaces have been studied by several authors. The equivalence of the conditions (a) and (c) in Theorem 3.10 was proved in the case $p = q$ in [105] and [12]; for similar statements, see also [246], [247]. In its full generality, Theorem 3.10 follows, of course, from Theorems 2.11 and 2.12. The Hölder condition (3.38) was considered in Lebesgue spaces by W.S. Wang [365]. In this connection, we point out also that other analytical properties of F in Lebesgue spaces are reflected in analogous properties of the corresponding function f. For instance, subadditive, subhomogeneous, and convex superposition operators F in L_p are characterized in [88] in terms of the generating function f. The paper [77] deals with applications of superposition operators which are *maximal monotone* in H. Brézis' sense (see [63]).

The equivalence of the Lipschitz condition (3.35) and the Darbo condition (3.41) was proved for Lebesgue spaces first in [12]. In fact, the construction proposed in [12] is the same as in the proof of Theorem 2.13. We remark that also a "global version" of Theorem 3.10 is true, i.e. with (3.35) replaced by the condition $\|Fx_1 - Fx_2\|_q \leq k\|x_1 - x_2\|_p$ $(x_1, x_2 \in L_p)$, see also (2.43). This global Lipschitz condition is in turn equivalent to the global Darbo condition $\alpha_q(FN) \leq k\alpha_p(N)$ $(N \subset L_p$ bounded), see also (2.49). Applications of this fact may be found in [13] and [20].

6. The first (sufficient) differentiability conditions for the superposition operator in Lebesgue spaces (with $\Omega_d = \emptyset$) go back to M.A. Krasnosel'skij

[166], M.M. Vajnberg [346], and P.P. Zabrejko [372]. The fact that every differentiable superposition operator "degenerates" in case $p \leq q$ (more precisely, is linear for $p = q$ and constant for $p < q$) was observed by many authors (e.g. [105], [166], [347]). The example after Theorem 3.12 may be found in [182]. The necessary and sufficient conditions given in Theorems 3.13, 3.14 and 3.15 are all taken from [14]; see also [365]. In this connection, we mention another classical fact: if the superposition operator F maps the space L_p into the space $L_{\tilde{p}}$ (with $1/p + 1/\tilde{p} = 1$) then F is a *potential operator*, namely the gradient of the *Golomb functional*

$$\Phi(x) = \int_\Omega \int_0^{x(s)} f(s, u)\,du\,ds \tag{3.56}$$

(see e.g. [347]). The first "pathologies" in the differentiability behaviour of superposition operators in Lebesgue spaces were discovered by M.A. Krasnosel'skij and M.M. Vajnberg. An example of a simple integral operator (involving nonlinear superpositions) which is Gâteaux differentiable but not Fréchet differentiable may be found in [95], Example 15.2.

The fact that any analytic superposition operator in Lebesgue spaces reduces to a polynomial (Theorem 3.16) is an immediate consequence of Theorem 2.20. This fact is rather disappointing, in view of the usefulness of Lebesgue spaces in applications, and may be the reason for the fact that analyticity properties of superposition operators in other than Lebesgue spaces have not been studied (except for [32] and [33]).

7. Many results mentioned in Section 3.7 are well-known "folklore". All proofs for the case $\Omega_d = \emptyset$ may be found in Chapter 10 of the book [380], for the case $\Omega_c = \emptyset$ in [93] or [94].

8. The first paper on the $\mathcal{L}$-characteristic of both linear and nonlinear operators in Lebesgue spaces is [381]. Much information on $\mathcal{L}$-characteristics of the superposition operator in Lebesgue spaces is contained in §17 of the book [182]. The "geometric" statement of Theorem 3.18, however, is new (see [37] for further details).

9. We point out that there exist also some literature on the superposition operator in Lebesgue spaces of *vector valued* (or even Banach space valued) functions. Continuity and boundedness are studied in the Banach space case, for instance, in [9], [160] and [200]; a very detailed survey is [257]. In [200] the authors show the following: if B_1 and B_2 are two Banach spaces, f is a Borel–measurable function from $\Omega \times B_1$ into B_2, and F is continuous between the *Bochner–Lebesgue spaces* $L_p(\Omega, B_1)$ and $L_q(\Omega, B_2)$, then f is in fact a Carathéodory function. The papers [59–61] are concerned with the continuity, boundedness, and absolute boundedness of "generalized" superposition operators in Bochner–Lebesgue spaces.

A more general notion of superposition operators in Bochner–Lebesgue spaces was considered by G. Bruckner in [65–67]. An operator A from $L_p(\Omega, B_1)$ into $L_q(\Omega, B_2)$ is called *generalized superposition operator* in [65] if, for each $s \in \Omega$, there is an operator $A(s)$ from B_1 into B_2 such that $Ax(s) = A(s)x(s)$ $(x \in L_p(\Omega, B_1))$. Using this concept, the author shows that certain "local" properties of such operators (Lipschitz continuity, Hölder continuity, monotonicity etc.) imply the corresponding "global" properties; for further results in this spirit, see also [232].

Finally, the survey [254] contains a detailed study of the superposition operator between Lebesgue spaces on C^∞ vector bundles over a compact manifold.

10. To conclude, we mention that there is also a vast literature on the integral functional (1.39) on subsets of Lebesgue spaces; the "interaction" of the properties of this functional and the superposition operator F on subsets of the space S was discussed in Chapter 1 (see e.g. Lemma 1.7 or [36]). A crucial point is here the representation problem for certain additive functionals Φ on Lebesgue spaces in the form (1.39). Some aspects of this problem are dealt with in [227] and, more generally, in [9] and [111]; for a recent survey, see Chapter 2 of [71]. For Banach space valued functions, this problem is studied from a very general point of view in [145]. The paper [212] deals with the problem of extending functionals like (1.39) from closed subspaces to the whole of L_p. Finally, boundedness and continuity properties of the related *integral operator*

$$\Psi x(t) = \int_\Omega f(t, x(s))ds$$

in Lebesgue spaces are discussed in [253].

Chapter 4

The superposition operator
in Orlicz spaces

Whenever one has to deal with problems involving rapidly increasing nonlinearities (e.g. of exponential type), Orlicz spaces are more appropriate than Lebesgue spaces. Since Orlicz spaces are ideal spaces, many statements of this section are just reformulations of the general results of Chapter 2, and therefore are cited mostly without proofs. However, in contrast to Lebesgue spaces, several new features occur in Orlicz spaces. For instance, the superposition operator may act from one Orlicz space into another and be bounded but not continuous, or continuous but not bounded.

It turns out again that F is weakly continuous between Orlicz spaces if and only if f is affine in u. Further, the Lipschitz and Darbo conditions for F coincide, as in Lebesgue spaces, and one gets the same "degeneracy" phenomenon if the second space is "essentially smaller" than the first one.

As in Lebesgue spaces, one can give conditions for differentiability and asymptotic linearity which are both necessary and sufficient. Again, f reduces to an affine function if F is differentiable from a "large" into a "small" space. On the other hand, the class of analytic superposition operators in Orlicz spaces is reasonably large; in contrast to Lebesgue spaces, a degeneration to polynomials occurs only under an additional restriction on the first space.

4.1 Orlicz spaces

As before, let Ω be an arbitrary set, $\mathcal{M}$ some σ-algebra of subsets of Ω, and μ a σ-finite and countably additive measure on $\mathcal{M}$; together with μ we shall sometimes consider an equivalent normalized countably additive measure λ (see Section 1.1).

Recall that an $(\mathcal{M} \otimes \mathcal{B}, \mathcal{B})$-measurable function $M : \Omega \times \mathbb{R} \to [0, \infty]$ is called a *Young function* if, for almost all $s \in \Omega$, $M(s, \cdot)$ is even, convex, and lower semi-continuous on $\mathbb{R}$ with $M(s, 0) = 0$; to exclude trivial situations, we suppose that $M(s,u) \not\equiv 0$ and $M(s,u) \not\equiv \infty$. Important examples are

$$M(u) = M_p(u) = \frac{1}{p}|u|^p \quad (1 \le p < \infty) \tag{4.1}$$

and

$$M(u) = M_\infty(u) = \begin{cases} 0 & \text{if } |u| \le 1, \\ \infty & \text{if } |u| > 1. \end{cases} \tag{4.2}$$

With each Young function M on $\Omega \times \mathbb{R}$ we associate the function

$$\delta(s) = \delta_M(s) = \sup\{|u| : M(s,u) < \infty\}, \tag{4.3}$$

taking values in $[0,\infty]$. For $u \in [-\delta(s), \delta(s)]$ the function M may be represented as integral

$$M(s,u) = \int_0^{|u|} m(s,t)dt, \tag{4.4}$$

where $m(s,\cdot)$ is increasing on $[0,\infty)$.

Given a measurable function x on Ω, the functional

$$M[x] = \int_\Omega M(s, x(s))ds \tag{4.5}$$

is called the *Orlicz modular* generated by the Young function M. For each $r > 0$ the set

$$\Sigma_M^r = \{x : x \in S, \ M[x/r] < \infty\} \tag{4.6}$$

is called the *Orlicz class* (of height r) generated by M. Observe that the set Σ_M^r is, in general, not a linear space; in fact, for $m(s,u) = e^u - 1$ and $x(s) = -\frac{1}{\alpha}\log s \ (0 < s < 1)$ we have $x \in \Sigma_M^r$ for $\alpha r > 1$ and $x \notin \Sigma_M^r$ for $\alpha r \le 1$. The set

$$L_M = L_M(\Omega, \mathcal{M}, \mu) = \bigcup_{r>0} \Sigma_M^r, \tag{4.7}$$

however, is a linear space. Equipped with either the *Luxemburg norm*

$$\|x\|_M = \inf\{\lambda : \lambda > 0, \ M[x/\lambda] \le 1\} \tag{4.8}$$

or the *Orlicz–Amemiya norm*

$$\||x|\|_M = \inf_{\ell>0} \frac{1}{\ell}(1 + M[\ell x]), \tag{4.9}$$

the set L_M becomes an ideal space, the *Orlicz space* generated by the Young function M. Both norms, (4.8) and (4.9), are equivalent, since $\|x\|_M \leq \|\|x\|\|_M \leq 2\|x\|_M$. In the sequel, however, we shall mostly consider the Luxemburg norm (4.8), because it has simpler properties than the norm (4.9). For instance, the unit ball $B_1(L_M)$ with respect to (4.8) is simply characterized by the condition $M[x] \leq 1$; for further results, see e.g. Lemma 4.1.

Orlicz spaces are natural generalizations of Lebesgue spaces: in fact, for M as in (4.1) (respectively (4.2)) we get $L_M = L_p$ (respectively $L_M = L_\infty$), and the three norms (3.3), (4.8) and (4.9) coincide up to multiplicative constants.

As in Section 2.1, we consider the subspace L_M^0 of all $x \in L_M$ with absolutely continuous norm; the elements $x \in L_M^0$ may be characterized equivalently by $\lim_{\lambda(D)\to 0} \|P_D x\| = 0$ (see (2.1)) or $\lim_{\lambda(D)\to 0} M[kP_D x] = 0$ $(k > 0)$. Moreover, L_M^0 is precisely the closure of the space $M(u_0)$ (see (2.5)) with respect to one of the norms (4.8) or (4.9), where u_0 is an arbitrary unit in L_M^0. Finally, the space L_M^0 coincides with the space

$$E_M = E_M(\Omega, \mathcal{M}, \mu) = \begin{cases} \bigcap_{r>0} \Sigma_M^r & \text{if } \Omega_d = \emptyset, \\ \bigcap_{r>0} \Sigma_{M,d}^r & \text{if } \Omega_c = \emptyset, \end{cases} \qquad (4.10)$$

where $\Sigma_{M,d}^r$ consists of all $x \in \Sigma_M^r$ such that $\int_{\lambda(s)<\varepsilon} M(s, x(s))ds < \infty$ for some $\varepsilon = \varepsilon(x) > 0$.

In order to describe the sets Σ, Π and Δ introduced in Section 2.2, we first remark that (see (2.11))

$$\delta(s) = \delta_M(s) = \sup\{|u| : M(s, u) < \frac{1}{\mu(s)}\}; \qquad (4.11)$$

in particular, in case $\Omega_d = \emptyset$ this gives again (4.3); moreover, L_M is quasi-regular if and only if $\delta_M(s) \equiv \infty$, and completely irregular if and only if $\delta_M(s) < \infty$ for almost all $s \in \Omega$. Further, for the function (2.13) we have

$$\pi(x) = \pi_M(x) = \inf\{r : r > 0, \ x \in \Sigma_M^r\}.$$

We conclude that

$$\Delta = \Delta_M = \begin{cases} L_M & \text{if } \Omega_d = \emptyset, \\ \{x : x \in L_M, \ \mu(s)M(s, x(s)) \leq 1\} & \text{if } \Omega_c = \emptyset, \end{cases} \qquad (4.12)$$

and

$$\Sigma = \Sigma_M = \{x : x \in S,\ M[x] < \infty\} \tag{4.13}$$

is just the class Σ_M^1 introduced in (4.6).

Concerning the split functional $H = H_M$ on the space L_M (see (2.15)), we remark that the two-sided estimates

$$M[x] \le H_M(x) \le 1 + M[x] \quad (\Omega_d = \emptyset,\ x \in L_M) \tag{4.14}$$

and

$$M[x] \le H_M(x) \le 1 + 2M[x] \quad (\Omega_c = \emptyset,\ x \in \Delta_M) \tag{4.15}$$

hold which correspond to the estimates (3.15) and (3.16), respectively, for Lebesgue spaces.

Given a Young function M, the function

$$\widetilde{M}(s,v) = \sup_{u \ge 0}\{u|v| - M(s,u)\} \tag{4.16}$$

is called the *associated Young function* to M. By definition, the *Young inequality*

$$uv \le M(s,u) + \widetilde{M}(s,v) \tag{4.17}$$

holds; moreover, it is not hard to see that $\widetilde{\widetilde{M}} = M$. If M is given by the relation (4.4), $\widetilde{M}$ may be given by

$$\widetilde{M}(s,v) = \int_0^{|v|} \widetilde{m}(s,t)dt\,,$$

where $\widetilde{m}(s,v) = \sup\{u : u \ge 0,\ m(s,u) < v\}$ is the left continuous inverse of the function m. In particular, in case $M = M_p$ (see (4.1) and (4.2)) we have $\widetilde{M} = M_{\widetilde{p}}$ with $p^{-1} + \widetilde{p}^{-1} = 1$ $(1 \le p \le \infty)$.

The associated Young function $\widetilde{M}$ allows us to give a precise description of the associate space $\widetilde{L}_M$ of L_M. Indeed, the relations $\|x\|_{\widetilde{M}} = \sup\{<x, y> : \|\|y\|\|_M \le 1\}$ and $\|\|x\|\|_{\widetilde{M}} = \sup\{<x, y> : \|y\|_M \le 1\}$ show that $\widetilde{L}_M$ with the norm (4.8) (respectively the norm (4.9)) is nothing else but the Orlicz space $L_{\widetilde{M}}$ with the norm (4.9) (respectively the norm (4.8)).

Now we are going to describe the multiplicator space L_N/L_M. To this end, we first suppose that $\Omega_d = \emptyset$ and $\mu(\Omega_c) < \infty$. Given two Young functions M and N, we write $M \sqsupset N$ if

$$\varlimsup_{u \to \infty} \frac{N(s, ku)}{M(s, u)} = 0 \tag{4.18}$$

uniformly in $s \in \Omega$ for every $k > 0$, and $M \sqsupseteq N$ if

$$\overline{\lim_{u \to \infty}} \frac{N(s, ku)}{M(s, u)} < \infty \qquad (4.19)$$

uniformly in $s \in \Omega$ for some $k > 0$.

In the first case, L_M is imbedded into L_N and the imbedding is absolutely bounded (see Section 2.4); in the second case, L_M is imbedded into L_N and the imbedding is only bounded, but not necessarily absolutely bounded. We point out that the condition $M \sqsupseteq N$ (the condition $M \sqsupset N$, respectively) is also necessary for the fact that L_M is imbedded (absolutely boundedly imbedded, respectively) into L_N if the Young functions M and N do not depend on s. For example, $M_p \sqsupseteq M_q$ (respectively $M_p \sqsupset M_q$) if and only if $p \geq q$ (respectively $p > q$).

Generally speaking, Orlicz spaces L_M with an *autonomous Young function* $M = M(u)$ are much easier to handle than those with $M = M(s, u)$. For instance, they are examples of so-called *symmetric spaces*, which will be described in detail in Chapter 5. An important property of such spaces is that the norm of the characteristic function χ_D of $D \in \mathcal{M}$ depends only on the measure $\mu(D)$ of D. In fact, an easy calculation shows that in case $M = M(u)$

$$\|\chi_D\|_M = \frac{1}{M^{-1}(1/\mu(D))}, \quad \|\|\chi_D\|\|_M = \mu(D)\tilde{M}^{-1}(1/\mu(D)). \qquad (4.20)$$

Now, the multiplicator space L_N/L_M (see (2.40)) may be described as follows (in case $\Omega_d = \emptyset$ and $\mu(\Omega_c) < \infty$):

$$L_N/L_M = \begin{cases} \{\theta\} & \text{if } N \sqsupset M, \\ L_\infty & \text{if } N \sqsupseteq M, \\ L_R & \text{if } M \sqsupset N, \end{cases} \qquad (4.21)$$

where L_R is the Orlicz space defined by the Young function

$$R(s, v) = \sup_{u \geq 0} \{N(s, uv) - M(s, u)\}. \qquad (4.22)$$

Observe again that in case $M = M_p$ and $N = M_q$ (see (4.1), (4.2)) the Young function (4.22) in the third case (i.e. $p > q$) becomes $R = M_r$ with $r^{-1} = q^{-1} - p^{-1}$. Moreover, the choice $N = M_1$ in (4.22) gives again the associated Young function (4.16), in agreement with the fact that $\tilde{L}_M = L_1/L_M = L_{\tilde{M}}$.

A similar result holds for the n-th order multiplicator space L_N/L_M^n (see (2.67)). For instance, if

$$\varlimsup_{u \to \infty} \frac{N(s, ku^n)}{M(s, u)} = 0$$

for all $k > 0$, i.e. the Young function $N_n(s, u) = N(s, u^n)$ satisfies $M \sqsupset N_n$, the space L_N/L_M^n is just the Orlicz space L_{R_n} generated by the Young function

$$R_n(s, v) = \sup_{u \geq 0} \{N(s, u^n v^n) - M(s, u)\}.$$

To conclude our summary on Orlicz spaces, we discuss an important special class of Young functions. It is evident that the Young function M determines the properties of the Orlicz space L_M. Generally speaking, the more restricted the rate of growth of M in u, the "nicer" the properties of the space L_M. The most important of such growth restrictions is the so-called Δ_2 *condition* which states that for any $c > 0$ one can find $b > 0$, $\varepsilon > 0$, $\delta > 0$, and $p \in L_1$ such that

$$M(s, cu) \leq bM(s, u) + p(s) \qquad (4.23)$$

for almost all $s \in \Omega$ with $\lambda(s) \leq \varepsilon$ and $\mu(s)M(s, u) \leq \delta$; here the problem of specifying the arguments $(s, u) \in \Omega \times \mathbb{R}$ for which (4.23) should hold is somewhat delicate. Loosely speaking, (4.23) is required for large $|u|$ if $\Omega_d = \emptyset$ and $\mu(\Omega_c) < \infty$, for small $|u|$ and large s if $\Omega_c = \emptyset$ and $\Omega_d \subseteq \mathbb{N}$, and for all u if $\mu(\Omega_c) = \infty$; in the sequel we will tacitly assume that the Δ_2 condition (4.23) is understood in the appropriate sense, depending on the underlying set Ω.

To give a simple example, the Young function $M(u) = \frac{1}{p}|u|^p$ satisfies the Δ_2-condition (4.23), while the Young function $M(u) = e^{|u|} - |u| - 1$ (see above) does not.

Roughly speaking, if M satisfies a Δ_2 condition (which we denote by $M \in \Delta_2$ in what follows), the corresponding Orlicz space L_M has properties similar to those of the Lebesgue space L_p. For example, the space L_M is regular if and only if $M \in \Delta_2$; consequently, the associate space $\tilde{L}_M$ coincides with the dual space L_M^* if and only if $M \in \Delta_2$. In case $\Omega_d = \emptyset$ the regular part L_M^0 of L_M (see (2.1)) is nothing else but the subspace E_M introduced in (4.10), and hence $L_M = E_M$ if and only if $M \in \Delta_2$. Moreover, we remark that $(L_M^0)^* = L_{\widetilde{M}}$, with $\widetilde{M}$ given by (4.16); this gives again the relation $L_M^* = \tilde{L}_M = L_{\widetilde{M}}$ in case $M \in \Delta_2$. Finally, the space L_M is reflexive if both $M \in \Delta_2$ and $\widetilde{M} \in \Delta_2$.

Let us return to the estimates (4.14) and (4.15). Let $\Omega_d = \emptyset$. Since $L_M = E_M$ if and only if $M \in \Delta_2$, the modular (4.5) is bounded on each ball $B_r(L_M)$ with $r > 1$ if and only if $M \in \Delta_2$; by (4.14) and (4.15), the same is true for H_M. In this way, we arrive at the following important lemma:

Lemma 4.1 Let $\Omega_d = \emptyset$. The Orlicz space L_M (with the Luxemburg norm (4.8)) is a Δ_2 space if and only if $M \in \Delta_2$.

Lemma 4.1 motivates the name "Δ_2 space" for ideal spaces satisfying one of the properties stated in Lemma 2.3. The question arises, whether or not an Orlicz space L_M with $M \notin \Delta_2$ belongs to some other class of ideal spaces introduced in Section 2.3. Here we have the following:

Lemma 4.2 Every Orlicz space L_M (with the Luxemburg norm (4.8)) is a δ_2 space.

The proof of Lemma 4.2 is easy: in fact, the modular $h(x) = M[x]$ has the properties (a) and (b) occuring in the definition of a δ_2 space. Thus, every Orlicz space L_M with $M \notin \Delta_2$ may serve as an example of a δ_2 space (and hence split space, by Lemma 2.5) which is not a Δ_2 space.

4.2 Acting conditions

The purpose of this section is to give a necessary and sufficient condition on f under which the corresponding superposition operator F acts between two Orlicz spaces L_M and L_N. This is easy only in the case when the Young function $M = M(s, u)$ has an inverse M^{-1} with respect to u for each $s \in \Omega$. In this case, the following observation reduces the acting problem to the L_1-case (compare Lemma 3.1):

Lemma 4.3 The superposition operator F generated by some sup-measurable function f maps the ball $B_r(L_M)$ into the ball $B_R(L_N)$ if and only if the superposition operator G generated by the function

$$g(s, u) = N\left\{s, \frac{1}{R}[f(s, rM^{-1}(s, u))]\right\} \tag{4.24}$$

maps the unit ball $B_1(L_1)$ into itself.

$\Rightarrow$ Suppose that G maps $B_1(L_1)$ into itself for some $r > 0$ and $R > 0$, and let $x \in B_r(L_M)$. By definition of the Luxemburg norm (4.8), this means that $M[x/r] \leq 1$, where $M[x]$ is the modular (4.5). Consequently, the function $\xi(s) = M(s, x(s)/r)$ belongs to $B_1(L_1)$, and thus the function $\eta(s) = G\xi(s) = N(s, f(s, x(s))/R)$ belongs to $B_1(L_1)$ as well, by assumption. But this means that $N[y/R] \leq 1$ for $y = Fx$, and hence $y \in B_R(L_N)$ as claimed.

The converse implication is proved analogously. $\Leftarrow$

Note that in case $M \in \Delta_2$ we have $\Sigma_M = L_M$, by (4.13), and hence, by Lemma 2.1, F is defined on the whole space L_M, whenever F is defined on the ball $B_1(L_M)$. In case $M \notin \Delta_2$, however, this is false. Consider, for example, the function

$$f(u) = M(u), \qquad (4.25)$$

where M is some (autonomous) Young function which does not satisfy a Δ_2 condition. The corresponding superposition operator F maps the ball $B_1(L_M)$ into the space L_1, since $\|Fx\|_1 = M[x] \leq 1$ for $\|x\|_M \leq 1$. On the other hand, for any $r > 1$ one may choose a function $x \in \Sigma_M^r \setminus \Sigma_M^1$, since $M \notin \Delta_2$; for this function x, we obviously have $Fx \notin L_1$.

Lemma 4.3 allows us to formulate necessary and sufficient acting conditions if the inverse M^{-1} of M with respect to u exists. In the general case, however, such acting conditions must be proved independently of Theorem 3.1 on the superposition operator in Lebesgue spaces. The following result is a generalization of Theorem 3.1, but much more difficult to prove.

Theorem 4.1   *The superposition operator F generated by f maps the ball $B_r(L_M)$ into the space L_N if and only if there exist $R > 0$, $\delta > 0$, $\varepsilon > 0$, $b \geq 0$ and $a \in L_1$ such that*

$$N\left(s, \frac{1}{R}f(s, u)\right) \preceq a(s) + bM\left(s, \frac{u}{r}\right) \qquad (4.26)$$

for all $(s, u) \in \Omega \times \mathbb{R}$ satisfying $\lambda(s) \leq \varepsilon$ and $\mu(s)M(s, u/r) \leq \delta$.

$\Rightarrow$ The sufficiency of (4.26) is proved by a straightforward calculation; let us prove the necessity of (4.26). To this end, we may assume that f is a Carathéodory function, and $f(s, 0) = 0$ almost everywhere on Ω. By the partial additivity of the operator F (see (1.10)), it suffices to consider F on the set

$$\hat{\Sigma}_M^r = \{x : x \in S, \ M[P_{\Omega_c}x/r] < \infty, \ \mu(s)M(s, x(s)/r) \leq 1\}. \qquad (4.27)$$

Suppose first that $\Omega_c = \emptyset$; we shall establish condition (4.26) in two steps.

First step.   We claim that for each $\gamma > 0$ there exist $R > 0$, $\varepsilon > 0$, and $\delta > 0$ such that $M[\frac{1}{r}P_{\Omega(\varepsilon)}x] \leq \delta$ implies that $N[\frac{1}{R}P_{\Omega(\varepsilon)}Fx] \leq \gamma$, where $\Omega(\varepsilon) = \{s : s \in \Omega, \lambda(s) \leq \varepsilon\}$. If this is false, we can find a sequence $x_n \in S$ such that $M[\frac{1}{r}P_{\Omega(1/n)}x_n] \leq 2^{-n}$ and $N[\frac{1}{2^n}P_{\Omega(1/n)}Fx_n] \geq \gamma_0 > 0$. Choosing appropriate indices $n' > n$, we get in particular $M[\frac{1}{r}P_{\Omega_{n,n'}}x_n] \leq 2^{-n}$ and $N[\frac{1}{2^n}P_{\Omega_{n,n'}}Fx_n] \geq \gamma_0 > 0$, where $\Omega_{n,n'} = \Omega(1/n) \setminus \Omega(1/n') = \{s : \frac{1}{n'} < \lambda(s) \leq \frac{1}{n}\}$. By induction, we construct a sequence $n_1 = 1$, $n_2 = n_1' + 1, \cdots, n_{j+1} = n_j' + 1, \cdots$ and set

$$x_*(s) = \sum_{j=1}^{\infty} P_{\Omega_{n_j,n_j'}} x_{n_j}(s) .$$

By construction, we have $x_* \in B_r(L_M)$, but $Fx_* \notin L_N$, contradicting our hypothesis.

Second step.   With the above notation, let

$$\Delta(s) = \{u : u \in \mathbb{R},\ \mu(s)M(s, u/r) \leq \delta\},$$

$$f_\gamma(s, u) = \max\{0, N(s, \frac{1}{R}f(s, u)) - \frac{2\gamma}{\delta}M(s, u/r)\},$$

and

$$a_\gamma(s) = \sup_{u \in \Delta(s)} f_\gamma(s, u).$$

By considering the set $\hat{\Sigma}_M^r$ given in (4.27), arguments similar to those in the proof of Theorem 3.1 show that then $a_\gamma \in L_1$ and $\|a_\gamma\|_1 \leq \gamma$. Consequently, the estimate (4.26) holds with $a = a_\gamma$ and $b = 2\gamma/\delta$. This proves the necessity of (4.26) in case $\Omega_c = \emptyset$.

Suppose now that $\Omega_d = \emptyset$; we shall establish condition (4.26) in three steps.

First step.   We consider the operator F on the set Σ_M^r defined in (4.6) and construct a special unit u_0 in L_M. To this end, let v_0 be a unit in L_N such that

$$\|y\|_N \leq \|y\|_{M(v_0)} , \qquad (4.28)$$

where $\| \cdot \|_{M(v_0)}$ denotes the norm (2.5). Denoting now $\Delta(s) = \{u : u \in \mathbb{R}, M(s, u/r) \leq \eta(s)\}$ where $\eta = d\lambda/d\mu$, the sequence

$$a_n(s) = \sup_{u \in \Delta(s)} \frac{1}{v_0(s)} |f(s, u/n)|$$

converges to zero almost everywhere on Ω. Let

$$\Omega_n = \{s : s \in \Omega,\ a_n(s) \le 1\}, \quad D_n = \Omega_n \setminus \bigcup_{j=1}^{n-1} \Omega_j\,;$$

then $\lambda(\Omega \setminus \bigcup_{n=1}^{\infty} \Omega_n) = \lambda(\Omega \setminus \bigcup_{n=1}^{\infty} D_n) = 0$. Now we define a unit u_0 by putting $u_0(s) = \frac{1}{n} \sup \Delta(s)$ for $s \in D_n$. By construction, $\|x\|_{M(u_0)} \le 1$ implies that $\|Fx\|_{M(v_0)} \le 1$, hence $\|Fx\|_N \le 1$, by (4.28).

Second step.   We claim that there exist $R > 0$, $\varepsilon > 0$, and $\delta > 0$ such that $M[x/r] \le \delta$ implies that $N[Fx/R] \le \varepsilon$. For the proof, we suppose in addition that $M(s, u) = 0$ $(s \in \Omega)$ holds only for $u = 0$. Assuming the contrary, we can find a sequence $x_n \in S$ such that $\lim_{n\to\infty} M[x_n/r] = 0$ and $\lim_{n\to\infty} N[2^{-n} Fx] = \infty$. By passing to a subsequence, if necessary, we may assume that $M[P_{T_n} x_n/r] \le 2^{-n}$ and $N[2^{-n} P_{T_n} Fx_n] \ge 2^{n+1}$, where T_1, T_2, $\cdots$, T_n, $\cdots$ is a decreasing sequence of sets $T_n \in \mathcal{M}$ with $\lambda(T_n) = 2^{-n}$; moreover, we may assume that $M(s, x_n(s)/r) \to 0$ almost everywhere on Ω as $n \to \infty$. By our additional assumption on M, $x_n \to \theta$ almost everywhere, hence also $\frac{x_n}{u_0} \to \theta$ almost everywhere, with u_0 being the unit constructed in the first step. Now the statement follows similarly as in the first step of the case $\Omega_c = \emptyset$.

If $M(s, u) = 0$ $(s \in \Omega)$ for some $u > 0$, we fix some unit w in the associate space $L_{\tilde{M}}$ and replace M by the Young function $\widehat{M}(s, u) = M(s, u) + |u| w(s)$, which is zero, of course, only for $u = 0$.

Third step.   With the above notation, let

$$\Delta(s) = \{u : u \in \mathbb{R},\ M(s, u/r) < \infty\}\,,$$
$$f_\varepsilon(s, u) = \max\{0, N(s, \tfrac{1}{R} f(s, u)) - \tfrac{\varepsilon}{\delta} M(s, u/r)\}\,,$$

and

$$a_\varepsilon(s) = \sup_{u \in \Delta(s)} f_\varepsilon(s, u)\,.$$

Then again $a_\varepsilon \in L_1$ and $\|a_\varepsilon\|_1 \le \varepsilon$. Consequently, the estimate (4.26) holds with $a = a_\varepsilon$ and $b = \varepsilon/\delta$. This proves the necessity of (4.26) in case $\Omega_d = \emptyset$.
$\Leftarrow$

4.3 Boundedness conditions

Now we pass to boundedness conditions for the superposition operator F between two Orlicz spaces. As already observed, every Orlicz space L_M is a

δ_2 space (and hence a split space, by Lemma 2.5). Consequently, Theorem 2.3 implies the following

Theorem 4.2   *Let f be a sup-measurable function, and suppose that the domain of definition $\mathcal{D}(F)$ of the superposition operator F, generated by f and considered as an operator between two Orlicz spaces L_M and L_N, has interior points. Then F is bounded on each ball contained in $\mathcal{D}(F)$ if and only if, for each $s \in \Omega_d$, the function $f(s, \cdot)$ is bounded on each bounded interval in $\mathbb{R}$.*

We are now in a position to give a simple example which shows that the operator F need not be bounded on large balls. Consider again the function (4.25), where $M \notin \Delta_2$. For $r > 1$, fix some function $x \in \Sigma_M^r \setminus \Sigma_M^1$ and let

$$x_n(s) = \begin{cases} x(s) & \text{if } |x(s)| \le n, \\ 0 & \text{if } |x(s)| > n. \end{cases}$$

Then we have $x_n \in B_r(E_M) \subseteq B_r(L_M)$, hence

$$\sup_{x \in B_r(L_M)} \|Fx\|_1 \ge \sup_n \int_\Omega |Fx_n(s)| ds = \sup_n \int_\Omega M(x_n(s)) ds = \infty;$$

indeed, if the last supremum were finite, the function x would belong to Σ_M^1, by the Fatou property of the integral.

Let μ_F be the growth function of the superposition operator F defined in (2.24). For Lebesgue spaces, we derived a two-sided estimate (3.22) for $\mu_F(r)$ in terms of the generating function f. A parallel result is given by the following theorem whose proof is similar to that of Theorem 4.1:

Theorem 4.3   *Let f be a sup-measurable function. The superposition operator F generated by f is bounded from the ball $B_r(L_M)$ into the space L_N (the space E_N, respectively) if and only if for some $R > 0$ (for all $R > 0$, respectively) the relation (4.26) holds for almost all $s \in \Omega$ with $\mu(s)M(s, u/r) \le 1$, and some $a \in L_1$ and $b \ge 0$. Moreover, the two-sided estimate (3.22) holds, where $\mu_F(r)$ is defined by (2.24), and $\nu_F(r)$ denotes the infimum of all $R > 0$ such that (4.26) holds with $\|a\|_1 \le 1$ and $b \le 1$.*

To conclude this section, we reformulate Theorem 2.4 for Orlicz spaces:

Theorem 4.4   *Let f be a sup-measurable function. Then the superposition operator F generated by f is absolutely bounded from the ball $B_r(L_M)$ into the space L_N if and only if there exists a monotonically increasing function Φ_r on $[0, \infty)$ such that (2.29) holds, and the superposition operator $\tilde{F}_r$*

generated by the function $\tilde{f}_r(s,u) = u_0(s) \, \Phi_r \, [|f(s,u)|u_0(s)^{-1}]$ *(with* u_0 *an arbitrary unit in* L_N*) is bounded on the ball* $B_r(L_M)$.

4.4 Continuity conditions

In this section, we shall be concerned with various continuity properties of the superposition operator between Orlicz spaces. We begin with the following fundamental theorem:

Theorem 4.5 *Let* f *be a sup-measurable function, and suppose that the interior* G *of the domain of definition* $\mathcal{D}(F)$ *of the superposition operator* F, *generated by* f *and considered as an operator between* L_M *and* L_N, *is non-empty. Then, if the operator* F *is continuous on* G, *the function* f *is sup-equivalent to some Carathéodory function on* $\delta(G)$ *(see (2.39)). Conversely, if the function* f *is sup-equivalent to some Carathéodory function on* $\delta(G)$ *and* $N \in \Delta_2$, *the operator* F *is continuous on* G.

$\Rightarrow$ The statement follows immediately from Theorem 2.6 and the fact that the Orlicz space L_N is regular if and only if $N \in \Delta_2$. $\Leftarrow$

At this point, we give a simple example which shows that the second statement of Theorem 4.5 is false if $N \notin \Delta_2$. As a "dual" example to (4.25), consider the function

$$f(u) = N^{-1}(u), \tag{4.29}$$

where N is some (autonomous) Young function which does not satisfy a Δ_2 condition. The corresponding superposition operator F maps the space L_1 into the space L_N, since $N[Fx] = \|x\|_1 < \infty$ for $x \in L_1$. However, F is discontinuous at $x_0 = \theta$. To see this, choose a function $y \in L_N \setminus E_N$, and let

$$y_n(s) = \begin{cases} y(s) & \text{if } |y(s)| \le n, \\ 0 & \text{if } |y(s)| > n. \end{cases}$$

Then the functions $x_n(s) = N(y(s) - y_n(s))$ belong to the space L_1 and converge to θ, since $\lim_{n \to \infty} \|x_n\|_1 = \lim_{n \to \infty} N[y(s) - y_n(s)]ds = 0$. On the other hand, the sequence $Fx_n = y - y_n$ does not converge to θ in L_N; otherwise the function y would be in the closure of L_∞ with respect to the Luxemburg norm (4.8), and hence belong to E_N, a contradiction.

For the sake of completeness, we also mention the following variant of Theorem 2.7 for Orlicz spaces:

Theorem 4.6 *Let* f *be a Carathéodory function, and let* x_0 *be an interior point of the domain of definition* $\mathcal{D}(F)$ *of the superposition operator* F, generated by f and considered as an operator between L_M and L_N. Then F is continuous at x_0 if and only if the superposition operator $\widetilde{F}$ generated by the function (1.11) maps the space L_M^0 into the space L_N^0. Moreover, if F is continuous at x_0, F is continuous at every point of the set $x_0 + L_M^0$.

In addition to Theorem 4.6, we point out that one may also give an example of a superposition operator F in an Orlicz space L_M which is continuous at θ, and hence on E_M, but discontinuous at every point $x_0 \in L_M \setminus E_M$.

Finally, we remark that one can give a criterion for the *uniform continuity* of F on bounded sets analogous to that given in Theorem 3.8 for Lebesgue spaces.

To conclude this section, we give the following analogue to Theorem 3.9 which is a consequence of the general Theorem 2.10 in ideal spaces:

Theorem 4.7   Let f be a sup-measurable function, and suppose that the superposition operator F generated by f acts from L_M into L_N. Then F is weakly continuous if and only if the restriction of f to $\Omega_c \times \mathbb{R}$ satisfies (3.34) with $a \in L_N$ and $b \in L_N/L_M$, and the restriction of f to $\Omega_d \times \mathbb{R}$ is a Carathéodory function.

4.5 Lipschitz and Darbo conditions

The results of this section will be parallel to those of Section 3.5. First, we mention the following consequence of Theorem 2.11:

Theorem 4.8   Let f be a Carathéodory function, and suppose that the superposition operator F generated by f acts from L_M into L_N. Then the following three conditions are equivalent:
(a) The operator F satisfies a Lipschitz condition

$$\|Fx_1 - Fx_2\|_N \le k(r)\|x_1 - x_2\|_M \quad (x_1, x_2 \in B_r(L_M)). \tag{4.30}$$

(b) Given two functions $x_1, x_2 \in B_r(L_M)$, one can find a function $\xi \in B_{k(r)}(L_N/L_M)$ such that (2.45) holds.
(c) The function f satisfies a Lipschitz condition

$$|f(s,u) - f(s,v)| \le g(s,w)|u - v| \quad (|u|, |v| \le w), \tag{4.31}$$

where the function g generates a superposition operator G which maps the ball $B_r(L_M)$ into the ball $B_{k(r)}(L_N/L_M)$.

In Theorem 3.10, a further equivalent condition (d) was given which may be regarded as a family of Lipschitz conditions for the function $f(s, \cdot)$. A similar fact may be proved here: if M and N are two Young functions with $M \sqsupset N$, all conditions of Theorem 4.8 are equivalent to the following one:

(d) For each $\lambda > 0$, there exists a function $a_\lambda \in L_1$ such that $c(\rho) = \sup\limits_{\lambda \geq \rho} \|a_\lambda\|_1 < \infty$ $(\rho > 0)$ and

$$N\left(s, \frac{\lambda}{k(r)}|f(s,u) - f(s,v)|\right) \leq a_\lambda(s) + M\left(s, \frac{u}{r}\right) + M(s, \lambda|u - v|). \quad (4.32)$$

Observe that then (3.37) is a special case of (4.32).

In order to analyze the "degeneracy" phenomena in Lemma 2.8, we give now a condition under which two Orlicz spaces L_M and L_N form a V-pair or a V_0-pair.

Lemma 4.4   Let $\Omega_d = \emptyset$, and let $M = M(u)$ and $N = N(u)$ be two autonomous Young functions. Then (L_M, L_N) is a V-pair if and only if

$$\varlimsup_{u \to 0} \frac{N^{-1}(1/u)}{M^{-1}(1/u)} < \infty. \quad (4.33)$$

Similarly, (L_M, L_N) is a V_0-pair if and only if

$$\lim_{u \to 0} \frac{N^{-1}(1/u)}{M^{-1}(1/u)} = 0. \quad (4.34)$$

$\Rightarrow$ Suppose that (L_M, L_N) is a V-pair, i.e. $\|P_D u_0\|_M \|P_D v_0\|_{\tilde{N}} = O(\mu(D))$ $(\mu(D) \to 0)$ for two units $u_0 \in L_M^0$ and $v_0 \in L_{\tilde{N}}$. Without loss of generality, we may assume that $c_0 \chi_D \leq u_0$ and $d_0 \chi_D \leq v_0$ for some $c_0, d_0 > 0$. By the Young inequality (4.17), we have $N^{-1}(t) \tilde{N}^{-1}(t) \leq 2t$, hence, by (4.20),

$$\frac{N^{-1}(1/\mu(D))}{M^{-1}(1/\mu(D))} \leq \frac{2}{\mu(D)M^{-1}(1/\mu(D))\tilde{N}^{-1}(1/\mu(D))} =$$

$$= \frac{2}{\mu(D)}\|\chi_D\|_M \|\chi_D\|_{\tilde{N}} \leq \frac{2}{c_0 d_0 \mu(D)}\|P_D u_0\|_M \|P_D v_0\|_{\tilde{N}} = O(1)$$

as $\mu(D) \to 0$ which implies (4.33). To prove the sufficiency of (4.33) we may take $u_0(s) = \sum_{n=1}^{\infty} c_n \chi_{D_n}(s)$ and $v_0(s) = \sum_{n=1}^{\infty} d_n \chi_{D_n}(s)$, where $\bigcup_{n=1}^{\infty} D_n = \Omega$, $c_n = 2^{-n} M^{-1}(1/\mu(D_n))$, and $d_n = 2^{-n} \tilde{N}^{-1}(1/\mu(D_n))$.

The second part is proved analogously. $\Leftarrow$

Lemma 4.4 and Theorem 2.12 imply the following:

Theorem 4.9   Let $\Omega_d = \emptyset$. Let f be a Carathéodory function, and suppose that the superposition operator F generated by f acts from L_M into L_N, where the Young functions M and N satisfy the relation (4.34). Then F satisfies the Lipschitz condition (4.30) if and only if F is constant, i.e. the function f does not depend on u.

Let us now analyze the Darbo condition

$$\alpha_N(FN) \leq k(r)\alpha_M(N) \quad (N \subseteq B_r(L_M)), \tag{4.35}$$

where α_M denotes the Hausdorff measure of noncompactness (2.32) in the space L_M; to this end, we need the following:

Lemma 4.5   The Orlicz space L_M with either the Luxemburg norm (4.8) or the Orlicz–Amemiya norm (4.9) is α-nondegenerate if $\Omega_d = \emptyset$.

$\Rightarrow$ We restrict ourselves to the Luxemburg norm (4.8), the proof for the norm (4.9) is similar. So, for fixed $u_0 \in L_M$, we have to show that

$$\alpha(R[-u_0, u_0]) \geq \|u_0\|_M, \tag{4.36}$$

with $R[-u_0, u_0]$ given by (2.51). Without loss of generality, let $\|u_0\|_M \geq 1$, hence $M[u_0] \geq 1$ (see (4.5)). Given a finite ε-net $\{z_1, \cdots, z_m\}$ for $R[-u_0, u_0]$ in L_M, we may suppose that u_0 and $z_1, \cdots, z_m$ have the form (2.53). Moreover, given $\eta > 0$, we may choose the partition $\omega = \{D_1, D_2, \cdots\}$ of Ω in such a way that $m_{k,n}^{\pm} \leq M(s, |c_n \pm d_{k,n}|) \leq m_{k,n}^{\pm} + \eta$ for some numbers $m_{k,n}^{\pm}$ $(k = 1, \cdots, m; \; n = 1, 2, \cdots)$. Since $\Omega_d = \emptyset$, we may divide each D_n into two sets A_n and B_n of equal measure. The function $x_* = \sum_{n=1}^{\infty} c_n(\chi_{A_n} - \chi_{B_n})$ belongs to $R[-u_0, u_0]$, hence $\|z_k - x_*\|_M \leq \varepsilon$ for some k. Further, we have

$$M(s, |z_k(s) - x_*(s)|) = \sum_{n=1}^{\infty} [M(s, |c_n - d_{k,n}|)\chi_{A_n} + M(s, |c_n + d_{k,n}|)\chi_{B_n}].$$

Let

$$\mu_k(s) = \sum_{n=1}^{\infty} [m_{k,n}^{-}\chi_{A_n}(s) + m_{k,n}^{+}\chi_{B_n}(s)] \quad (k = 1, \cdots, m),$$

and denote by P_ω the averaging operator (2.52) associated with the partition ω. Since the space L_1 is average stable (see Section 2.6), we have $\|P_\omega \mu_k\|_1 \leq \|\mu_k\|_1 \leq M[z_k - x_*]$. On the other hand,

$$P_\omega \mu_k(s) = \frac{1}{2} \sum_{n=1}^{\infty} (m_{k,n}^{-} + m_{k,n}^{+})\chi_{D_n}(s)$$

and

$$m_{k,n}^- + m_{k,n}^+ \geq M(s, |c_n - d_{k,n}|) + M(s, |c_n + d_{k,n}|) - 2\eta \geq M(s, |c_n|) - 2\eta \,;$$

consequently, $|P_\omega \mu_k(s)| \geq M(s, u_0(s)) - \eta$, hence $M[z_k - x_*] \geq M[u_0] \geq 1$, since $\eta > 0$ was arbitrary. This shows that $\varepsilon \geq \|z_k - x_*\|_M \geq 1$, and hence (4.36) holds. $\Leftarrow$

By Theorem 2.13, we get from Lemma 4.5 the following:

Theorem 4.10   *Let f be a Carathéodory function, and suppose that the superposition operator F generated by f acts from L_M into L_N. Then the Lipschitz condition (4.30) and the Darbo condition (4.35) are equivalent.*

4.6 Differentiability conditions and analyticity

In this section we shall reformulate the differentiability results of Section 2.7 and 2.8 for the superposition operator between Orlicz spaces. As for Lebesgue spaces, more precise statements than those given in Chapter 2 may be given. We begin with the following analogue of Theorem 3.12:

Theorem 4.11   *Let f be a Carathéodory function, and suppose that the superposition operator F generated by f acts from L_M into L_N. If F is differentiable at $x \in L_M$, the derivative has the form (2.55), where the function a given by (2.56) belongs to the multiplicator space L_N/L_M. In particular, if (4.33) holds, the function f has the form (2.59) for $s \in \Omega_c$ with $a \in L_N$ and $b \in L_\infty$; if (4.34) holds, the function $f(s, \cdot)$ is constant for $s \in \Omega_c$.*

Conversely, if the superposition operator G generated by the function

$$g(s, u) = \begin{cases} \frac{1}{u}[f(s, x(s) + u) - f(s, x(s))] & \text{if } u \neq 0, \\ a(s) & \text{if } u = 0, \end{cases}$$

acts from L_M into L_N/L_M and is continuous at θ, then F is differentiable at x and (2.55) holds.

$\Rightarrow$ The statements on the differentiability of F follow immediately from Theorem 2.14. The fact that the function $f(s, \cdot)$ degenerates for $s \in \Omega_c$ if (4.33) or (4.34) hold follows from Theorem 2.15 and Lemma 4.4. $\Leftarrow$

In order to generalize Theorem 3.13, we assume that $M \sqsupset N$, and hence L_N/L_M is the Orlicz space L_R generated by the Young function (4.22).

Theorem 4.12   *Let f be a Carathéodory function, and suppose that the superposition operator F generated by f acts from L_M into L_N $(M \sqsupset N)$. Then F is differentiable at $x \in L_M$ if and only if the limit (2.56) exists, belongs to L_R (with R given in (4.22)), and satisfies the following condition: for each $\lambda > 0$, there exists a function $a_\lambda \in L_1$ such that $\|a_\lambda\|_1 \to 0$ $(\lambda \to \infty)$ and*

$$N(s, \lambda | f(s, x(s) + u) - f(s, x(s)) - a(s)u|) \leq a_\lambda(s) + M(s, \lambda u). \quad (4.37)$$

Similarly, Theorem 3.14 reads now as follows:

Theorem 4.13   *Let f be a Carathéodory function, and suppose that the superposition operator F generated by f acts from L_M into L_N $(M \sqsupset N)$. Then F is asymptotically linear if and only if the limit (2.65) exists, belongs to L_R (with R given in (4.22)), and satisfies the following condition: for each $\lambda > 0$, there exists a function $a_\lambda \in L_1$ such that $\|a_\lambda\|_1 \to 0$ $(\lambda \to 0)$ and*

$$N(s, \lambda | f(s, u) - a_\infty(s)u|) \leq a_\lambda(s) + M(s, \lambda u). \quad (4.38)$$

Observe that (4.37) and (4.38) generalize (3.43) and (3.44), respectively.

Finally, we remark that an analogous result for higher differentiability holds as that given in Theorem 3.15 for Lebesgue spaces. We do not go into the details, but just remark that, if the operator F is n-times differentiable from L_M into L_N at some point $x \in L_M$, and (L_M, L_N) is a V^n-pair, by Theorem 2.20 the function f has the form (2.71), i.e. is a polynomial of degree n. A necessary and sufficient condition for (L_M, L_N) to be a V^n-pair, in turn, is that in case $M = M(u)$ and $N = N(u)$

$$\varlimsup_{u \to 0} \frac{N^{-1}(1/u)}{[M^{-1}(1/u)]^n} < \infty; \quad (4.39)$$

compare this with (4.33).

The results of Section 2.8 on analyticity conditions may also be sharpened for Orlicz spaces. First of all, we recall that, given an ideal space Y and a Young function $M = M(s, u)$, the inverse M-transform $M^{-1}[Y]$ of Y is defined by means of the norm (2.86). A comparison with (4.8) and (2.82) shows that $M^{-1}[L_1]$ is just the Orlicz space L_M with the Luxemburg norm.

Suppose now that $\Omega_d = \emptyset$, and f is a Carathéodory function over $\Omega \times \mathbf{C}$ such that $f(s, 0) = 0$ and $f(s, \cdot)$ is entire analytic on $\mathbf{C}$ for almost all $s \in \Omega$, i.e. the expansion

$$f(s, u) = \sum_{n=1}^{\infty} a_n(s) u^n$$

holds for all $u \in \mathbb{C}$. Let

$$M(s, u) = \max_{|v|=u} |f(s, v)|,$$

and consider the Orlicz space $L_M = M^{-1}[L_1]$. Since all hypotheses of Theorem 2.23 are satisfied, we have the following:

Theorem 4.14   *Under the above hypotheses, the superposition operator F, generated by f and considered as an operator from L_M into L_1, is analytic at θ.*

Theorem 4.14 allows us to associate to each entire analytic function f an Orlicz space L_M such that the superposition operator F generated by f becomes analytic between L_M and L_1. However, the operator F need not be entire analytic in this case. Consider, for example, the function $f(u) = e^u - 1$. This function is entire analytic and generates an analytic superposition operator F from L_M (with $M(u) = e^{|u|} - 1$) into L_1. The domain of analyticity of F, however, is not the whole space L_M, but just the class $\Sigma_M^1 = \{x : x \in S, M[x] < \infty\}$ which is a proper subset of L_M since $M \notin \Delta_2$.

The above reasoning shows also that, at least in case $M \notin \Delta_2$, the class of analytic superposition operators between the Orlicz spaces L_M and L_N is reasonably large. On the other hand, Theorem 2.20 and Lemma 4.1 together imply the following:

Theorem 4.15   *Let $\Omega_d = \emptyset$, let f be a sup-measurable function, and suppose that the superposition operator F generated by f acts between two Orlicz spaces L_M and L_N, where $M \in \Delta_2$. Assume that F is analytic in a neighbourhood of some point $x \in L_M$. Then F is a polynomial operator on L_M.*

4.7 Notes, remarks and references

1.   Orlicz spaces were introduced in the thirties by W. Orlicz [251], [252], who discovered that to each increasing convex function $M = M(u)$ there corresponds a Banach space L_M just as the Lebesgue space L_p corresponds to the function $M(u) = |u|^p$ $(1 < p < \infty)$. Afterwards, this concept was generalized in various directions. First, Orlicz' original assumption that

$M(u)u^{-1} \to 0$ and ∞ as $u \to 0$ and ∞, respectively, was weakened by A.C. Zaanen [368] in order to include the spaces L_1 and L_∞ as well. *Variable* Young functions $M = M(s, u)$ seem to have been considered first in [146] (without L_1 and L_∞), and then in [305] and [306] (with L_1 and L_∞). *Vector-valued* Young functions were introduced by M.S. Skaff [323], [324] and have been studied since by many authors, e.g. J. Chatelain, A.D. Ioffe, A. Kozek, and others. In the meantime, Orlicz spaces have become such a well established part of functional analysis that the literature is vast; loosely speaking, whenever a concept is introduced in the general theory of Banach spaces, there are some papers concerned with this concept in the framework of Orlicz spaces.

Most of the literature deals with Orlicz spaces $L_M = L_M(\Omega, \mathcal{M}, \mu)$, where Ω is a bounded domain in Euclidean space, $M = M(u)$, and μ is the Lebesgue measure on Ω; most bibliographical references in the sequel refer first to this situation, which we shall call the "standard case". Fundamental work on Young functions and Orlicz spaces has been done by M.A. Krasnosel'skij and Ja.B. Rutitskij; we mention, in particular, the monograph [178]. A classical reference is also the book [369]; a brief recent survey may be found in [135]. We remark that Young functions are also called *Orlicz functions* or *N-functions* by some authors (e.g. [178]). Orlicz spaces are special examples of the more general class of so-called *modular spaces*; see e.g. the monographs [163] and [234].

The imbedding conditions and the formula (4.21) for the multiplicator space L_N/L_M may be found for the standard case in [11], [364] and [371]. The estimate (4.14) is contained in [32].

As already mentioned, the Δ_2 condition (4.23) is extremely important in both the theory and applications of Orlicz spaces. In the standard case, the Δ_2 condition (4.23) simply reads

$$M(cu) \leq bM(u) \quad (|u| \geq u_0), \quad\quad\quad (4.40)$$

of course. From this one could conjecture that any Young function which has at most polynomial growth satisfies the Δ_2 condition; a counterexample in [178] shows that this is false.

Lemma 4.1 is proved in [23], Lemma 4.2 in [35]; the statement of Lemma 4.2 shows, in particular, that there exist δ_2 spaces which are not Δ_2 spaces. **2.** The superposition operator in Orlicz spaces was extensively studied, especially in the standard case, by M.A. Krasnosel'skij, Ja.B. Rutitskij, I.V. Shragin, and M.M. Vajnberg [176–178], [277–279], [290], [291], [293], [295–301], [348–351]. The "transformation lemma" 4.3 has been used implicitly by many authors in case $M = M(u)$, $N = N(u)$. Theorem 4.1 may be found, even for vector valued functions, in [242], [243]. Of course, in the standard case the acting condition (4.26) becomes, for f a Carathéodory function,

$$N(\frac{f(s,u)}{R}) \leq a(s) + bM(\frac{u}{r})$$

and may be obtained with Lemma 4.3 [349]. In the vector-valued case, a general acting theorem was given in [394]. As already observed, any acting condition for F from L_M into L_N implies, by the particular choice $f(u) = u$, an *imbedding condition* for $L_M \subseteq L_N$. In this way, it follows from Theorem 4.1 that L_M is imbedded in L_N if and only if (4.26) holds for $f(u) = u$. In particular, in the standard case this is equivalent to $M \sqsupseteq N$.

3. The boundedness result Theorem 4.3 is due to M.A. Krasnosel'skij and Ja.B. Rutitskij in the standard case [176–178], [277] (see also [371]), in the general case to P.P. Zabrejko and H.T. Nguyên [243], [382]. We remark that in the case $\Omega_d = \emptyset$ the upper estimate in (3.28) holds with 2 instead of 3. The counterexample following Theorem 4.2 may be found in [178]. Some auxiliary results on bounded superposition operators in Orlicz spaces are contained in [294]. In [299] it is shown that every superposition operator which maps the Orlicz class Σ_M^r of height r into L_N is bounded on the ball $B_r(L_M)$.

Theorem 4.4 is a consequence of the general Theorem 2.4; see also [134], [136], and [371]. Absolutely bounded superposition operators in Orlicz spaces are considered from a different viewpoint in [23]. Similar conditions for F between various Orlicz spaces which ensure that the composition KF of F with the linear integral operator (2.83) is compact are discussed in [391].

In the special case $f(u) = u$, Theorem 4.4 gives a necessary and sufficient condition on M and N under which the imbedding $L_M \subseteq L_N$ is absolutely bounded. In the standard case, this means again that $M \sqsupseteq N$.

4. Continuity properties for the superposition operator in Orlicz spaces have been studied from the very beginning of the theory. In [178] and [290] one can find acting and continuity conditions for the simplest operator $Fx(s) = a(s)x(s)$; the paper [290] contains also a condition for the *weak continuity* of this operator. Theorem 4.5 may be found in §17 of [178], [296] and [297] in the standard case. A brief survey concerning conditions for boundedness, continuity and weak continuity of F in Orlicz spaces is [295]; weak continuity of F from E_M into L_N is treated in [293]. The counterexample after Theorem 4.5 is again mentioned in [178]. Theorem 4.6 is first mentioned in [297] and [298], all proofs of these two papers may be found in [300]. Of course, the ideas contained in [300] carry over to general ideal spaces (see [249]). In [301] the author gives an example of a superposition operator F in an Orlicz space L_M which is continuous at $x_0 = \theta$ (and hence on E_M), but discontinuous at each $x_0 \in L_M \setminus E_M$.

The paper [387] contains a necessary and sufficient condition for the uniform continuity of superposition operators between Orlicz spaces. Another special type of continuity of the superposition operator in Orlicz spaces is mentioned in Lemmas 4.3 and 4.17 of [122] in view of applications to nonlinear *elliptic boundary value problems*. Acting, boundedness, and continuity properties of the operator F from the space C of continuous functions into some Orlicz space L_M may be found in [188], [292], and [302]. Finally, we remark that Theorem 4.7 was proved in the standard case in [293].

5. All results presented in Section 4.5 follow from the corresponding results in Section 2.7. Lemma 4.4 is implicitly contained in [34], Theorem 4.8 and Lemma 4.5 in [27]. The additional condition (d) after Theorem 4.8 may be found in [20] or [365]; the latter paper contains also a criterion for the *Hölder continuity* of F in Orlicz spaces, i.e.

$$\|Fx_1 - Fx_2\|_N \le k(r)\|x_1 - x_2\|_M^\alpha \quad (x_1, x_2 \in B_r(L_M)), \qquad (4.41)$$

which is an analogue to (3.38) for Lebesgue spaces. In this connection, we also mention the paper [68], which shows that, if certain properties of F (Hölder continuity, monotonicity etc.) hold on $B_r(L_M)$ for some $r > 1$, they also hold on any ball; another typical example of such a "local-global" phenomenon is Lemma 2.3, see also [148].

The degeneracy condition given in Theorem 4.9 is of course a direct consequence of Lemma 4.4. Theorem 4.10 is taken from [27].

6. The first differentiability conditions for F, in part sufficient, in part necessary, in Orlicz spaces may be found in [176], [363], [371] and [373]. The necessary and sufficient condition in Theorem 4.12 was proved in [365] and, independently, in [19], Theorem 4.13 and a corresponding result for higher differentiability also in [19] in the standard case. The *potential* (3.56) was studied in Orlicz spaces in [349].

The fact that L_M is a Δ_2 space if and only if M satisfies a Δ_2 condition is first proved in [23]. The degeneracy result given in Theorem 4.15 is taken from [32]. In our discussion concerning analytic superposition operators in Orlicz spaces, especially Theorem 4.14, we followed [32].

7. So far, most of our bibliographical remarks mainly referred to the standard case. In the case $\Omega_c = \emptyset$ some acting theorems have been obtained by I.V. Shragin [311], [313]; the continuity of F under the acting condition $F(L_M) \subseteq L_N^0$ is proved in [276], a general continuity theorem in [273]. The results of [305] carry over also to the discrete case $\Omega_c = \emptyset$ (see [307]).

There exists also some literature now on the vector-valued case for non-autonomous Young functions $M = M(s,u)$ and $N = N(s,u)$. Acting, boundedness, and continuity conditions for F between L_M and L_1 may be found in [148], much general information for F between L_M and L_N in [62], [83–85], [112], [113], [258], [259], [320], [360], and [361]. In [260–262] the author studies superposition operators between generalized Orlicz spaces by means of the theory of *modular spaces* of vector valued functions. Representation theorems for the functional (1.39) on generalized Orlicz spaces are dealt with in [263]. The most complete presentation of all important properties of the superposition operator F in Orlicz spaces of vector functions with non-autonomous Young functions is the thesis [243].

Chapter 5

The superposition operator
in symmetric spaces

Symmetric spaces are ideal spaces whose norm may be defined by means of the decreasing rearrangement of measurable functions. Thus, all general results discussed in Chapter 2 carry over to such spaces, but some results may be sharpened. For instance, the main statements on the boundedness, Lipschitz continuity, or differentiability of the superposition operator between symmetric spaces can be formulated more explicitly in terms of the so-called fundamental function.

The most important examples of symmetric spaces, apart from those discussed in Chapters 3 and 4, are the Lorentz space Λ_ϕ and the Marcinkiewicz space M_ϕ. These spaces play a fundamental role, for example, in interpolation theory of linear operators.

After recalling the notions and properties of symmetric spaces, in general, and Lorentz or Marcinkiewicz spaces, in particular, we formulate some elementary results on the superposition operator between such spaces. Unfortunately, the theory is here much less advanced than in, say, Lebesgue and Orlicz spaces. The results presented here are mainly combinations of special properties of symmetric spaces and general results obtained in Chapter 2.

5.1 Symmetric spaces

Let Ω be an arbitrary set, $\mathcal{M}$ some σ-algebra of subsets of Ω, and μ a σ-finite and countably additive measure on $\mathcal{M}$; as before, by λ we denote some equivalent normalized measure on $\mathcal{M}$. Since the spaces and their properties which we are going to study below are practically unknown in case of a discrete measure, we shall assume throughout that the measure μ is atomic-free on Ω, i.e. $\Omega = \Omega_c$. By $S^0 = S^0(\Omega, \mathcal{M}, \mu)$ we denote the set of all *completely measurable* functions x over Ω, which means that the *distribution function* (see also (1.19))

$$\mu(x, h) = \mu(\{s : s \in \Omega, |x(s)| > h\}) \tag{5.1}$$

is finite for any $h \in [0, \infty)$. In case $\mu(\Omega) < \infty$ the set S^0 coincides, of course, with the whole space S; as in Chapter 1, we consider S^0 as a metric space with the metric (see also (1.1))

$$\rho(x, y) = \inf_{0 < h < \infty} \{h + \mu(x - y, h)\}. \tag{5.2}$$

In what follows, we shall only deal with functions in S^0.

For further reference, we collect some properties of the distribution function in the following:

Lemma 5.1   *The functional (5.1) has the following properties:*
(a) For any $x \in S^0$, the function $\mu(x, \cdot)$ is monotonically decreasing and right-continuous on $[0, \infty)$.
(b) If $x_1 \in S$, $x_2 \in S^0$, and $|x_1| \le |x_2|$, then also $x_1 \in S^0$ and $\mu(x_1, h) \le \mu(x_2, h)$.
(c) For all $x_1, x_2 \in S^0$,

$$\mu(x_1 + x_2, h_1 + h_2) \le \mu(x_1, h_1) + \mu(x_2, h_2).$$

(d) For all $x \in S^0$,

$$\int_\Omega |x(s)| ds = \int_0^\infty \mu(x, h) dh. \tag{5.3}$$

(e) $\mu(\{s : s \in \Omega, x_1(s) \ne x_2(s)\}) \le \delta$ implies that $|\mu(x_1, h) - \mu(x_2, h)| \le \delta$.
(f) If a sequence $x_n \in S^0$ converges (in the metric (5.2)) to $x \in S$, the sequence $\mu(x_n, h)$ converges (in $\mathbb{R}$) to $\mu(x, h)$ at each point of continuity h of $\mu(x, \cdot)$.

$\Rightarrow$ All properties follow immediately from the definition (5.1). Observe, in particular, that (d) is just the classical "Cavalieri principle". The property (c) is a weak form of subadditivity; simple examples show that the function $\mu(\cdot, h)$ is, in general, not subadditive in the usual sense. $\Leftarrow$

By Lemma 5.1 (a), the function $\mu(x, \cdot)$ admits a (unique) right-continuous monotonically decreasing inverse

$$x^*(t) = \inf\{h : h > 0, \mu(x, h) < t\} \quad (0 < t < \infty) \tag{5.4}$$

which is usually called the *decreasing rearrangement* of x. Observe that then

$$\mu(x, x^*(t)) \leq t, \quad x^*(\mu(x, h)) \geq h,$$

and equality occurs at each point of continuity.

Lemma 5.2   *The function (5.4) has the following properties:*
(a) For any $x \in S^0$, the function x^* is monotonically decreasing and right-continuous on $(0, \infty)$.
(b) If $x_1 \in S$, $x_2 \in S^0$, and $|x_1| \leq |x_2|$, then also $x_1 \in S^0$ and $x_1^* \leq x_2^*$.
(c) For all $x_1, x_2 \in S^0$,

$$(x_1 + x_2)^*(t_1 + t_2) \leq x_1^*(t_1) + x_2^*(t_2).$$

(d) For all $x \in S^0$,

$$\begin{aligned}
\int_0^t x^*(\tau)d\tau &= \sup_{\mu(D)=t} \int_D |x(s)|ds, \\
\int_t^\infty x^*(\tau)d\tau &= \inf_{\mu(D)=t} \int_{\Omega \backslash D} |x(s)|ds.
\end{aligned} \tag{5.5}$$

(e) $\mu(\{s : s \in \Omega, x_1(s) \neq x_2(s)\}) \leq \delta$ implies that $x_1^*(t + \delta) \leq x_2^*(t) \leq x_1^*(t - \delta)$.
(f) If a sequence $x_n \in S^0$ converges (in the metric (5.2)) to $x \in S$, the sequence $x_n^*(t)$ converges (in $\mathbb{R}$) to $x^*(t)$ at each point of continuity t of x^*.

$\Rightarrow$ The proof follows immediately from the definition (5.4). $\Leftarrow$

Finally, we introduce the *average rearrangement* of x,

$$x^{**}(t) = \frac{1}{t} \int_0^t x^*(\tau)d\tau \quad (0 < t < \infty), \tag{5.6}$$

which will be important in the definition of a large class of spaces considered in the following section.

Lemma 5.3   *The function (5.6) has the following properties:*
(a) For any $x \in S^0$, the function x^{**} is continuous on $(0, \infty)$ and satisfies $x^{**} \geq x^*$.
(b) If $x_1 \in S$, $x_2 \in S^0$, and $|x_1| \leq |x_2|$, then also $x_1 \in S^0$ and $x_1^{**} \leq x_2^{**}$.
(c) For all $x_1, x_2 \in S^0$,

$$(x_1 + x_2)^{**}(t) \leq x_1^{**}(t) + x_2^{**}(t).$$

(d) For all $x \in S^0$,

$$x^{**}(t) = \frac{1}{t} \sup_{\mu(D)=t} \int_D |x(s)| ds. \tag{5.7}$$

(e) If a sequence $x_n \in S^0$ converges (in the metric (5.2)) to $x \in S$, the squence x_n^{**} converges pointwise (in $\mathbb{R}$) to x^{**}.

$\Rightarrow$ The proof follows again from the definition (5.6). Observe, in particular, that x^{**} has the usual subadditivity property (c); this follows from (d). $\Leftarrow$

To give a trivial example, for a characteristic function $x = \chi_D$ $(D \in \mathcal{M})$ we get

$$\mu(\chi_D, h) = \begin{cases} \mu(D) & \text{if } 0 < h < 1, \\ 0 & \text{if } 1 \le h < \infty, \end{cases}$$

$$\chi_D^*(t) = \chi_{(0,\mu(D))}(t),$$

and

$$\chi_D^{**}(t) = \begin{cases} 1 & \text{if } 0 < t \le \mu(D), \\ \frac{1}{t}\mu(D) & \text{if } \mu(D) < t < \infty. \end{cases}$$

Recall that two functions x_1 and x_2 on Ω are called *equi-measurable* if $\mu(x_1, h) = \mu(x_2, h)$ for all $h \ge 0$. The advantage of introducing the decreasing rearrangement (5.4) consists in the fact that passing from x to x^* one gets an equi-measurable function which is defined on the positive half-axis and preserves the L_1-norm, by (5.5).

By means of the rearrangement x^*, we introduce now an important class of ideal spaces. A space X is called *symmetric* if, for $x \in X$, every measurable function which is equi-measurable with x belongs also to X and has the same norm. In other words, X is symmetric if and only if the relations $x_1 \in S$, $x_2 \in X$ and $x_1^* \le x_2^*$ imply that $x_1 \in X$ and $\|x_1\|_X \le \|x_2\|_X$. Likewise, we call a space X *supersymmetric* if the relations $x_1 \in S$, $x_2 \in X$ and $x_1^{**} \le x_2^{**}$ imply that $x_1 \in X$ and $\|x_1\|_X \le \|x_2\|_X$. Obviously, every supersymmetric space is symmetric, and every symmetric space is ideal.

Examples of supersymmetric spaces are the Lebesgue spaces L_p ($1 \le p \le \infty$). The Orlicz space L_M (see Section 4.1) is symmetric if and only if the generating Young function $M = M(u)$ does not depend on s. We give now a third example which is particularly useful in interpolation theory. Given $p, q \in [1, \infty]$, denote by $L_{p \wedge q}$ the space $L_p \cap L_q$ equipped with norm (see (3.3))

$$\|x\|_{p \wedge q} = \max\{\|x\|_p, \|x\|_q\}, \tag{5.8}$$

and by $L_{p \vee q}$ the space of all functions $x = u + v$ with $u \in L_p$ and $v \in L_q$, equipped with the norm

$$\|x\|_{p \vee q} = \inf\{\|u\|_p + \|v\|_q : x = u + v\}, \qquad (5.9)$$

where the infimum is taken over all pairs $(u, v) \in L_p \times L_q$ with $u + v = x$. It is not hard to see that $L_{p \wedge q}$ and $L_{p \vee q}$ are symmetric spaces; a particularly important case is $p = 1$ and $q = \infty$.

A few more notions are in order. Given a symmetric space X, by X^σ we denote the ideal space of all functions $\xi \in S([0, \infty))$ for which the norm

$$\|\xi\|_{X^\sigma} = \|x\|_X \qquad (5.10)$$

if finite, where $x \in X$ with $x^* = \xi^*$ is arbitrary; the correctness of this definition, i.e. the independence of x, follows from the symmetry of X. The space X^σ which is often more convenient to study than the space X itself is called the *rearrangement* of X.

Let X be a symmetric space. Since the norm $\|\chi_D\|_X$ of a characteristic function depends only on the measure $\mu(D)$ of $D \in \mathcal{M}$, but not on the explicit structure of the set D, we may introduce the *fundamental function*

$$\varphi_X(\lambda) = \|\chi_D\|_X \quad (\mu(D) = \lambda) \qquad (5.11)$$

of the space X. With X^σ given by (5.10), we have of course $\varphi_X(\lambda) = \|\chi_{(0,\lambda)}\|_{X^\sigma}$. The fundamental function is a *quasi-concave function* on $[0, \infty)$; this means that $\varphi_X(0) = 0$ and both functions $\varphi_X(\lambda)$ and $\lambda/\varphi_X(\lambda)$ are positive and increasing on $(0, \infty)$.

Given a symmetric space X and a number $\lambda \in (0, \infty)$, the *dilatation operator* σ_λ is defined in the space X^σ by

$$\sigma_\lambda \xi(t) = \xi(t/\lambda) \quad (\xi \in X^\sigma) \qquad (5.12)$$

Observe that this operator commutes with the rearrangements (5.4) and (5.6), i.e. $\sigma_\lambda(\xi^*) = (\sigma_\lambda \xi)^*$ and $\sigma_\lambda(\xi^{**}) = (\sigma_\lambda \xi)^{**}$. Moreover, the operator σ_λ is bounded in X^σ with

$$\min\{1, \lambda\} \leq \|\sigma_\lambda\|_{X^\sigma} \leq \max\{1, \lambda\}. \qquad (5.13)$$

The norm of the dilatation operator (5.12) in X,

$$\sigma_X(\lambda) = \|\sigma_\lambda\|_{X^\sigma} \quad (\lambda > 0) \qquad (5.14)$$

is called the *dilatation function* of the space X; this is also a quasi-concave function on $(0, \infty)$.

A relation between the fundamental function (5.11) and the dilatation function (5.14) is given by

$$\sup_{0 < \mu < \infty} \frac{\varphi_X(\lambda\mu)}{\varphi_X(\mu)} \leq \sigma_X(\lambda); \qquad (5.15)$$

there exist symmetric spaces X such that strict inequality occurs in (5.15).

Several important properties of a symmetric space may be expressed in terms of its fundamental function or dilatation function. For instance, a symmetric space X is quasi-regular if and only if

$$\lim_{\lambda \to 0} \varphi_X(\lambda) = 0, \qquad (5.16)$$

and regular if and only if (5.16) holds and the set of simple functions is dense in X (see Section 2.1).

In order to illustrate the functions (5.11) and (5.14) we consider again the spaces $X = L_p$, $X = L_M$, and $X = L_{p \wedge q}$ and $X = L_{p \vee q}$ given above; other examples will be given in subsequent sections.

A trivial calculation shows that $\varphi_{L_p}(\lambda) = \sigma_{L_p}(\lambda) = \lambda^{1/p}$ $(1 \leq p \leq \infty)$; thus, in case $X = L_p$ equality holds in (5.15). More generally, in the Orlicz space L_M with Luxemburg norm (4.8) we have

$$\varphi_{L_M}(\lambda) = \frac{1}{M^{-1}(1/\lambda)}, \qquad (5.17)$$

while the Orlicz–Amemiya norm (4.9) gives

$$\varphi_{L_M}(\lambda) = \lambda \tilde{M}^{-1}(1/\lambda) \qquad (5.18)$$

(see (4.20)), where $\tilde{M}$ is the associated Young function (4.16). Explicit formulas for the dilatation function (5.14) in L_M are not known, but estimates may be given using (5.15).

Finally, a straightforward calculation shows that

$$\varphi_{L_{p \wedge q}}(\lambda) = \begin{cases} \lambda^{\min\{1/p, 1/q\}} & \text{if } \lambda \leq 1, \\ \lambda^{\max\{1/p, 1/q\}} & \text{if } \lambda > 1, \end{cases}$$

and

$$\varphi_{L_{p \vee q}}(\lambda) = \begin{cases} \lambda^{\max\{1/p, 1/q\}} & \text{if } \lambda \leq 1, \\ \lambda^{\min\{1/p, 1/q\}} & \text{if } \lambda > 1; \end{cases}$$

the same formulas hold for $\sigma_{L_{p \wedge q}}$ and $\sigma_{L_{p \vee q}}$.

Recall (see Section 2.1) that the associate space $\tilde{X}$ of an ideal space is defined by the norm (2.8). The associate space $\tilde{X}$ to a symmetric space X is again symmetric. If an ideal space Z is representable as associate space $\tilde{X}$ of some symmetric space X, or, equivalently, if $\tilde{\tilde{Z}} = Z$ (isometric isomorphism), then Z is supersymmetric. In fact, if $x_1^{**} \leq x_2^{**}$ for some $x_1 \in S^0$ and $x_2 \in Z$, then

$$\int_0^\infty x_1^*(\tau)y^*(\tau)d\tau \leq \int_0^\infty x_2^*(\tau)y^*(\tau)d\tau \qquad (5.19)$$

for all $y \in S^0$; taking the supremum over all y with $\|y\|_X \leq 1$ in (5.19) gives $\|x_1\|_Z \leq \|x_2\|_Z$.

For example, the relations $\tilde{L}_{p\vee q} = L_{\tilde{p}\wedge\tilde{q}}$ and $\tilde{L}_{p\wedge q} = L_{\tilde{p}\vee\tilde{q}}$ hold ($p^{-1} + \tilde{p}^{-1} = q^{-1} + \tilde{q}^{-1} = 1$), as may be seen by applying (2.8) directly to (5.8) and (5.9). Recall that, given two ideal spaces X and Y, the multiplicator space of X with respect to Y is defined by the norm (2.40); in particular, $\tilde{X} = L_1/X$. If X and Y are symmetric, then so is Y/X.

Lemma 5.4   *The fundamental and dilatation functions of $\tilde{X}$ satisfy*

$$\varphi_{\tilde{X}}(\lambda) = \frac{\lambda}{\varphi_X(\lambda)}, \quad \sigma_{\tilde{X}}(\lambda) \leq \lambda\sigma_X(1/\lambda). \qquad (5.20)$$

The fundamental and dilatation functions of Y/X satisfy

$$\varphi_{Y/X}(\lambda) = \frac{\varphi_Y(\lambda)}{\varphi_X(\lambda)}, \quad \sigma_{Y/X}(\lambda) \leq \sigma_Y(\lambda)\sigma_X(1/\lambda). \qquad (5.21)$$

$\Rightarrow$ On the one hand,

$$\varphi_{\tilde{X}}(\lambda) = \|\chi_{(0,\lambda)}\|_{\tilde{X}^\sigma} \leq \sup_{\|x\|_X \leq 1} \int_0^\infty x^*(t)\chi_{(0,\lambda)}(t)dt$$

$$= \sup_{\|x\|_X \leq 1} \int_0^\lambda x^*(t)dt \leq \frac{\lambda}{\varphi_X(\lambda)};$$

on the other, the choice $y(s) = \chi_D(s)/\varphi_X(\lambda)$ ($\mu(D) = \lambda$) gives

$$\varphi_{\tilde{X}}(\lambda) \geq \int_0^\infty y^*(t)\chi_{(0,\lambda)}(t)dt = \frac{\lambda}{\varphi_X(\lambda)}.$$

The second formula in (5.20) follows from

$$\|\sigma_\lambda x^*\|_{\tilde{X}^\sigma} = \sup_{\|y\|_X \leq 1} \lambda < \sigma_{1/\lambda} x^*, y^* > \leq \lambda \sigma_X(1/\lambda)\|x^*\|_{\tilde{X}^\sigma} .$$

The relations (5.21) are proved similarly. $\Leftarrow$

We remark that in case $\tilde{\tilde{X}} = X$ the equality $\sigma_X(\lambda) = \lambda\sigma_{\tilde{X}}(1/\lambda) = \sigma_{\approx\atop X}(\lambda)$
holds. For example, the formulas for σ_X in the spaces $X = L_{p\wedge q}$ and
$X = L_{p\vee q}$ may be obtained in this way. Among these spaces, the case $p = 1$
and $q = \infty$ is particularly important. In fact, the spaces $L_{1\wedge\infty}$ and $L_{1\vee\infty}$
are in a certain sense *extremal* in the class of symmetric spaces, inasmuch
as

$$L_{1\wedge\infty} \subseteq X \subseteq L_{1\vee\infty} \tag{5.22}$$

(continuous imbeddings) for any symmetric space X. This fact explains the
importance of the spaces $L_{1\wedge\infty}$ and $L_{1\vee\infty}$ in interpolation theory.

We pass now to the study of V-pairs and V_0-pairs introduced in Section
2.6. It turns out that V-pairs and V_0-pairs (X,Y) of symmetric spaces may
be characterized by a very simple condition on their fundamental functions.

Lemma 5.5   *The following statements are equivalent for two quasi-
regular symmetric spaces X and Y:*
(a) The fundamental functions φ_X and φ_Y satisfy

$$\overline{\lim_{\lambda\to 0}} \frac{\varphi_X(\lambda)}{\varphi_Y(\lambda)} < \infty. \tag{5.23}$$

(b) (X,Y) is a V-pair.
(c) $(Y/X)^0 = \{\theta\}$.
(d) $Y/X \subseteq L_\infty$.

$\Rightarrow$ Let (a) hold. The fact that X and Y are quasi-regular implies that one
may choose $u_0 = c_0\chi_D$ and $v_0 = d_0\chi_D$ in the definition of a V-pair. But
then (b) follows immediately from the formula (5.20) for the fundamental
function $\varphi_{\tilde{Y}}$ of $\tilde{Y}$. The fact that (b) implies (c) was already proved in
Lemma 2.8, and the fact that (c) implies (d) and (d) implies (a) is obvious.
$\Leftarrow$

Analogously, one may prove the following:

Lemma 5.6   *The following statements are equivalent for two quasi-
regular symmetric spaces X and Y:*

(a) *The fundamental functions φ_X and φ_Y satisfy*

$$\lim_{\lambda \to 0} \frac{\varphi_X(\lambda)}{\varphi_Y(\lambda)} = 0.\tag{5.24}$$

(b) *(X, Y) is a V_0-pair.*
(c) *$Y/X = \{\theta\}$.*

Observe that, by the formula (5.17) for the fundamental function of the Orlicz space L_M, Lemma 4.4 is a special case of Lemmas 5.5 and 5.6.

We remark that similar criteria may be formulated for V^n-pairs and V_0^n-pairs (see Section 2.8) in terms of the fundamental function. For instance, (X, Y) is a V^n-pair (V_0^n-pair, respectively) if and only if $\overline{\lim}_{\lambda \to 0} \varphi_X^n(\lambda)/\varphi_Y(\lambda)$ $< \infty$ (if and only if $\lim_{\lambda \to 0} \varphi_X^n(\lambda)/\varphi_Y(\lambda) = 0$, respectively); compare this again with (4.39).

As we have seen in Sections 2.2 and 2.3, several important properties of ideal spaces (and the superposition operator between them) may be deduced from the asymptotic behaviour of the split functional H defined in (2.15). We say that the fundamental function φ_X of a symmetric space X satisfies a Δ_2 *condition* if there exist $\delta, \varepsilon \in (0, 1)$ such that

$$\varphi_X(\delta\lambda) \leq \varepsilon\varphi_X(\lambda) \quad (\lambda > 0),\tag{5.25}$$

or, equivalently, for all $\varepsilon > 0$ there exists a $\delta \in (0, 1)$ satisfying (5.25). In case $X = L_M$ with $M = M(u)$, this amounts, by (5.17), to $M(u/\varepsilon) \leq M(u)/\delta$, which is the usual Δ_2 condition for Young functions (4.40) in the special case $\Omega_d = \emptyset$. The Δ_2 condition for the dilatation function σ_X is defined analogously.

Lemma 5.7 *Given a symmetric space X, let*

$$H_\sigma(x) = \inf\{\lambda : \|\sigma_\lambda x^*\|_{X^\sigma} < 1\}.\tag{5.26}$$

Then the split functional (2.15) of X satisfies the estimate

$$H(x) \leq H_\sigma(x) + 1.\tag{5.27}$$

In particular, if the dilatation function (5.14) satisfies a Δ_2 condition, then X is a Δ_2 space.

$\Rightarrow$ Fix $x \in X$ with $H_\sigma(x) < \infty$, and let $n = \lceil H_\sigma(x) \rceil$. By (5.26) and the continuity of the measure μ (recall that $\Omega_d = \emptyset$!), we can find a partition

$\{D_1, \cdots, D_n\}$ of Ω such that $(P_{D_i}x)^*(t) \le x^*(t/n)$ $(i = 1, \cdots, n)$. This implies that $\|P_{D_i}x\|_X \le 1$, hence

$$H(x) \le \sum_{i=1}^{n} \|P_{D_i}x\|_X \le n \le H_\sigma(x) + 1$$

as claimed. To prove the second assertion, choose $\lambda = 1/\delta$ and $r \in (1, \frac{1}{\varepsilon})$, where δ and ε are given in (5.25). For any $x \in B_r(X)$ we have then

$$\|\sigma_{1/\lambda}x^*\|_{X^\sigma} \le r\sigma_X(1/\lambda) \le \varepsilon r\sigma_X(1) = \varepsilon r < 1,$$

which shows that H_σ is bounded on the ball $B_r(X)$; the statement follows now from (5.27) and Lemma 2.3. $\Leftarrow$

Recall that by P_ω we denote the averaging operator (2.52) associated with some countable partition $\{D_1, D_2, \cdots\}$ of Ω. As before (see Section 2.6), we call a symmetric space X average stable if every operator P_ω acts from X into the second associate space $\overset{\approx}{X}$ and has norm $\|P_\omega\| = 1$. The following condition for average-stability, which we cite without proof, follows from Mitjagin's interpolation theorem.

Lemma 5.8 *Every supersymmetric space is average-stable. In particular, every symmetric space X which is isometrically isomorphic to $\overset{\approx}{X}$ is average-stable.*

5.2 Lorentz and Marcinkiewicz spaces

We are now going to study certain classes of symmetric spaces which are of great importance in the general interpolation theory of linear operators, but have not yet been given much attention form the viewpoint of nonlinear superposition operators.

We call a quasi-concave function $\phi : [0, \infty) \to [0, \infty)$ a *Lorentz–Marcinkiewicz function*, or *LM-function* for short, if $\lim_{\lambda \to 0} \phi(\lambda) = \lim_{\lambda \to 0} \lambda/\phi(\lambda) = 0$. The function

$$\tilde{\phi}(\lambda) = \frac{\lambda}{\phi(\lambda)} \tag{5.28}$$

will be called *associated LM-function* to ϕ.

Let ϕ be an LM-function. Obviously, the set of all $x \in S^0$ for which the functional

$$[x]_\phi = \int_0^\infty \phi(\mu(x, h))dh = \int_0^\infty x^*(t)d\phi(t) \tag{5.29}$$

is finite, is a linear space Λ_ϕ. Equipped with the norm

$$\|x\|_{\Lambda_\phi} = \sup\{< x, y >: \int_D |y(s)|ds \leq \phi(\mu(D))\}, \tag{5.30}$$

Λ_ϕ becomes a symmetric perfect ideal space, the *Lorentz space* generated by ϕ.

If the LM-function ϕ is concave, the norm (5.30) coincides with the functional (5.29), while in general only the equivalence $\|x\|_{\Lambda_\phi} \leq [x]_\phi \leq 2\|x\|_{\Lambda_\phi}$ holds. Similarly, the set of all $x \in S^0$ for which the norm

$$\|x\|_{M_\phi} = \sup_{D \subseteq \Omega} \frac{\phi(\mu(D))}{\mu(D)} \int_D |x(s)|ds \tag{5.31}$$

is defined and finite is a symmetric perfect ideal space, the *Marcinkiewicz space* M_ϕ generated by ϕ. The norm (5.31) may also be defined by

$$\|x\|_{M_\phi} = \sup_{0<h<\infty} \frac{\phi(\mu(x,h))}{\mu(x,h)} \int_h^\infty k d\mu(x,k) \tag{5.32}$$

or

$$\|x\|_{M_\phi} = \sup_{0<t<\infty} \frac{\phi(t)}{t} \int_0^t x^*(\tau)d\tau = \sup_{0<t<\infty} \phi(t)x^{**}(t). \tag{5.33}$$

For some purposes it is useful to consider also the functional

$$< x >_\phi = \sup_{0<t<\infty} \phi(t)x^*(t)$$

instead of (5.33); by Lemma 5.3 (a), the estimate $< x >_\phi \leq \|x\|_{M_\phi}$ holds. A bilateral estimate as for $[x]_\phi$, however, holds only if the function $\alpha(t) = 1/\phi(t)$ belongs to the space M_ϕ^σ (see (5.10)); in this case, $< x >_\phi \leq \|x\|_{M_\phi} \leq \|\alpha\|_{M_\phi^\sigma} < x >_\phi$.

Let us now calculate the functions (5.11) and (5.14) in the spaces Λ_ϕ and M_ϕ. First, we have for $D \in \mathcal{M}$ with $\mu(D) = \lambda$

$$\varphi_{\Lambda_\phi}(\lambda) = \sup\{< \chi_D, y >: \int_D |y(s)|ds \leq \phi(\mu(D))\} = \phi(\lambda)$$

and

$$\varphi_{M_\phi}(\lambda) = \sup_{0<t<\infty} \phi(t)\chi_D^{**}(t) = \phi(\lambda).$$

This shows, in particular, that both Λ_ϕ and M_ϕ are quasi-regular spaces. The Lorentz space Λ_ϕ is regular if and only if either $\mu(\Omega) < \infty$, or $\mu(\Omega) = \infty$ and $\phi(\infty) = \lim_{\lambda\to\infty} \phi(\lambda) = \infty$. On the other hand, the Marcinkiewicz space M_ϕ is never regular, since the function $\beta(t) = \frac{d}{dt}\tilde{\phi}(t)$ always belongs to M_ϕ^σ, but has not an absolutely continuous norm in M_ϕ. In fact, the space M_ϕ^0 consists precisely of all functions $x \in M_\phi$ for which

$$\lim_{\lambda(D)\to 0} \frac{\phi(\mu(D))}{\mu(D)} \int_D |x(s)|ds = \lim_{t\to 0} \frac{\phi(t)}{t} \int_0^t x^*(\tau)d\tau = 0\,,$$

but, by the definition (5.28) of $\tilde{\phi}$,

$$\lim_{t\to 0} \frac{\phi(t)}{t} \int_0^t \beta(\tau)d\tau = 1\,.$$

Concerning the dilation function (5.14), we just remark that equality holds in (5.15) for $X = \Lambda_\phi$ or $X = M_\phi$.

From Lemma 5.7 it follows that both Λ_ϕ and M_ϕ are Δ_2 space (hence split spaces; see Lemma 2.4) if $\phi \in \Delta_2$. This implies natural boundedness results for the superposition operator which will be discussed in Section 5.4.

It follows from the definition of the norms (5.30) and (5.31) that the spaces Λ_ϕ and $M_{\tilde{\phi}}$ are associate to each other, i.e. $\|x\|_{\Lambda_\phi} = \sup\{< x, y >:$ $\|y\|_{M_{\tilde{\phi}}} \le 1\}$ and $\|x\|_{M_\phi} = \sup\{< x, y >: \|y\|_{\Lambda_{\tilde{\phi}}} \le 1\}$. Moreover, if either $\mu(\Omega) < \infty$, or $\mu(\Omega) = \infty$ and $\phi(\infty) = \infty$, then $M_{\tilde{\phi}} = \tilde{\Lambda}_\phi = \Lambda_\phi^*$; on the other hand, $\Lambda_{\tilde{\phi}} = \tilde{M}_\phi = (M_\phi^0)^*$.

In Section 5.1 we mentioned the fact that the spaces $L_{1\wedge\infty}$ and $L_{1\vee\infty}$ are extremal in the class of all symmetric spaces (see (5.22)). If we consider only spaces X with fixed fundamental function $\varphi_X = \phi$, then the corresponding Lorentz and Marcinkiewicz spaces play the role of extremal spaces:

Lemma 5.9   *Let X be a symmetric space whose fundamental function $\varphi_X = \phi$ satisfies $\phi(\infty) = \infty$. Then*

$$\Lambda_\phi \subseteq X \subseteq M_\phi \qquad\qquad (5.34)$$

(continuous imbeddings).

⇒ The first imbedding follows from the fact that the simple functions are dense in Λ_ϕ in case $\phi(\infty) = \infty$, and the equality $\|\chi_D\|_X = \|\chi_D\|_{\Lambda_\phi}$. The second imbedding follows from the estimates

$$\|x\|_X = \|x^*\|_{X^\sigma} \geq \|P_{(0,t)}x^*\|_{X^\sigma} \geq \frac{\phi(t)}{t}\int_0^t x^*(\tau)d\tau$$

and the definition (5.33) of the norm in M_ϕ. $\Leftarrow$

At this point, an important example is in order. For $1 < p < \infty$, the function $\phi(t) = t^{1/p}$ is obviously a (concave) LM-function; the associated function (5.28) is of course $\tilde{\phi}(t) = t^{1/\tilde{p}}$ with $p^{-1}+\tilde{p}^{-1} = 1$. Here the Lorentz space Λ_p consists of all $x \in S^0$ for which

$$\|x\|_{\Lambda_p} = \int_0^\infty \mu(x,h)^{1/p}dh = \frac{1}{p}\int_0^\infty x^*(t)t^{1/p}\frac{dt}{t} < \infty. \qquad (5.35)$$

Likewise, the Marcinkiewicz space M_p consists of all functions $x \in S^0$ for which

$$\|x\|_{M_p} = \sup_{0<h<\infty} \mu(x,h)^{(1-p)/p}\int_h^\infty k\, d\mu(x,k)$$
$$= \sup_{0<t<\infty} t^{(1-p)/p}\int_0^t x^*(\tau)d\tau < \infty, \qquad (5.36)$$

while its regular part M_p^0 is characterized by the condition

$$\lim_{t\to 0} t^{(1-p)/p}\int_0^t x^*(\tau)d\tau = 0.$$

It is easily seen that $M_{\tilde{p}} = \tilde{\Lambda}_p = \Lambda_p^*$ and $\Lambda_{\tilde{p}} = \tilde{M}_p = (M_p^0)^*$; moreover, $\Lambda_p \subseteq X \subseteq M_p$ for each symmetric space X with $\varphi_X(\lambda) = \lambda^{1/p}$ (for example, $X = L_p$).

The Lorentz and Marcinkiewicz spaces may be generalized in various ways. For instance, given an LM-function ϕ and $q \in [1,\infty]$, one may consider the generalized Lorentz space $\Lambda_{\phi,q}$ of all functions $x \in S^0$ for which the norm

$$\|x\|_{\Lambda_{\phi,q}} = \begin{cases} \left\{\int_0^\infty \phi(t)^{q-1}x^{**}(t)^q d\phi(t)\right\}^{1/q} & \text{if } 1 \leq q < \infty, \\ \sup_{0<t<\infty} \phi(t)x^{**}(t) & \text{if } q = \infty, \end{cases} \qquad (5.37)$$

is finite. A comparison of (5.37) with (5.29) and (5.33) shows that $\Lambda_{\phi,1} = \Lambda_\phi$ and $\Lambda_{\phi,\infty} = M_\phi$. Since $\varphi_{\Lambda_{\phi,q}}(\lambda) = \phi(\lambda)$ for any q, we have

$$\|x\|_{\Lambda_{\phi,\infty}} \leq \|x\|_{\Lambda_{\phi,q}} \leq \|x\|_{\Lambda_{\phi,1}} \quad (1 < q < \infty)$$

by Lemma 5.9. Choosing again $\phi(t) = t^{1/p}$ $(1 < p < \infty)$, we obtain the classical interpolation space $\Lambda_{p,q}$ of all functions $x \in S^0$ satisfying

$$\|x\|_{\Lambda_{p,q}} = \left\{ \int_0^\infty t^{q/p} x^{**}(t)^q \frac{dt}{t} \right\}^{1/q} < \infty \quad (1 \leq q < \infty)$$

and

$$\|x\|_{\Lambda_{p,\infty}} = \sup_{0 < t < \infty} t^{1/p} x^{**}(t) < \infty \quad (q = \infty),$$

respectively. In particular, $\Lambda_{p,1} = \Lambda_p$, $\Lambda_{p,p} = L_p$, and $\Lambda_{p,\infty} = M_p$.

Another possibility of generalizing the Lorentz space Λ_ϕ consists in the following. Let ϕ be a Lorentz–Marcinkiewicz function and M a Young function (see Section 4.1). The *Lorentz–Orlicz space* $\Lambda_{\phi,M}$ contains, by definition, all functions $x \in S^0$ for which the norm

$$\|x\|_{\Lambda_{\phi,M}} = \inf\left\{ k : k > 0, \int_0^\infty M\left[\frac{x^*(t)}{k}\right] d\phi(t) \leq 1 \right\} \quad (5.38)$$

is finite. In particular, the "extremal" choice $M(u) = |u|$ or $\phi(t) = t$ gives the Lorentz space Λ_ϕ and the Orlicz space L_M, respectively. In the space $\Lambda_{\phi,M}$ the fundamental function is given by

$$\varphi_{\Lambda_{\phi,M}}(\lambda) = \frac{1}{M^{-1}(1/\phi(\lambda))}.$$

Finally, we remark that another symmetric space is the space $\Lambda_{M,N}$ of all functions $x \in S^0$ for which the norm

$$\|x\|_{\Lambda_{M,N}} = \inf\left\{ k : k > 0, \int_0^\infty N\left[\frac{1}{M^{-1}(1/t)}\right] N\left[\frac{x^*(t)}{k}\right] \frac{dt}{t} \leq 1 \right\} \quad (5.39)$$

is finite, where M and N are two Young functions. In particular, $\Lambda_{M,M}$ is the Orlicz space L_M, while $\Lambda_{M,\infty}$ coincides with the "weak Orlicz space" of all functions $x \in S$ for which

$$\|x\|_{\Lambda_{M,\infty}} = \sup_{0 < t < \infty} \frac{x^{**}(t)}{M^{-1}(1/t)} < \infty. \quad (5.40)$$

In the case $M(u) = |u|^p/p$ and $N(u) = |u|^q/q$, of course, we get again the spaces $\Lambda_{p,q}$ and $\Lambda_{p,\infty} = M_p$ described above.

5.3 Acting conditions in symmetric spaces

The investigation of the superposition operator in general symmetric spaces is far from being complete. Nevertheless, since symmetric spaces are special ideal spaces, many results proved in Chapter 2 may be made more precise.

Let X and Y be two symmetric spaces with fundamental functions φ_X and φ_Y, respectively. Suppose that the function $f = f(u)$ does not depend on s, that $f(0) = 0$, and that the corresponding superposition operator F acts between X and Y and is bounded on some ball $B_r(X)$. Since all functions of the form $(D \in \mathcal{M}, r > 0)$

$$x(s) = \frac{r}{\varphi_X(\mu(D))}[\chi_{D_0}(s) - \chi_{D \setminus D_0}(s)] \quad (D_0 \subseteq D)$$

belong to the ball $B_r(X)$, their images

$$Fx(s) = f\left(\frac{r}{\varphi_X(\mu(D))}\right)[\chi_{D_0}(s) - \chi_{D \setminus D_0}(s)]$$

belong to some ball $B_{m(r)}(Y)$. This implies that $\left|f\left(\frac{r}{\varphi_X(\mu(D))}\right)\right|\varphi_Y(\mu(D)) \leq m(r)$, hence $|f(r/\varphi_X(\lambda))| \leq m(r)/\varphi_Y(\lambda) \quad (0 \leq \lambda \leq \mu(\Omega))$. Putting $r/\varphi_X(\lambda) = u$ we thus arrive at the following elementary theorem:

Theorem 5.1   Assume that the function $f = f(u)$ does not depend on s, and that $f(0) = 0$. Suppose that the superposition operator F generated by f acts between two symmetric spaces X and Y and is bounded on some ball $B_r(X)$. Then f satisfies the estimate

$$|f(u)| \leq m(r)[\varphi_Y(\varphi_X^{-1}(r|u|^{-1}))]^{-1} \tag{5.41}$$

for $c_-(r) \leq |u| \leq c_+(r)$, where $c_-(r) = r\varphi_X(\mu(\Omega))^{-1}$ and $c_+(r) = r(\varphi_X(0+))^{-1}$.

The problem arises whether the condition (5.41) is also sufficient for F to be bounded between X and Y. The simplest case when the answer to this problem is affirmative is when $X = Y$; hence $\varphi_X = \varphi_Y$. In this case (5.41) becomes

$$|f(u)| \leq \frac{m(r)}{r}|u| \quad (c_-(r) \leq |u| \leq c_+(r)),$$

which is trivially sufficient for $F(B_r(X)) \subseteq B_{m(r)}(Y)$. A simple generalization can be obtained if the fundamental functions φ_X and φ_Y are related by $\varphi_Y(\lambda) = \gamma(\varphi_X(\lambda))$, where the superposition operator $\Gamma x(s) = \gamma(x(s))$ acts from X into Y. Apart from the trivial case $\gamma(t) = t$, the case $\gamma(t) = t^\kappa$ is rather important; here the operator Γ acts from X into the space $Y = X^\kappa$ ($\kappa > 0$) of all functions $x \in S$ for which the norm

$$\|x\|_{X^\kappa} = \| \, |x|^{1/\kappa} \|_X^\kappa$$

makes sense and is finite. For example, $(L_p)^\kappa = L_q$ with $\kappa q = p$, and $(L_M)^\kappa = L_N$ with $N(u) = M(u^{1/\kappa})$. Observe that the classical acting condition (3.19) for F between L_p and L_q may be deduced in this way.

Another type of sufficient acting conditions may be obtained in the following manner. Since any symmetric space X with fundamental function $\phi = \varphi_X$ is imbedded in the Marcinkiewicz space M_ϕ (Lemma 5.9), the acting condition $F(M_\phi) \subseteq Y$ implies trivially the acting condition $F(X) \subseteq Y$. For $F(M_\phi) \subseteq Y$ it is in turn sufficient that the function $\tilde{y}(s) = f(r/\phi(s))$ belongs to the space Y for some $r > 0$, because all functions $x \in B_r(X)$ satisfy the estimate $x^*(t) \leq r/\phi(t)$, by the definition of the norm (5.33). We summarize with the following:

Theorem 5.2   *Suppose that the function $f = f(u)$ does not depend on s, and that $f(0) = 0$. Given two symmetric spaces X and Y, assume that the function $\tilde{y}(s) = f(r/\varphi_X(s))$ belongs to Y for some $r > 0$. Then the superposition operator F generated by f maps the ball $B_r(X)$ into the space Y.*

Unfortunately, the requirement that $\tilde{y} \in Y$ is, in general, very restrictive. The only case when the acting condition $F(X) \subseteq Y$ implies that $\tilde{y} \in Y$ is that of the Marcinkiewicz space M_ϕ with $\phi \in \Delta_2$. In this way, we arrive at the following acting condition, which is, in contrast with Theorems 5.1 and 5.2, both necessary and sufficient.

Theorem 5.3   *Suppose that the function $f = f(u)$ is independent of s, continuous, and monotonically increasing with $f(0) = 0$. Let ϕ and ψ be two LM-functions. Then the superposition operator F generated by f maps the ball $B_r(M_\phi)$ into the space M_ψ if and only if the estimate*

$$|f(u)|\psi(\phi^{-1}(r/|u|)) \leq m(r) \tag{5.42}$$

holds for $c_-(r) \leq |u| \leq c_+(r)$ and some $m(r) > 0$.

$\Rightarrow$ The sufficiency of (5.42) follows from a trivial calculation and the fact that $(Fx)^* = F(x^*)$, by the monotonicity and continuity of f. To prove the necessity, one has to consider the functions $u = r/\phi(s)$ and apply Theorem 5.2. $\Leftarrow$

We remark that for the classical Marcinkiewicz spaces M_p and M_q (see (5.36)) condition (5.42) takes the form

$$|f(u)| \le b|u|^{p/q} \quad (r(\mu(\Omega))^{1/p} \le |u| < \infty)$$

with $b = b(r) = m(r)r^{-p/q}$, which is similar to the classical Krasnosel'skij condition (3.19) in the case $f = f(u)$ and $f(0) = 0$.

In general symmetric spaces, necessary and sufficient acting conditions are not known. Since symmetric spaces are defined in terms of the decreasing rearrangement (5.4), a "good" condition for handling the acting problem in such spaces would be the condition $(Fx)^* = F(x^*)$ used in Theorem 5.3. Unfortunately, this condition is very difficult to verify, except for very restrictive classes of nonlinearities f, as those considered in Theorem 5.3.

In order to obtain general information about the superposition operator F in case of a function $f = f(s, u)$ depending on both s and u, it would be useful to find the explicit form of the *rearrangement* of the operator F

$$F^*\xi(t) = \sup_{x^* \le \xi} (Fx)^*(t) \quad (\xi \in X^\sigma), \tag{5.43}$$

where the supremum is taken over all functions $x \in X$ with $x^*(t) \le \xi(t)$ almost everywhere on $(0, \infty)$. Under the hypotheses of Theorem 5.3 we have $F^* = F$, of course; in general, an explicit description of F^* is extremely difficult. A good knowledge of the operator F^* would enable us to give conditions (at least sufficient) for the operator F to act between two symmetric spaces X and Y. For instance, if F^* given in (5.43) acts in the space L_1, then F acts in both the Lorentz space Λ_ϕ and the Marcinkiewicz space M_ϕ, as well as in the interpolation spaces defined in (5.37).

The foregoing discussion shows that very little is known on the acting problem for the superposition operator in general symmetric spaces, or, in particular, in Lorentz and Marcinkiewicz spaces. If we know a priori, however, that F acts in such spaces, some analytical and topological properties of F hold automatically which may be obtained on the base of the general results of Chapter 2. A description of such properties will be carried out in the following section.

5.4 Some properties in symmetric spaces

Combining the general results obtained in Chapter 2 with the special properties of symmetric spaces (especially Lorentz and Marcinkiewicz spaces) discussed in Section 5.2, we arrive at several statements on the superposition operator between such spaces. Here we assume as before that the generating function f is a Carathéodory function, and the underlying measure is atomic-free (i.e. $\Omega_d = \emptyset$).

Theorem 5.4 *Suppose that the superposition operator F generated by f acts between two symmetric spaces X and Y, where the dilatation function σ_X of X (see (5.14)) satisfies a Δ_2 condition. Then F is bounded on bounded sets. The assertion is true, in particular, if $X = \Lambda_\phi$ or $X = M_\phi$ with $\phi \in \Delta_2$.*

$\Rightarrow$ The proof follows from Theorem 2.3 and Lemma 5.7. $\Leftarrow$

Theorem 5.5 *Suppose that the superposition operator F generated by f acts between two symmetric spaces X and Y, where the simple functions are dense in Y and the fundamental function of Y satisfies condition (5.16). Then F is continuous. The assertion is true, in particular, if $Y = \Lambda_\phi$ with $\phi(\infty) = \infty$ or $Y = M_\phi^0$.*

$\Rightarrow$ The proof follows from Theorem 2.6 and the fact that (5.16) implies the regularity of Y. $\Leftarrow$

Theorem 5.6 *Suppose that the superposition operator F generated by f acts between two symmetric spaces X and Y. Then the following three conditions are equivalent:*
(a) The operator F satisfies the Lipschitz condition (2.44).
(b) Given any two functions $x_1, x_2 \in B_r(X)$, one can find a function $\xi \in B_{k(r)}(Y/X)$ such that (2.45) holds.
(c) The function f satisfies a Lipschitz condition

$$|f(s,u) - f(s,v)| \leq g(s,w)|u - v| \quad (|u|,|v| \leq w), \tag{5.44}$$

where the function g generates a superposition operator G which maps the ball $B_r(X)$ into the ball $B_{k(r)}(Y/X)$.
In particular, if (5.23) holds, then the function g in (5.44) is bounded; similarly, if (5.24) holds, then F satisfies a Lipschitz condition if and only if f does not depend on u. The assertion is true, in particular, if X and Y are Lorentz or Marcinkiewicz spaces.

$\Rightarrow$ The proof follows from Theorem 2.11 and Lemmas 5.5 and 5.6. $\Leftarrow$

Theorem 5.7 *Suppose that the superposition operator F generated by f acts between two symmetric spaces X and Y, where Y is supersymmetric. Then the Lipschitz condition (2.44) and the Darbo condition (2.50) are equivalent. The assertion is true, in particular, if X and Y are Lorentz or Marcinkiewicz spaces.*

$\Rightarrow$ The proof follows from Theorem 2.13. $\Leftarrow$

Theorem 5.8   *Suppose that the superposition operator F generated by f acts between two symmetric spaces X and Y and is differentiable at some point $x \in X$. Then f is of the form (2.59) with $a \in Y$ and $b \in L_\infty$ if (5.23) holds, and f is independent of u if (5.24) holds. The assertion is true, in particular, if X and Y are Lorentz or Marcinkiewicz spaces.*
$\Rightarrow$ The proof follows from Theorem 2.15. $\Leftarrow$

5.5 Notes, remarks and references

1.   Symmetric spaces are, apart from Orlicz spaces and their generalizations, the most important class of ideal spaces of measurable functions. We do not pretend to give all the original papers where the concepts discussed above appear for the first time, but have mainly followed the presentation in [184]; see also [69]. Symmetric spaces are often called *rearrangement-invariant spaces* in a great deal of the literature, since any two equi-measurable functions have the same norm in such a space. Much information on decreasing rearrangements can be found in the classical monographs [75], [144], and [196].

Supersymmetric spaces are extremely important in interpolation theory, since they are precisely the *interpolation spaces* between $L_{1\wedge\infty}$ and $L_{1\vee\infty}$ with interpolation constant 1; see Theorem 4.3 in [184].

The notion of the *fundamental function* of a symmetric space is due to Je.M. Semjonov [285]. Almost all the properties of symmetric spaces which are described in terms of the fundamental function in Section 5.1 (in particular, Lemma 5.4) may be found in the book [184]. The description of V-pairs and V_0-pairs of symmetric spaces by means of their fundamental functions (Lemmas 5.5 and 5.6) is taken from [34].

Lemma 5.7 is proved in [32]; Mitjagin's interpolation theorem, which is referred to in Lemma 5.8, may be found in [226].

2.   Lorentz spaces have been introduced in [195], where also the relation $\Lambda_\phi^* = M_{\tilde\phi}$ is established. The Marcinkiewicz subspaces M_ϕ^0 have been considered by Je.M. Semjonov in [284], where also the relation $(M_\phi^0)^* = \Lambda_{\tilde\phi}$ is established. The imbedding result given in Lemma 5.9 is taken from [285]. The observation that, given two symmetric spaces X and Y with fundamental functions φ_X and φ_Y, respectively, the imbedding $M_{\varphi_X} \subseteq \Lambda_{\varphi_Y}$ implies the imbedding $X \subseteq Y$, leads to the following sufficient condition [285]: the space X is *imbedded* in the space Y if

$$\int_0^{\mu(\Omega)} \frac{d}{d\lambda}\tilde\varphi_X(\lambda)d\varphi_Y(\lambda) < \infty,$$

where $\tilde\varphi_X$ is the associated function to φ_X (see (5.28)).

The Δ_2 condition (5.25) for an LM-Function ϕ plays a similar role for the spaces Λ_ϕ and M_ϕ as the Δ_2 condition (4.40) for Orlicz spaces. For example, $\phi \in \Delta_2$ implies that both the Lorentz space Λ_ϕ and the Marcinkiewicz space M_ϕ are Δ_2-spaces. Conversely, if Λ_ϕ is a Δ_2-space, then $\phi \in \Delta_2$. In fact, in case $\phi \notin \Delta_2$ the superposition operator $Fx(s) = 1/\phi(1/|x(s)|)$ acts from Λ_ϕ into the Lebesgue space L_1, but is unbounded on each ball $B_r(\Lambda_\phi)$ with radius $r > 1$; thus Λ_ϕ is not a Δ_2-space, by Theorem 2.3. We do not know whether the condition $\phi \in \Delta_2$ is necessary for M_ϕ to be a Δ_2-space.

The spaces $\Lambda_{\phi,M}$ (see (5.38)) are described in [207], the spaces $\Lambda_{M,N}$ (see (5.39)) in [17], and the spaces $\Lambda_{\phi,q}$ (see (5.37)) in [23]. The spaces $\Lambda_{p,q}$, $\Lambda_p = \Lambda_{p,1}$, and $M_p = \Lambda_{p,\infty}$ are of course classical (see e.g. [75], [147], or [250]).

3. Very little is known about the superposition operator in symmetric spaces. Apart from the Lebesgue and Orlicz space case, acting conditions which are both necessary and sufficient may be given only for Marcinkiewicz spaces (see Theorem 5.3, which was proved in the special case $M_\phi = M_p$ and $M_\psi = M_q$ in [37]).

Almost nothing is known about the operator (5.43) either. Nevertheless, the *integral functional*

$$\Psi\xi = \sup_{x^* \leq \xi} \int_\Omega f(s, x(s))ds \quad (\xi \in X^\sigma)$$

is somewhat easier to deal with. For instance, a classical result (due to G.G. Lorentz) states that, if the function $f = f(s, u)$ is nonnegative, decreasing in s, and increasing in u, then

$$\Psi\xi = \int_0^\infty f(t, \xi(t))dt.$$

In the paper [149], the authors study the function

$$\psi\xi = \inf \int_\Omega f(s, x(s))ds \quad (\xi \in \mathbb{R}),$$

where the infimum is taken over all functions $x \in L_1$ with integral ξ. Finally, some relations between $(Fx)^*$ and $F(x^*)$ in symmetric spaces, especially Orlicz spaces, are established in [41].

4. All results presented in Section 5.4 are, of course, immediate consequences of the more general results given in Chapter 2. We remark, however, that more statements on the superposition operator between symmetric spaces may be obtained than those following from Chapter 2. For example, if F acts in a symmetric space X and is absolutely bounded (see Section 2.4), then F maps the Marcinkiewicz space M_ϕ into its regular part M_ϕ^0, where $\phi = \varphi_X$; the converse is not true [23].

Chapter 6

The superposition operator
in the spaces C and BV

Apart from the space $S = S(\Omega)$, which was studied in detail in Chapter 1, the space $C = C(\Omega)$ of all continuous functions on a complete metric space Ω plays an important role in both linear and nonlinear analysis. If Ω is a compact subset of Euclidean space without isolated points, most results on the superposition operator in $C(\Omega)$ are of course well-known "folklore". For instance, in this case F maps C into itself if and only if f is continuous on $\Omega \times \mathbb{R}$, and F is always bounded and continuous. A somewhat more careful analysis is required, however, if Ω has isolated points.

Before studying the superposition operator in the space C, we discuss a certain continuity property "up to small sets" of functions $f = f(s, u)$ which is usually called the Scorza–Dragoni property. It turns out that the functions having this property are precisely the Carathéodory functions.

The main sections of this chapter are devoted to the study of the superposition operator from C into C, from C into S, and from S into C. Since C is a thick set in S, there is no essential difference between the cases $F(S) \subseteq S$ and $F(C) \subseteq S$. On the other hand, the requirement that $F(S) \subseteq C$ leads to a strong degeneracy, as will be shown at the end of Section 6.4.

The last section will be concerned with the superposition operator in the space BV of functions of bounded variation. Here one can give a necessary and sufficient acting condition only in the autonomous case $f = f(u)$, while in the general case $f = f(s, u)$ only sufficient conditions are known.

6.1 The space C

Let Ω be a complete metric space with metric d. As usual, by $C = C(\Omega)$ we denote the set of all continuous (real or complex) functions x on Ω; with the natural algebraic operations, the set C becomes a linear space.

In the most important case when Ω is compact, the space $C(\Omega)$ is a separable Banach space, equipped with the norm

$$\|x\|_C = \max_{s \in \Omega} |x(s)| . \tag{6.1}$$

Moreover, the trivial inequality $\|xy\|_C \leq \|x\|_C \|y\|_C$ shows that the space C is also an algebra.

We point out that only the set Ω' of all accumulation points of Ω is of interest, since any function is trivially continuous at each isolated point.

We recall briefly some classical facts about the space C. First, a subset $N \subset C$ is precompact if and only if N is bounded and equicontinuous (Arzelà–Ascoli's theorem). If the space C is considered with the norm (6.1), its dual space C^* is the space of all Borel measures μ on Ω with norm $\|\mu\|_{C^*} = \text{var}\,(\mu, \Omega)$, where $\text{var}\,(\mu, \Omega)$ denotes the total variation of μ on Ω. Finally, a subalgebra $A \subseteq C$ is dense in C if A separates the points of Ω, is stable with respect to complex conjugation, and contains all constants (Stone–Weierstrass' theorem).

If Ω is not compact it is not possible to introduce a norm on $C(\Omega)$, but only a family of seminorms p_K given by $p_K(x) = \max\limits_{s \in K} |x(s)|$, where K runs over all compact subsets of Ω. In this way, $C(\Omega)$ becomes a locally convex linear space. If Ω is countably compact (i.e. if Ω is a countable union of compact sets, say $K_1, K_2, \cdots$), the space $C(\Omega)$ is even a separable metrizable locally convex space, where the metric may be given by

$$\rho(x, y) = \sum_{n=1}^{\infty} 2^{-n} \frac{p_{K_n}(x - y)}{1 + p_{K_n}(x - y)} .$$

In general, the space $C(\Omega)$ is not metrizable.

For each function $x \in C$ we denote by

$$\omega(x, \sigma) = \sup_{d(s,t) \leq \sigma} |x(s) - x(t)| \tag{6.2}$$

its *modulus of continuity*; obviously, $\omega(x, \cdot)$ is decreasing in σ and

$$\lim_{\sigma \to 0} \omega(x, \sigma) = 0 \tag{6.3}$$

if and only if x is uniformly continuous on Ω.

Usually the set Ω is not only a metric space, but also a measure space in the sense mentioned at the beginning of Chapter 1, i.e. Ω is equipped with a σ-algebra $\mathcal{M}$ of subsets and a countably additive and σ-finite measure μ. In this case, the following compatibility condition is usually assumed between the measure and the metric structure of Ω which we called *regularity* of μ in Section 1.1: given $D \in \mathcal{M}$ and $\varepsilon > 0$, there exists a compact subset D_ε of Ω such that $\lambda(D \triangle D_\varepsilon) < \varepsilon$, where λ denotes an equivalent normalized measure. In this situation, the σ-algebra $\mathcal{M}$ contains all compact (and hence all Borel) subsets of Ω.

The fact that Ω carries both a measure and a metric structure enables us to consider the spaces $S = S(\Omega)$ and $C = C(\Omega)$ simultaneously. Since the closed and bounded subsets of Ω belong to $\mathcal{M}$, the space C is always imbedded in the space S. Conversely (Luzin's theorem), given $x \in S$ and $\varepsilon > 0$, one can find a subset $\Omega_\varepsilon \subset \Omega$ such that $\lambda(\Omega_\varepsilon) < \varepsilon$ and the function x is continuous on $\Omega \setminus \Omega_\varepsilon$. A certain analogue to this for functions $f = f(s, u)$ will be considered below (see Theorem 6.1).

Luzin's theorem implies, in particular, that C is a thick set in S (see the remark after Lemma 1.2).

6.2 Some properties of Carathéodory functions

This section is devoted to some special properties of functions $f = f(s, u)$ of two variables which are useful in applications to nonlinear equations involving the superposition operator or related nonlinear operators.

We shall say that a function $f = f(s, u)$ on $\Omega \times \mathbb{R}$ is a *Scorza–Dragoni function* (or function which satisfies a Scorza–Dragoni condition) if for each $\varepsilon > 0$ one can find an open subset $\Omega_\varepsilon \subset \Omega$ such that $\lambda(\Omega_\varepsilon) < \varepsilon$ and the function f is continuous on the product $(\Omega \setminus \Omega_\varepsilon) \times \mathbb{R}$. Note that the Scorza–Dragoni property plays a similar role for functions on $\Omega \times \mathbb{R}$ as the Luzin property for functions on Ω. The following result, which is usually referred to as the *Scorza–Dragoni lemma*, establishes an important relation to Carathéodory functions (see Section 1.4).

Theorem 6.1 *Every Scorza–Dragoni function is a Carathéodory function, and vice versa.*

$\Rightarrow$ The fact that every Scorza-Dragoni function is a Carathéodory function is obvious; therefore we shall prove only the converse.

Let n be a natural number, and denote by $\gamma_n(s)$ the supremum of all $\gamma \in \mathbb{R}$ with the property that, whenever $u_1, u_2 \in \mathbb{Q}$ satisfy $|u_1 - u_2| < \gamma$ and $|u_1|, |u_2| \leq n$, we have $|f(s, u_1) - f(s, u_2)| \leq \frac{1}{n}$. Obviously, all functions γ_n are positive almost everywhere on Ω, and converge to 0 almost everywhere as $n \to \infty$. The functions γ_n are also measurable: in fact, we have $\{s : \gamma_n(s) \geq h\} = \{s : \varphi_{n,h}(s) \leq \frac{1}{n}\}$ $(0 < h < \infty)$, where

$$\varphi_{n,h}(s) = \sup\{|f(s,u_1)-f(s,u_2)| : u_1, u_2 \in \mathbb{Q};\ |u_1-u_2| \le h;\ |u_1|, |u_2| \le n\},$$

and the functions $\varphi_{n,h}$ are measurable as least upper bounds of measurable functions.

Let $\varepsilon > 0$ be given. From the measurability of γ_n it follows that there exists a natural number $k(n)$ such that the set $D_n^0 = \{s : \gamma_n(s) < \frac{1}{k(n)}\}$ has λ-measure at most $2^{-(n+1)}\varepsilon$. Now let Q_n be the set of all $u_{n,j} = -n + \frac{j}{k(n)}$ $(j = 0, 1, \cdots, N(n) = 2nk(n))$ and denote by $D_{n,j}$ and $f_{n,j}$ $(j = 0, 1, \cdots, N(n))$ subsets of Ω and continuous functions, respectively, with the property that $\lambda(D_{n,j}) \le \varepsilon 2^{-(n+1)}[N(n)+1]^{-1}$ and $f(s, u_{n,j}) = f_{n,j}(s)$ $(s \in \Omega \backslash D_{n,j})$, which is possible by Luzin's theorem. Finally, consider the functions

$$f_n(s, u) =$$

$$\begin{cases} f_{n,0}(s) & \text{if } u \le -n, \\[2mm] \dfrac{u_{n,j} - u}{u_{n,j} - u_{n,j-1}} f_{n,j-1}(s) + \dfrac{u - u_{n,j-1}}{u_{n,j} - u_{n,j-1}} f_{n,j}(s) & \text{if } u_{n,j-1} \le u \le u_{n,j}, \\[2mm] f_{n,N(n)}(s) & \text{if } u \ge n. \end{cases}$$

By construction, the functions f_n are continuous on $\Omega \times \mathbb{R}$ and satisfy the estimate

$$|f(s, u) - f_n(s, u)| \le \frac{1}{n} \quad (s \in \Omega \setminus D_n,\ |u| \le n), \tag{6.4}$$

where $D_n = D_n^0 \cup D_{n,1} \cup D_{n,2} \cup \cdots \cup D_{n,N(n)}$. The set $\Omega_\varepsilon = \bigcup_{n=1}^\infty D_n$ has λ-measure at most ε, by construction. Moreover, without loss of generality we may assume that Ω_ε is open, since the measure μ was assumed to be regular. Now, (6.4) implies that the sequence f_n converges uniformly on $(\Omega \setminus \Omega_\varepsilon) \times [-m, m]$ to f, for each $m \in \mathbb{N}$; consequently, the function f is continuous on $(\Omega \setminus \Omega_\varepsilon) \times \mathbb{R}$ as claimed. $\Leftarrow$

The following result, which is usually called the *Krasnosel'skij–Ladyzhenskij lemma*, has been used already earlier (see e.g. the proof of Theorem 2.6).

Theorem 6.2   *Let f be a Carathéodory function, and let v and w be two functions in S (not necessarily finite almost everywhere) such that $v \le w$. Then the function*

$$M(s) = \sup\{f(s, u) : v(s) \le u \le w(s)\} \tag{6.5}$$

is measurable. Moreover, if the supremum in (6.5) is attained for almost all
$s \in \Omega$, *there exists* $x_* \in S$ *such that* $v \leq x_* \leq w$ *and* $M(s) = f(s, x_*(s))$.

$\Rightarrow$ The measurability of M follows from the fact that f is a Carathéodory
function. Moreover, the functions $M_h(s) = \sup\{f(s,u) : v(s) \leq u \leq \min\{h, w(s)\}\}$ are also measurable. Denote by $x_*(s)$ the minimal value
of u such that $f(s, u) = M(s)$; we claim that x_* is a measurable function.
In fact, this follows from the equality $\{s : x_*(s) > h\} = \{s : M(s) > M_h(s)\}$
and the measurability of M and M_h. This completes the proof. $\Leftarrow$

6.3 The superposition operator in the space C

Throughout the remaining part of this chapter we assume that the set Ω is
compact. In this case, the problem of characterizing the acting condition
$F(C) \subseteq C$ becomes rather easy.

Theorem 6.3   *The superposition operator F generated by f maps the
space C into itself if and only if the function f is continuous on $\Omega' \times \mathbb{R}$,
where Ω' denotes the set of all accumulation points of Ω.*

$\Rightarrow$ The fact that the continuity of f on $\Omega' \times \mathbb{R}$ implies the acting condition
$F(C) \subseteq C$ is obvious. To prove the converse implication, we observe first
that $F(C) \subseteq C$ implies the continuity of $f(\cdot, u)$ on Ω' for all $u \in \mathbb{R}$. Moreover, if f is discontinuous at some point $(s_0, u_0) \in \Omega' \times \mathbb{R}$, we may find a
sequence (s_n, u_n) converging to (s_0, u_0) such that

$$|f(s_n, u_n) - f(s_0, u_0)| > \varepsilon_0 > 0. \qquad (6.6)$$

Here it suffices to distinguish the two cases when either all s_n belong to
Ω', or all s_n belong to $\Omega \setminus \Omega'$. In the first case we may find a sequence
of different points $\bar{s}_n$ such that $d(s_n, \bar{s}_n) < \frac{1}{n}$ and $|f(\bar{s}_n, u_n) - f(s_n, u_n)| < |f(s_n, u_n) - f(s_0, u_0)| - \varepsilon_0$; thus we may assume that all s_n in (6.6) are
mutually different. In the second case, we may always suppose, without
loss of generality, that the s_n are different.

By the classical Tietze–Uryson theorem there exists a continuous function
x_* such that $x_*(s_n) = u_n$ for $n = 0, 1, \cdots$; but (6.6) shows that then Fx_* is
discontinuous at $s = s_0$, contradicting the hypothesis $F(C) \subseteq C$. $\Leftarrow$

In the case when the set Ω is *perfect* (i.e. $\Omega' = \Omega$), Theorem 6.3 amounts
to the fact that $F(C) \subseteq C$ if and only if f is continuous on $\Omega \times \mathbb{R}$; in this
version, Theorem 6.3 was used by many people in applications of nonlinear
analysis.

Theorem 6.4   *Suppose that the superposition operator F generated by f maps the space C into itself. Then F is bounded if and only if $f(s, \cdot)$ is bounded for each $s \in \Omega \setminus \Omega'$.*

$\Rightarrow$ The fact that the boundedness of F implies that of the functions $f(s, \cdot)$ for $s \in \Omega \setminus \Omega'$ is obvious. By Theorem 6.3, f is continuous on $\Omega \times \mathbb{R}$. By Weierstrass' theorem, for each $r > 0$ we may choose $m(r) > 0$ and $\delta(r) > 0$ such that $|f(s, u)| \leq m(r)$ for $|u| \leq r$ and dist $(s, \Omega') \leq \delta(r)$. Now, let $N \subset C$ be bounded, say $N \subseteq B_r(C)$, and let $m(r)$ and $\delta(r)$ be chosen as above. Since Ω is compact, by our general assumption, the estimate dist $(s, \Omega') > \delta(r)$ holds only for finitely many $s \in \Omega$. This means that

$$\tilde{m}(r) = \sup\{|f(s, u)| : |u| \leq r, \text{ dist } (s, \Omega') > \delta(r)\} < \infty.$$

Consequently, for $x \in N$ we have

$$|f(s, x(s))| \leq \begin{cases} m(r) & \text{if dist } (s, \Omega') \leq \delta(r), \\ \tilde{m}(r) & \text{if dist } (s, \Omega') > \delta(r), \end{cases}$$

and hence $FN \subset C$ is also bounded. $\Leftarrow$

The proof of Theorem 6.4 shows that the growth function (2.24) in the space C satisfies the estimate

$$\mu_F(r) \leq \max\{m(r), \tilde{m}(r)\}. \tag{6.7}$$

Of course, in case $\Omega = \Omega'$ this simplifies to $\mu_F(r) = \nu_f(r)$, where

$$\nu_f(r) = \max_{\substack{|u| \leq r \\ s \in \Omega}} |f(s, u)|; \tag{6.8}$$

compare this with the corresponding formula for L_∞ (see (3.50) and (3.51)).

Theorem 6.5   *Suppose that the superposition operator F generated by f maps the space C into itself. Then F is continuous if and only if $f(s, \cdot)$ is continuous for each $s \in \Omega \setminus \Omega'$.*

$\Rightarrow$ The statement follows from the definition of the norm (6.1) and classical theorems of calculus. In particular, the continuity of $f(s, \cdot)$ ($s \in \Omega \setminus \Omega'$) follows trivially from the continuity of F. $\Leftarrow$

We remark that, under the hypotheses of Theorem 6.5, F is also *uniformly continuous* (on bounded sets). Moreover, its modulus of continuity is given by

$$\omega_F(r,\delta) = \max\{|f(s,u) - f(s,v)| : s \in \Omega; \ |u|, |v| \le r; \ |u - v| \le \delta\};$$

compare this with (3.31).

To conclude this section we give a comparison between the Lipschitz and Darbo conditions for the superposition operator F in the space C. It is remarkable that, although C is not an ideal space, the following analogue to Theorems 2.12 and 2.14 holds in case $\Omega = [0,1]$:

Theorem 6.6   *Suppose that the superposition operator F generated by f maps the space C into itself (over $\Omega = [0,1]$). Then the following three conditions are equivalent:*
(a) The function f satisfies a Lipschitz condition

$$|f(s,u) - f(s,v)| \le g(s,w)|u - v| \quad (|u|, |v| \le w), \tag{6.9}$$

where the function g generates a superposition operator G which maps the ball $B_r(C)$ into the ball $B_{k(r)}(C)$.
(b) The operator F satisfies a Lipschitz condition

$$\|Fx_1 - Fx_2\|_C \le k(r)\|x_1 - x_2\|_C \quad (x_1, x_2 \in B_r(C)). \tag{6.10}$$

(c) The operator F satisfies a Darbo condition

$$\alpha_C(FN) \le k(r)\alpha_C(N) \quad (N \subseteq B_r(C)), \tag{6.11}$$

where α_C denotes the Hausdorff measure of noncompactness (2.32) in the space C.

$\Rightarrow$ The fact that (a) implies (b) and (b) implies (c) is trivial. Assume that (a) is false, i.e. there exist $s_0 \in [0,1]$ and $u_0, v_0 \in [-r, r]$ such that

$$|f(s_0, u_0) - f(s_0, v_0)| > k(r)|u_0 - v_0|; \tag{6.12}$$

without loss of generality we may assume that $0 < s_0 < 1$ and $u_0 < v_0$. Let N_0 be the set of all functions x_n $(n = 1, 2, \cdots)$ defined by

$$x_n(s) = \begin{cases} v_0 & \text{if } 0 \le s \le s_0, \\ (v_0 - u_0)\left(1 - \frac{s - s_0}{\delta_n}\right) + u_0 & \text{if } s_0 < s < s_0 + \delta_n, \\ u_0 & \text{if } s_0 + \delta_n \le s \le 1, \end{cases}$$

where $\delta_n = \frac{1}{n}(1-s_0)$. Then clearly $\alpha_C(N_0) \leq \frac{1}{2}(v_0 - u_0)$ since the singleton $z_0(s) \equiv \frac{1}{2}(u_0 + v_0)$ is a $\frac{1}{2}(v_0 - u_0)$-net for N_0 in C. Moreover, observe that $N_0 \subseteq B_r(C)$. We claim that

$$\alpha_C(FN_0) \geq \frac{1}{2}|f(s_0, u_0) - f(s_0, v_0)|. \tag{6.13}$$

In fact, write γ for the right-hand side of (6.13), and assume that $\{z_1, \cdots , z_m\}$ is a finite η-net for FN_0 in C, where $\eta < \gamma$. Given $x_n \in N_0$, we have $|f(s, x_n(s)) - z_j(s)| \leq \eta$ for some $j \in \{1, \cdots, m\}$. Since z_j and f are continuous functions, we have $|z_j(s_0 + \delta_n) - z_j(s_0)| \leq \frac{1}{2}(\gamma - \eta)$ and $|f(s_0 + \delta_n, u_0) - f(s_0, u_0)| \leq \frac{1}{2}(\gamma - \eta)$ for sufficiently large n, hence $|f(s_0, u_0) - f(s_0, v_0)| \leq |f(s_0, u_0) - f(s_0 + \delta_n, u_0)| + |f(s_0 + \delta_n, u_0) - z_j(s_0 + \delta_n)| + |z_j(s_0 + \delta_n) - z_j(s_0)| \leq \gamma + \eta < 2\gamma$, contradicting the definition of γ. By (6.12) and (6.13), we get $\alpha_C(FN_0) > \frac{1}{2}k(r)|u_0 - v_0| \geq k(r)\alpha_C(N_0)$, i.e. (c) fails. $\Leftarrow$

When dealing with the the Darbo condition (6.11), we applied directly the definition of the Hausdorff measure of noncompactness (2.32). Alternatively, by the Arzelà–Ascoli theorem, we could have used the bilateral estimate

$$\frac{1}{2}\alpha_C(N) \leq \tilde{\alpha}_C(N) \leq 2\alpha_C(N), \tag{6.14}$$

where

$$\tilde{\alpha}_C(N) = \lim_{\sigma \to 0} \sup_{x \in N} \omega(x, \sigma) \tag{6.15}$$

with $\omega(x, \sigma)$ denoting the modulus of continuity (6.2); in particular, $N \subset C$ is precompact if and only if (6.3) holds uniformly in $x \in N$.

Observe that, by Theorem 6.6, the superposition operator F is compact in the space C if and only if F is constant; thus from the viewpoint of compactness the same degeneracy occurs as in ideal spaces, see Theorem 2.5.

We close with two results on the differentiability and analyticity of the superposition operator in the space C, where $\Omega = \Omega'$. First of all, it is easy to see that the derivative $F'(x)$ of F at $x \in C$, or the asymptotic derivative of F (if it exists!), has necessarily the form

$$F'(x)h(s) = a(s)h(s) \tag{6.16}$$

or

$$F'(\infty)h(s) = a_\infty(s)h(s), \tag{6.17}$$

respectively, where

$$a(s) = \lim_{u \to 0} \frac{1}{u}[f(s, x(s) + u) - f(s, x(s))] \qquad (6.18)$$

and

$$a_\infty(s) = \lim_{u \to \infty} \frac{1}{u} f(s, u). \qquad (6.19)$$

Theorem 6.7   *The superposition operator F generated by f is differentiable at $x \in C$ (respectively asymptotically linear) if and only if the limit (6.18) (respectively the limit (6.19)) exists as a continuous function and satisfies*

$$\sup_{\substack{s \in \Omega \\ |u| \le r}} |f(s, x(s) + u) - f(s, x(s)) - a(s)u| = o(r) \quad (r \to 0)$$

(respectively

$$\sup_{\substack{s \in \Omega \\ |u| \le r}} |f(s, u) - a_\infty(s)u| = o(r) \quad (r \to \infty)).$$

$\Rightarrow$ The proof follows immediately from the definition of the derivatives $F'(x)$ and $F'(\infty)$ and the fact that the growth function (2.24) of F in the space C coincides with the function (6.8) in case $\Omega' = \Omega$. $\Leftarrow$

The following theorem on the analyticity of F is what one should expect in "reasonable" spaces; this is in sharp contrast, of course, with Theorem 2.22 and, especially, Theorems 3.16 and 4.15.

Theorem 6.8   *The superposition operator F is analytic in the space C if and only if the corresponding function f is analytic.*

6.4 The superposition operator between C and S

In some situations it may happen that one has to consider the superposition operator not just in the space C or in the space S, but between C and S, or S and C. In this section we shall briefly discuss this situation. We begin with the following:

Theorem 6.9   *The superposition operator F generated by f maps the space C into the space S if and only if f is sup-measurable.*

$\Rightarrow$ If f is sup-measurable, F maps S into S (by definition, see Section 1.3), and hence also C into S. Conversely, if F maps C into S, then also S into S, since the set C is thick in S. $\Leftarrow$

If, in addition, we require continuity of F from C into S, we arrive at the following analogue of Theorem 1.4:

Theorem 6.10   *The superposition operator F generated by f maps the space C into the space S and is continuous if and only if f is sup-equivalent to some Carathéodory function.*

$\Rightarrow$ Without loss of generality, suppose that $f(s,0) = 0$. We show that the continuity of F from C into S implies its continuity from S into S; the statement follows then from Theorem 1.4. If x_n converges in S to x, we may find, for each $\varepsilon > 0$, a subset $D \subset \Omega$ such that $\lambda(D) < \varepsilon$, the functions $P_{\Omega \setminus D} x_n$ and $P_{\Omega \setminus D} x$ are continuous, and $P_{\Omega \setminus D} x_n$ converges to $P_{\Omega \setminus D} x$, by Luzin's and Jegorov's theorems. By assumption and by (1.10), $P_{\Omega \setminus D} F x_n$ converges in S to $P_{\Omega \setminus D} F x$. Since $F x_n - F x = P_{\Omega \setminus D} F x_n - P_{\Omega \setminus D} F x + P_D (F x_n - F x)$ and $\lambda(D)$ is arbitrarily small, $F x_n$ converges in S to $F x$ as well. $\Leftarrow$

The preceding two theorems show that there is no difference between the behaviour of the superposition operator F from S into S or from C into S. On the other hand, the fact that C is essentially smaller than S has the consequence that F degenerates as an operator from S into C:

Theorem 6.11   *The superposition operator F generated by f maps the space S into the space C if and only if F is constant, i.e. the function f does not depend on u.*

$\Rightarrow$ Suppose that F maps S into C, and let $x \in S$ be fixed. We have to show that the function $y(s) = f(s, x(s)) - f(s, 0)$ is identically zero. If this is not so, we may find a function $y_* \in S$ such that $|y_*| \leq |y|$ and $y_* \notin C$. By Theorem 6.2, we may in turn find a function $x_* \in S$ such that $|x_*| \leq |x|$ and $f(s, x_*(s)) - f(s, 0) = y_*(s)$. By the assumption $F(S) \subseteq C$, the function y_* belongs to the space C, a contradiction. $\Leftarrow$

6.5 The superposition operator in the space BV

In this final section we shall be concerned with the superposition operator in the space $BV = BV(\Omega)$ of all functions x of bounded variation on Ω. Since one knows only some facts about the superposition operator on functions over intervals, we shall assume throughout this section that $\Omega = [0, 1]$.

Recall that the number

$$\text{var } (x; 0, 1) = \sup \sum_{i=1}^{n} |x(s_i) - x(s_{i-1})|,$$

where the supremum is taken over all (finite) partitions $\{s_0, s_1, \cdots, s_n\}$ of $[0, 1]$ (with n variable), is called the *total variation* of x on $[0, 1]$. By $BV = BV([0, 1])$ we denote the space of all functions x on $[0, 1]$ for which the norm

$$\|x\|_{BV} = |x(0)| + \text{var } (x; 0, 1) \tag{6.20}$$

is finite. In this way, BV becomes not only a Banach space, but also an algebra, since

$$\text{var } (xy; 0, 1) \leq \sup_{0 \leq s \leq 1} |x(s)| \text{var } (y; 0, 1) + \sup_{0 \leq s \leq 1} |y(s)| \text{var } (x; 0, 1). \tag{6.21}$$

Although neither of the spaces BV and C is contained in the other, there is some relation between them. In fact, the space $NBV = NBV([0, 1])$ of all *normalized functions x of bounded variation* (i.e. x is left-continuous and satisfies $x(0) = 0$) may be interpreted as dual space of $C = C([0, 1])$ by means of the representation formula

$$< x, y >= \int_0^1 y(t) dx(t) \quad (x \in NBV, \; y \in C), \tag{6.22}$$

where the integral in (6.22) is the classical Riemann–Stieltjes integral.

We mention now some facts about the superposition operator in the space BV. First of all, we point out that the fact that the function f itself is of bounded variation does not imply the acting condition $F(BV) \subseteq BV$. Consider, for example, the function

$$f(u) = \begin{cases} +1 & -\infty < u < -1, \\ \sqrt{|u|} & -1 \leq u \leq 1, \\ +1 & 1 < u < \infty; \end{cases}$$

then $x(s) = s^2 \sin^2 1/s$ belongs to the space BV, but $Fx(s) = s \sin 1/s$ does not. The point is here that f does not satisfy a Lipschitz condition at $u = 0$ (see Theorem 6.12 below). An obvious acting condition for F in the space BV is given in the following:

Theorem 6.12 *Suppose that the function $f(s, \cdot)$ is Lipschitz continuous (uniformly in $s \in [0, 1]$) on $\mathbb{R}$, and the function $f(\cdot, u)$ is of bounded variation (uniformly in $u \in \mathbb{R}$) on $[0, 1]$. Then the superposition operator F generated by f maps the space BV into itself and is bounded.*

$\Rightarrow$ The fact that F maps the space BV into itself follows from a straightforward calculation. Moreover, it follows from the definition of the norm (6.20) that the growth function (2.24) of F satisfies the estimate

$$\mu_f(r) \leq 2Lr + \sup_{u \in \mathbb{R}} \operatorname{var}\,(f(\cdot, u); 0, 1) + |f(0, 0)|,$$

where L is the global Lipschitz constant of $f(s, \cdot)$ on $\mathbb{R}$. $\Leftarrow$

It is an open problem whether the converse of Theorem 6.12 is also true. In the autonomous case $f = f(u)$, however, this problem is solved affirmatively:

Theorem 6.13 *Suppose that the superposition operator F generated by $f = f(u)$ acts in the space BV over $[0, 1]$. Then f satisfies a local Lipschitz condition*

$$|f(u) - f(v)| \leq k(r)|u - v| \quad (|u|, |v| \leq r). \tag{6.23}$$

$\Rightarrow$ Suppose that f does not satisfy a Lipschitz condition on some interval $[-r, r]$. Since the function $x(s) = s$ is of bounded variation, the function f is bounded on $[-r, r]$, say $|f(u)| \leq M$. Let u_n and v_n be sequences in $[-r, r]$ such that

$$|f(u_n) - f(v_n)| \geq (n^2 + n)|u_n - v_n|. \tag{6.24}$$

By passing to subsequences, if necessary, we may assume that $u_n \to u$ and $|u_n - u| \leq 2M(n + 1)^{-2}$. By (6.24), we have $\delta_n \leq 2M(n^2 + n)^{-1}$, where $\delta_n = |u_n - v_n|$.

Now let x_* be defined by

$$x_*(s) = \begin{cases} 0 & \text{if } s = 0, \\ u_n & \text{if } s = \frac{1}{n+1} + k|u_n - v_n| \ (k \in \mathbb{N}), \\ v_n & \text{otherwise on } [\frac{1}{n+1}, \frac{1}{n}), \\ u_1 & \text{if } s = 1. \end{cases}$$

We claim that $x_* \in BV$, but $Fx_* \notin BV$. To see this, let $m = \lceil 2M(n^2 + n)^{-1}\delta_n^{-1} \rceil$ and consider the partition $\omega = \omega_n = \{\frac{1}{n+1}, \frac{1}{n+1} + \frac{1}{2}\delta_n, \frac{1}{n+1} + \delta_n, \cdots, \frac{1}{n+1} + \frac{2m-1}{2}\delta_n, \frac{1}{n+1} + m\delta_n, \frac{1}{n}\}$ of the interval $[\frac{1}{n+1}, \frac{1}{n}]$. For the total variation of x_* on $[\frac{1}{n+1}, \frac{1}{n}]$ we get then

$$\text{var }(x_*; \tfrac{1}{n+1}, \tfrac{1}{n}) \le 2m\delta_n + |u_{n-1} - u| + |u - u_n| + \delta_n$$

$$\le 2\frac{2M}{n^2 + n} + \frac{2M}{n^2} + \frac{2M}{(n+1)^2} + \frac{2M}{n^2 + n} \le \frac{10M}{n^2} \, ;$$

hence $x_* \in BV$, since $\sum_{n=1}^{\infty} n^{-2}$ converges. On the other hand, by (6.24),

$$\text{var }(Fx_*; \tfrac{1}{n+1}, \tfrac{1}{n}) \ge 2m|f(u_n) - f(v_n)| \ge 2m(n^2 + n)\delta_n \ge 1/M \, ,$$

and thus $Fx_* \notin BV$. The proof is complete. $\Leftarrow$

As already mentioned, no general results on the acting, boundedness, or continuity of the superposition operator F are known in the non-autonomous case $f = f(s, u)$ (apart from trivial sufficient conditions, of course). Surprisingly, one may characterize the global Lipschitz condition (2.43) in the space $X = BV$ explicitly. To state this, we denote by

$$f^-(s, u) = \lim_{t \uparrow s} f(t, u) \tag{6.25}$$

the *left regularization* of the function $f(\cdot, u)$.

Theorem 6.14   *Suppose that the superposition operator F generated by f acts in the space BV over $[0, 1]$, and assume that F satisfies a global Lipschitz condition*

$$\|Fx_1 - Fx_2\|_{BV} \le k\|x_1 - x_2\|_{BV} \quad (x_1, x_2 \in BV). \tag{6.26}$$

Then the limits (6.25) exists for all $u \in \mathbb{R}$, and the function (6.25) has the form

$$f^-(s, u) = a(s) + b(s)u \, , \tag{6.27}$$

with $a, b \in NBV$.

$\Rightarrow$ Since the constant function $x_u(s) \equiv u$ is of bounded variation, the limits (6.25) exist for all $u \in \mathbb{R}$; moreover, $f^-(\cdot, u)$ is left-continuous by construction.

Now fix $u_1, u_2, v_1, v_2 \in \mathbb{R}$ and $\sigma \in (0, 1]$, and let $\{s_0, s_1, \cdots, s_{2n}, s_{2n+1}\}$ be a partition of $[0, \sigma]$. Define two functions $x_1, x_2 \in BV$ by

$$x_i(s) = \begin{cases} u_i & \text{if } s \notin \{s_2, s_4, \cdots, s_{2n}\}, \\ v_i & \text{if } s \in \{s_2, s_4, \cdots, s_{2n}\} \end{cases}$$

$(i = 1, 2)$. Obviously,

$$\|x_1 - x_2\|_{BV} = |u_1 - u_2| + 2n|v_1 - v_2 - u_1 + u_2|. \tag{6.28}$$

By (6.26) and (6.28) we have

$$\sum_{i=1}^{n} |f(s_{2i}, v_1) - f(s_{2i}, v_2) - f(s_{2i-1}, u_1) + f(s_{2i-1}, u_2)|$$

$$\leq \text{var } (Fx_1 - Fx_2; 0, 1) \leq k\|x_1 - x_2\|_{BV}$$

$$\leq k|u_1 - u_2| + 2kn|v_1 - v_2 - u_1 + u_2|.$$

Letting in this inequality s_1 tend to σ yields $n|f^-(\sigma, v_1) - f^-(\sigma, v_2) - f^-(\sigma, u_1) + f^-(\sigma, u_2)| \leq k|u_1 - u_2| + 2kn|v_1 - v_2 - u_1 + u_2|$. Setting $v_1 = \xi + \eta$, $v_2 = \xi$, $u_1 = \eta$, and $u_2 = 0$ gives $|f^-(\sigma, \xi + \eta) - f^-(\sigma, \xi) - f^-(\sigma, \eta) + f^-(\sigma, 0)| \leq \frac{k}{n}|\eta|$. Finally, letting $n \to \infty$ we obtain $f^-(\sigma, \xi + \eta) - f^-(\sigma, \xi) - f^-(\sigma, \eta) + f^-(\sigma, 0) = 0$. This shows that, for fixed $\sigma \in (0, 1]$, the function $\ell_\sigma(t) = f^-(\sigma, t) - f^-(\sigma, 0)$ $(t \in \mathbb{R})$ is additive; moreover, since $|f^-(\sigma, u) - f^-(\sigma, v)| \leq k|u - v|$, ℓ_σ is also continuous, hence linear. Consequently, we have $\ell_\sigma(t) = b(\sigma)t$ for some function b. Finally, setting $a(s) = f^-(s, 0)$ gives the representation (6.27), where both functions $a = f^-(\cdot, 0)$ and $b = f^-(\cdot, 1) - f^-(\cdot, 0)$ belong to NBV. $\Leftarrow$

The question arises whether or not, under the hypotheses of Theorem 6.14, the generating function f itself is affine in u, rather than its regularization (6.25). As the following example shows, the answer is negative.

Let $\{r_0, r_1, \cdots\}$ be the set of all rational numbers in $[0, 1]$ $(r_0 = 0)$, and let

$$f(s, u) = \begin{cases} 2^{-k} \sin u & \text{if } s = r_k, \ u \in \mathbb{R}, \\ 0 & \text{otherwise.} \end{cases}$$

For any partition $\{s_0, s_1, \cdots, s_n\}$ of $[0, 1]$ and $x \in BV$ we have then

$$\sum_{i=1}^{n} |Fx(s_i) - Fx(s_{i-1})| \leq 2 \sum_{i=0}^{n} |f(s_i, x(s_i))|$$

$$\leq 2 \sum_{k=0}^{\infty} |f(r_k, x(r_k))| = 2 \sum_{k=0}^{\infty} 2^{-k} |\sin x(r_k)| \leq 2,$$

which shows that F maps the space BV into itself. We claim that F satisfies a Lipschitz condition (6.26). To see this, fix $x_1, x_2 \in BV$, and let

$\{s_0, s_1, \cdots, s_n\}$ be again an arbitrary partition of $[0, 1]$. Since

$$\sum_{i=1}^{n} |(Fx_1 - Fx_2)(s_i) - (Fx_1 - Fx_2)(s_{i-1})|$$

$$= \sum_{i=1}^{n} |f(s_i, x_1(s_i)) - f(s_i, x_2(s_i)) - f(s_{i-1}, x_1(s_{i-1})) + f(s_{i-1}, x_2(s_{i-1}))|$$

$$\leq 2 \sum_{i=0}^{n} |f(s_i, x_1(s_i)) - f(s_i, x_2(s_i))|$$

$$\leq 2 \sum_{k=0}^{\infty} |f(r_k, x_1(r_k)) - f(r_k, x_2(r_k))|$$

$$\leq 2 \sum_{k=0}^{\infty} 2^{-k} |\sin x_1(r_k) - \sin x_2(r_k)|$$

$$\leq 2 \sum_{k=0}^{\infty} 2^{-k} |x_1(r_k) - x_2(r_k)| \leq 2\|x_1 - x_2\|_{BV},$$

we have that var $(Fx_1 - Fx_2; 0, 1) \leq 2\|x_1 - x_2\|_{BV}$ and

$$|Fx_1(0) - Fx_2(0)| = |\sin x_1(0) - \sin x_2(0)| \leq |x_1(0) - x_2(0)|,$$

i.e. (6.26) holds with $k = 2$. By Theorem 6.14, the left regularization (6.25) has the form (6.27); here it is of course easy to see directly that $f^-(s, u) \equiv 0$.

If we assume a priori that $f(\cdot, u) \in NBV$ for each $u \in \mathbb{R}$, we get the following:

Theorem 6.15 *Suppose that the superposition operator F generated by f acts in the space BV over $[0, 1]$, and assume that the function $f(\cdot, u)$ is left-continuous for all $u \in \mathbb{R}$. Then F satisfies the global Lipschitz condition (6.26) if and only if the function f has the form (2.59) (i.e. is an affine function in u), where $a, b \in NBV$.*

$\Rightarrow$ The "if" part follows from the fact that BV is an algebra (see (6.21)), while the "only if" part follows from (6.27) and the fact that $f^- = f$. $\Leftarrow$

6.6 Notes, remarks and references

1. The properties of the space $C(\Omega)$ for various Ω may be found in every textbook on functional analysis; see e.g. [104]. Luzin's theorem which we had used already in Chapter 1 is discussed from different points of view in [192]; for relations with other classical results on the space C see [318].

There are many compatibility conditions between the measure and metric structure of Ω which are similar to (but different from) the regularity we assumed in Section 6.1; see [104] or [256].

There is a certain pecularity when studying the "disposition" of the space C in the space S. On the one hand, one may consider C as space of continuous *functions*; trivially, every member of C then also belongs to S. On the other hand, one may consider C as a set of *classes of functions*, where the equivalence is, as usual, equality almost everywhere, and hence consider each such class as a member of S. Throughout this chapter, we followed the first interpretation.

2. The Scorza–Dragoni property of Carathéodory functions was first observed in [282]. Afterwards this property was studied in much more general situations and generalized by various authors; see e.g. [39], [271], [272] and, first of all, [57] and [395]. Many generalizations have been obtained also to multivalued maps.

Theorem 6.2 is due to M.A. Krasnosel'skij and L.A. Ladyzhenskij [169], see also [303]. We remark that the Theorems 6.1 and 6.2 have been motivated by the L_p theory of nonlinear superposition and integral operators; see e.g. [182]. Some more properties of Carathéodory functions, together with several illuminating examples, may be found in [80].

3. In the case when Ω is a compact perfect subset of Euclidean space, all results presented in Section 6.3 are, of course, well-known "folklore". The first elementary results in this direction seem to be due to I.V. Misjurkejev [225]. As Theorems 6.3, 6.4 and 6.5 show, one has to be careful at isolated points of Ω. As a matter of fact, the results in [225] are not quite correct, since the author tacitly assumes that $\Omega = \Omega'$, as was pointed out by N.V. Shuman [321]. The results given here may be found in [292], see also [187]; it is also shown there that, if F maps the space C into itself, F is always *weakly continuous*. This is of course in sharp contrast with Theorem 2.10. Observe that the basic results of Section 6.3 are valid for countably compact $\Omega = \bigcup_{n=1}^{\infty} K_n$, simply by restricting f to $K_n \times \mathbb{R}$.

The fact that the growth function $\mu_F(r)$ in the space C coincides with the function (6.8) (in case $\Omega = \Omega'$) makes it possible to describe the analytical properties of F in the space C quite explicitly; observe that actually the same is true in the space L_∞; see (3.50) and (3.51).

Continuous superposition operators which leave some cone in the space C invariant are considered in [150] and [151]. The paper [255] contains a necessary and sufficient continuity condition for F in the space $C(X)$, with X being an arbitrary topological space.

Theorem 6.6 shows that, from the viewpoint of the Lipschitz and Darbo conditions, the superposition operator behaves in rather the same way in the spaces C, L_p and L_M (see Theorems 3.11 and 4.10). We point out that a *global* version of Theorem 6.6 is also true: the Lipschitz condition

$$\|Fx_1 - Fx_2\|_C \le k\|x_1 - x_2\|_C \quad (x_1, x_2 \in C)$$

and the Darbo condition

$$\alpha_C(FN) \le k\alpha(N) \quad (N \subset C \text{ bounded})$$

for the operator F, as well as the Lipschitz condition

$$|f(s, u) - f(s, v)| \le k|u - v| \quad (u, v \in \mathbb{R})$$

for the function $f(s, \cdot)$, are all equivalent [12].

The differentiability theorems 6.7 and 6.8 are immediate consequences of the formula (6.8) for the growth function of F in the space C; compare the proof of Theorem 3.13.

We remark that there exists also some literature in the case when Ω is not compact. For example, in [358], [359] one may find some results on the superposition operator in the *scale of Banach spaces* $BC_\eta(\mathbb{R})$ of all continuous functions x on $\mathbb{R}$ for which the norm

$$\|x\|_\eta = \sup_{t \in \mathbb{R}} e^{-\eta|t|}|x(t)| \quad (\eta > 0)$$

is finite. Finally, the paper [109] is concerned with superposition operators of the form $Fx(\tau) = f(\tau, x(e^{i\tau}))$ $(\tau \in \mathbb{R})$, defined on the space of all functions $x = x(z)$ which are continuous on $|z| \le 1$ and holomorphic on $|z| < 1$.

4. The Theorems 6.9 and 6.10 are essentially "by-products" of the theory of the superposition operator in the space S developed in Chapter 1. We point out again that one must not replace C by smaller spaces which are not thick in S (e.g. C^1).

Theorem 6.11 is taken from [54]. In fact, it follows from the proof that the superposition operator F is necessarily constant if F maps the space S into some *completely non-ideal space* $X \subseteq S$, where X is called completely non-ideal (see [54]) if, given $x \in X$, $x \ne \theta$, one may find a function $y \in S$ such that $|y| \le |x|$ and $y \notin X$. In other words, a set $N \subset S$ is completely non-ideal if and only if, for each $u \in N \setminus \{\theta\}$, the ideal hull $id\{u\}$ (see (2.84)) is not contained in N.

Apart from studying the behaviour of F between the spaces C and S, one may also consider the superposition operator from C into some ideal space X. For instance, the operator F maps C into an ideal space X if and only if the corresponding function f satisfies the growth condition

$$|f(s,u)| \le a_r(s) \quad (|u| \le r)$$

with some function $a_r \in X$ which depends on $r > 0$. For $X = L_1$ this follows from Lemma 1.3; for arbitrary X, one has to choose some unit $v \in \tilde{X}$ (see (2.8)) and to pass to the auxiliary function $f_v(s,u) = f(s,u)v(s)$, which generates a superposition operator F_v from C into L_1. In the case $X = L_p$, for a similar result see Theorem 3.17; in the case of an Orlicz space $X = L_M$, the above acting criterion was proved by I.V. Shragin [292], see also [188] and [302].

In Chapter 1 we mentioned the problem of characterizing the *integral functional*

$$\Phi x = \int_\Omega f(s, x(s))ds \tag{6.29}$$

on the space S, which may be represented by a Carathéodory function (or more general functions). There is a parallel result in the space C which states that, if Φ is a nonlinear continuous bounded disjointly additive functional on C, then Φ has a representation (6.29) with some Carathéodory function f [81].

5. Very little is known about the superposition operator in the spaces BV and NBV. Theorem 6.12 is contained in [193], Theorem 6.13 (which is, of course, the nontrivial part) in [154].

Both Theorems 6.14 and 6.15, as well as the counterexample in between, are taken from [224]. Of course, analogous statements hold for right-continuous instead of left-continuous functions, with (6.25) replaced by the *right regularization*

$$f^+(s,u) = \lim_{t \downarrow s} f(t,u). \tag{6.30}$$

Finally we remark that necessary and sufficient acting conditions (for $f = f(u)$), as well as Lipschitz-type conditions, are known for the space HBV of all functions of harmonic bounded variation [82], for the space BV_φ of all functions of bounded φ-variation [86], and for the space A_ω of all integrable functions whose Fourier–Haar coefficients statisfy a summability condition with weight ω [336–339]. Interestingly, these are also local Lipschitz conditions for f, like that given in (6.23).

Chapter 7

The superposition operator
in Hölder spaces

The literature on the superposition operator in Hölder spaces is almost as vast as that in Lebesgue spaces. However, the "behaviour" of the superposition operator in Hölder spaces is quite different from that in spaces of measurable functions, or in the space C.

The topics discussed in the present chapter are the following: after formulating a (quite technical) necessary and sufficient acting condition, we discuss the boundedness of F. Interestingly, acting implies boundedness only in case $f = f(u)$, but not in case $f = f(s,u)$. Quite amazing is the fact that the operator F may act in a Hölder space, although the generating function f is not continuous (and thus F does not act in the space C!).

As far as the continuity of F is concerned, things are even worse: even in the autonomous case $f = f(u)$ the operator F may act between two Hölder spaces (and hence be bounded), but not continuous. On the other hand, it is possible to describe the "points of continuity", and to give conditions for both continuity and uniform continuity on bounded sets.

Another surprising fact concerns the Lipschitz and Darbo conditions, which in all spaces considered so far turned out to be equivalent. In Hölder spaces, the operator F may satisfy a Darbo condition for a reasonably large class of nonlinearities f, while a global Lipschitz condition for F leads, roughly speaking, necessarily to affine functions f (in u). This shows that the Banach contraction mapping principle may not be applied (globally) to problems involving superposition operators in Hölder spaces.

When dealing with boundedness properties of the superposition operator, it is possible to derive a bilateral estimate for its growth function between Hölder spaces. This enables us to give conditions for the differentiability and asymptotic linearity of F, in terms of f, which are both necessary and sufficient.

7.1 Hölder spaces

Throughout this chapter, by a *Hölder function* we mean a real function ϕ such that $\phi(0) = 0$, $\phi(1) = 1$, and both functions $\phi(t)$ and $t/\phi(t)$ are positive and increasing. A standard example is $\phi(t) = t^\alpha$ with $0 < \alpha \leq 1$. For further reference, we summarize some elementary properties of such functions with the following:

Lemma 7.1   *Every Hölder function ϕ has the following properties:*

$$\phi(\sigma + \tau) \leq \phi(\sigma) + \phi(\tau) \quad (\sigma, \tau \geq 0) ; \tag{7.1}$$

$$\frac{\phi(\tau)}{\tau} \leq 2\frac{\phi(\sigma)}{\sigma} \quad (0 < \sigma \leq \tau) ; \tag{7.2}$$

$$\frac{1}{2}\phi(\sigma) \leq \frac{1}{\sigma}\int_0^\sigma \phi(t)dt \leq \phi(\sigma) . \tag{7.3}$$

$\Rightarrow$ The inequality (7.1) follows from the monotonicity of the function $t/\phi(t)$.

To prove (7.2), we choose $n \in \mathbb{N}$ such that $(n-1)\sigma \leq \tau - \sigma \leq n\sigma$ and get, by repeated application of (7.1), $\frac{\phi(\tau)}{\tau} \leq \frac{\phi(\sigma)}{\tau} + \frac{\phi(\tau-\sigma)}{\tau} \leq \frac{\phi(\sigma)}{\sigma} + \frac{\phi(n\sigma)}{\sigma+(n-1)\sigma}$
$\leq \frac{\phi(\sigma)}{\sigma} + \frac{n\phi(\sigma)}{n\sigma} = 2\frac{\phi(\sigma)}{\sigma}$.

The left inequality in (7.3) follows from the relation

$$\phi(\sigma) - \frac{1}{\sigma}\int_0^\sigma \phi(t)dt = \frac{1}{\sigma}\int_0^\sigma [\phi(\sigma) - \phi(t)]dt \leq$$

$$\leq \frac{1}{\sigma}\int_0^\sigma \phi(\sigma - t)dt = \frac{1}{\sigma}\int_0^\sigma \phi(t)dt ,$$

while the right inequality follows from

$$\int_0^\sigma \phi(t)dt \leq \int_0^\sigma \phi(\sigma)dt = \sigma\phi(\sigma) . \quad \Leftarrow$$

Let Ω be a compact domain in Euclidean space without isolated points. Without loss of generality we assume that diam $\Omega = 1$. Given a Hölder function ϕ, the *Hölder space* $H_\phi = H_\phi(\Omega)$ consists, by definition, of all continuous functions x on Ω for which

$$h_\phi(x) = \sup_{\sigma > 0} \frac{\omega(x, \sigma)}{\phi(\sigma)} < \infty , \tag{7.4}$$

where

$$\omega(x,\sigma) = \sup\{|x(s) - x(t)| : s,t \in \Omega;\ |s - t| \le \sigma\} \qquad (7.5)$$

denotes the *modulus of continuity* of the function x. In case $\phi(t) = t^\alpha$ one usually writes H_α instead of H_ϕ; the space H_1 is sometimes denoted by *Lip*. Equipped with the norm

$$\|x\|_\phi = \max\{\|x\|_C, h_\phi(x)\}, \qquad (7.6)$$

the set H_ϕ becomes a Banach space. Apart from this space, the closed subspace H_ϕ^0 of all functions $x \in H_\phi$ with

$$\omega(x,\sigma) = o(\phi(\sigma)) \quad (\sigma \to 0) \qquad (7.7)$$

will be important in what follows. In some respect, the subspace H_ϕ^0 is much nicer than the whole space H_ϕ. For instance, H_ϕ^0 is always separable, while H_ϕ is not, and compact subsets of H_ϕ^0 are easily described (see Lemma 7.3).

In order to compare the "size" of different Hölder functions and spaces, we introduce the following notation: given two Hölder functions ϕ and ψ, we write $\phi < \psi$ (respectively $\phi << \psi$) if $\phi(t) = O(\psi(t))$ (respectively $\phi(t) = o(\psi(t))$) as $t \to 0$. The condition $\phi < \psi$ is then equivalent to each of the inclusions $H_\phi \subseteq H_\psi$, $H_\phi^0 \subseteq H_\psi$, and $H_\phi^0 \subseteq H_\psi^0$, while the condition $\phi << \psi$ is equivalent to $H_\phi \subseteq H_\psi^0$. In the special case $\phi(t) = t^\alpha$ and $\psi(t) = t^\beta$, we have of course $\phi < \psi$ (respectively $\phi << \psi$) if and only if $\alpha \ge \beta$ (respectively $\alpha > \beta$).

Although the space H_ϕ^0 is much easier to deal with than the space H_ϕ, it is sometimes convenient to pass to another space which is still smaller. Let ϕ be a fixed Hölder function, and let $1 \le p < \infty$. By $\mathcal{J}_{\phi,p} = \mathcal{J}_{\phi,p}(\Omega)$ we denote the set of all continuous functions x on Ω for which

$$j_{\phi,p}(x) = \int_0^1 \frac{\omega(x,t)^p}{\phi(t)^p} \frac{dt}{t} < \infty. \qquad (7.8)$$

Equipped with the norm

$$\|x\|_{\phi,p} = \max\{\|x\|_C, j_{\phi,p}(x)^{1/p}\}, \qquad (7.9)$$

the set $\mathcal{J}_{\phi,p}$ becomes a Banach space. A relation with the Hölder space

H_ϕ^0 is given by the continuous imbedding

$$\mathcal{J}_{\phi,p} \subseteq H_\phi^0 \quad (1 \le p < \infty). \tag{7.10}$$

In fact, since $\phi(2\sigma) \le 2\phi(\sigma)$, by (7.1), we get

$$\int_\sigma^{2\sigma} \phi(t)^{-p} \omega(x,t)^p \frac{dt}{t} \ge \frac{\omega(x,\sigma)^p}{2\phi(2\sigma)^p} \ge \frac{1}{4} \left[\frac{\omega(x,\sigma)}{\phi(\sigma)} \right]^p ,$$

and the left integral tends to zero, as $\sigma \to 0$, for each $x \in \mathcal{J}_{\phi,p}$.

An important special case of the space $\mathcal{J}_{\phi,p}$ is the following. Given $0 < \alpha \le 1$ and $0 < \beta < \infty$, we denote by $\mathcal{J}_{\alpha,\beta} = \mathcal{J}_{\alpha,\beta}(\Omega)$ the space of all continuous functions x on Ω for which

$$j_{\alpha,\beta}(x) = \int_0^1 t^{-(\beta+1)} \omega(x,t)^{\beta/\alpha} dt < \infty , \tag{7.11}$$

equipped with the norm

$$\|x\|_{\alpha,\beta} = \max\{\|x\|_C, j_{\alpha,\beta}(x)^{\alpha/\beta}\} .$$

In other words, we have $\mathcal{J}_{\alpha,\beta} = \mathcal{J}_{\phi,p}$, where $\phi(t) = t^\alpha$ and $p = \beta/\alpha$. In this case, (7.7) can be sharpened to

$$H_{\alpha+\varepsilon} \subseteq \mathcal{J}_{\alpha,\beta} \subseteq H_\alpha^0 \subseteq H_\alpha \quad (\varepsilon > 0) .$$

In fact, for $x \in H_{\alpha+\varepsilon}$ we have $j_{\alpha,\beta}(x) \le C \int_0^1 t^{-(\beta+1)} t^{(\alpha+\varepsilon)\beta/\alpha} dt < \infty$, hence $x \in \mathcal{J}_{\alpha,\beta}$.

We remark that all the spaces H_ϕ, H_ϕ^0 and $\mathcal{J}_{\phi,p}$ are algebras as well; this follows from the elementary inequality

$$\omega(xy,\sigma) \le \|x\|_C \omega(y,\sigma) + \|y\|_C \omega(x,\sigma) . \tag{7.12}$$

7.2 Acting conditions

In this section we give necessary and sufficient conditions under which the superposition operator acts between two Hölder spaces H_ϕ and H_ψ. Unfortunately, there is a certain "unsymmetry" in such conditions: *sufficient* conditions (for acting, boundedness, continuity etc.) may be established for arbitrary compact domains, while *necessary* conditions build on the construction of special functions on intervals. Therefore we shall always assume in the sequel that Ω is the unit interval $[0,1]$. The reader will find no difficulty in formulating the sufficient conditions for general (compact) Ω.

In case $\Omega = [0,1]$ we shall also use the norm

$$\|x\|_\phi = |x(0)| + h_\phi(x) \,, \tag{7.13}$$

which is of course equivalent to (7.6).

In what follows, we shall need special sets of points $(s,u) \in [0,1] \times \mathbb{R}$. Given $s_0 \in [0,1]$, $u_0 \in \mathbb{R}$, $r > 0$, $\delta > 0$, and a Hölder function ϕ, we denote by $W_\phi(s_0, u_0, r, \delta)$ the set of all $(s,u) \in [0,1] \times \mathbb{R}$ such that $|s - s_0| \leq \delta$ and $|u - u_0| \leq r\phi(|s - s_0|)$; geometrically, $W_\phi(s_0, u_0, r, \delta)$ is a "bow-tie" with centre at (s_0, u_0), width 2δ, and height $2r\phi(\delta)$. For stating the main result of this section, we need the following technical lemma:

Lemma 7.2   *Let* $(s_n, u_n), (t_n, u_n) \in W_\phi(s_0, u_0, r, \frac{1}{n})$ *such that* $|u_n - v_n| \leq \phi(|s_n - t_n|)$. *Then there exists a function* $x_* \in H_\phi$ *such that*

$$x_*(s_n) = u_n \,, \quad x_*(t_n) = v_n \tag{7.14}$$

for infinitely many indices n.

$\Rightarrow$ Starting with $n_1 = 1$, we construct by induction a strictly increasing sequence of natural numbers n_k such that

$$|u_{n_k} - u_{n_{k+1}}| \leq (1+\varepsilon)r\phi(|s_{n_k} - s_{n_{k+1}}|) \,,$$
$$|v_{n_k} - v_{n_{k+1}}| \leq (1+\varepsilon)r\phi(|t_{n_k} - t_{n_{k+1}}|) \,,$$
$$|u_{n_k} - v_{n_{k+1}}| \leq (1+\varepsilon)r\phi(|s_{n_k} - t_{n_{k+1}}|) \,,$$
$$|v_{n_k} - u_{n_{k+1}}| \leq (1+\varepsilon)r\phi(|t_{n_k} - s_{n_{k+1}}|) \,,$$

where $\varepsilon > 0$ is given. This is possible, as the strict inequalitities $|u_{n_k} - u_0| < (1+\varepsilon)\, r\phi(|s_{n_k} - s_0|)$ and $|v_{n_k} - u_0| < (1+\varepsilon)r\phi(|t_{n_k} - s_0|)$ hold for every k, and both sequences (s_n, u_n) and (t_n, v_n) converge to (s_0, u_0). Without loss of generality we may assume that $s_{n_{k+1}} \leq t_{n_{k+1}} \leq s_{n_k} \leq t_{n_k}$. If we set

$$x_*(s) = \begin{cases} u_{n_k} & \text{if } s = s_{n_k} \,, \\ v_{n_k} & \text{if } s = t_{n_k} \,, \\ \text{linear} & \text{otherwise}, \end{cases}$$

then x_* has the required properties. $\Leftarrow$

We are now in a position to state a necessary and sufficient acting condition.

Theorem 7.1   *The superposition operator F generated by f maps the space H_ϕ into the space H_ψ if and only if for all $(s_0, u_0) \in [0,1] \times \mathbb{R}$ and all $r > 0$ one may find $\delta > 0$ and $m > 0$ such that*

$$|f(s,u) - f(t,v)| \le m\{\psi(|s-t|) + \psi[\phi^{-1}(\frac{|u-v|}{r})]\} \qquad (7.15)$$

for $(s,u), (t,v) \in W_\phi(s_0, u_0, r, \delta)$.

$\Rightarrow$ We show first the sufficiency of (7.15). Let $x \in H_\phi$ with $\|x\|_\phi \le r$. By assumption, to each $s_0 \in [0,1]$ we may associate $m(s_0) > 0$ and $\delta(s_0) > 0$ such that

$$|f(s,u) - f(t,v)| \le m(s_0)\{\psi(|s-t|) + \psi[\phi^{-1}(\frac{|u-v|}{r})]\}$$

for $(s,u), (t,v) \in W_\phi(s_0, x(s_0), r, \delta(s_0))$. Since $\Omega = [0,1]$ is compact, we may cover Ω by a finite number of intervals $(s_j - (s_j), s_j + \delta(s_j))$ $(j = 1, \cdots, k)$. Thus, for $|s - s_j|, |t - s_j| \le \delta(s_j)$ we have $|Fx(s) - Fx(t)| \le 2m(s_j)\psi(|s-t|)$, which shows that $Fx \in H_\psi$.

The proof of the necessity of (7.15) is much harder. If (7.15) fails, we may choose two sequences (s_n, u_n) and (t_n, v_n) such that $|s_n - s_0| \le \frac{1}{n}$, $|u_n - u_0| \le r\phi(|s_n - s_0|)$, $|t_n - s_0| \le \frac{1}{n}$, $|v_n - u_0| \le r\phi(|t_n - s_0|)$, and

$$|f(s_n, u_n) - f(t_n, v_n)| \ge 3n\{\psi(|s_n - t_n|) + \psi[\phi^{-1}(\frac{|u_n - v_n|}{r})]\}. \qquad (7.16)$$

Now we have to distinguish several cases.

<u>Case 1</u>   Suppose that $(s_n - s_0)(t_n - s_0) > 0$ for infinitely many n; then we may even assume that this holds for all n, and that, moreover, $s_n \le t_n$.

<u>Case 1.1</u>   For infinitely many n,

$$|u_n - v_n| \le r\phi(|s_n - t_n|).$$

By Lemma 7.2, there exists a function $x_* \in H_\phi$ such that (7.14) holds for these n. But this implies that $|f(s_n, u_n) - f(t_n, v_n)| = |Fx_*(s_n) - Fx_*(t_n)| \le c\psi(|s_n - t_n|)$ for some $c > 0$, contradicting (7.16).

<u>Case 1.2</u>   For infinitely many n,

$$|u_n - v_n| > r\phi(|s_n - t_n|).$$

For these n we set $\tau_n = t_n - \phi^{-1}(\frac{|u_n - v_n|}{2r})$ and $w_n = \frac{u_n + v_n}{2}$. Then $(\tau_n, w_n) \in W_\phi(s_0, u_0, r, 1/n)$, since $\tau_n \ge t_n - \phi^{-1}[\frac{1}{2}(\phi(s_n - s_0) + \phi(t_n -$

$s_0))] \geq s_0$ and $|w_n - u_0| \leq r\phi(|\tau_n - s_0|)$; moreover, $|w_n - u_n| \leq r\phi(|\tau_n - s_n|)$ and $|w_n - v_n| \leq r\phi(|\tau_n - t_n|)$. Since $|f(s_n, u_n) - f(t_n, v_n)| \leq |f(s_n, u_n) - f(t_n, u_n)| + |f(t_n, u_n) - f(\tau_n, w_n)| + |f(\tau_n, w_n) - f(t_n, v_n)|$, (7.16) implies that at least one of the three estimates

$$|f(s_n, u_n) - f(t_n, u_n)| \geq n\{\psi(|s_n - t_n|) + \psi[\phi^{-1}(\frac{|u_n - v_n|}{r})]\} , \text{(7.17)}$$

$$|f(t_n, u_n) - f(\tau_n, w_n)| \geq n\{\psi(|s_n - t_n|) + \psi[\phi^{-1}(\frac{|u_n - v_n|}{r})]\} \text{ (7.18)}$$

or

$$|f(\tau_n, w_n) - f(t_n, v_n)| \geq n\{\psi(|s_n - t_n|) + \psi[\phi^{-1}(\frac{|u_n - v_n|}{r})]\} \text{ (7.19)}$$

holds true. Each case has to be considered again separately.

Case 1.2.1   If (7.17) holds for infinitely many n, by Lemma 7.2 we may construct a function $x_* \in H_\phi$ such that $x_*(s_n) = x_*(t_n) = u_n$; this leads to the same contradiction as before.

Case 1.2.2   If (7.18) holds for infinitely many n, by Lemma 7.2 we may construct a function $x_* \in H_\phi$ such that $x_*(\tau_n) = w_n$, $x_*(t_n) = u_n$. On the one hand, we have then $|Fx_*(\tau_n) - Fx_*(t_n)| \leq c\psi(|\tau_n - t_n|)$ for some $c > 0$, and on the other hand, by (7.18), we get that $|Fx_*(\tau_n) - Fx_*(t_n)| = |f(\tau_n, w_n) - f(t_n, u_n)| \geq n\psi[\phi^{-1}(\frac{|u_n - v_n|}{2r})] \geq n\psi[\phi^{-1}(\phi \ (|t_n - \tau_n|)] = n\psi(|t_n - \tau_n|)$, a contradiction.

Case 1.2.3   If (7.19) holds for infinitely many n, we arrive at the same contradiction, replacing u_n by v_n.

Case 2   Suppose now that $(s_n - s_0)(t_n - s_0) < 0$ for infinitely many n; in this case we may assume that $s_n < s_0 < t_n$. Since $|f(s_n, u_n) - f(t_n, v_n)| \leq |f(s_n, u_n) - f(s_0, u_0)| + |f(s_0, u_0) - f(t_n, v_n)|$, (7.16) implies again that at least one of the two estimates

$$|f(s_n, u_n) - f(s_0, u_0)| \geq n\{\psi(|s_n - t_n|) + \psi[\phi^{-1}(\frac{|u_n - v_n|}{r})]\} \text{ (7.20)}$$

or

$$|f(t_n, v_n) - f(s_0, u_0)| \geq n\{\psi(|s_n - t_n|) + \psi[\phi^{-1}(\frac{|u_n - v_n|}{r})]\} \text{ (7.21)}$$

holds.

Case 2.1   If (7.20) holds for infinitely many n, by Lemma 7.2 we may find a function $x_* \in H_\phi$ such that $x_*(s_n) = u_n$ and $x_*(s_0) = u_0$. On the one hand, we have then $|Fx_*(s_n) - Fx_*(s_0)| \leq c\psi(|s_n - s_0|)$, and, on the other, $|Fx_*(s_n) - Fx_*(s_0)| = |f(s_n, u_n) - f(s_0, u_0)| \geq n\psi(|s_n - t_n|) \geq n\psi(|s_n - s_0|)$, a contradiction.

<u>Case 2.2</u> If (7.21) holds for infinitely many n, we arrive at the same contradiction, replacing (s_n, u_n) by (t_n, v_n).

The proof is complete. $\Leftarrow$

As already observed, the "if" part of Theorem 7.1 carries over to arbitrary compact domains Ω, as the proof shows, while the "only if" part relies heavily on special properties of the interval $[0, 1]$. Theorem 7.1 gives a complete characterization of the acting condition $F(H_\phi) \subseteq H_\psi$. Similar results hold for the spaces H_ϕ^0 and H_ψ^0. We do not prove this in detail; the proof is parallel to that of Theorem 7.1.

Theorem 7.2   *The superposition operator F generated by f maps H_ϕ^0 into H_ψ (H_ϕ into H_ψ^0; H_ϕ^0 into H_ψ^0, respectively) if and only if for all $(s_0, u_0) \in [0, 1] \times \mathbb{R}$ there exist $r > 0$, $\delta > 0$ and $m > 0$ (if and only if for all $(s_0, u_0) \in [0, 1] \times \mathbb{R}$ and all $r > 0$ and $m > 0$ there exists $\delta > 0$; if and only if for all $(s_0, u_0) \in [0, 1] \times \mathbb{R}$ and all $m > 0$ there exist $r > 0$ and $\delta > 0$, respectively) such that (7.15) holds for $(s, u), (t, v) \in W_\phi(s_0, u_0, r, \delta)$.*

The relations between the various parameters occurring in Theorems 7.1 and 7.2 is sketched formally in the following scheme:

$$F(H_\phi) \subseteq H_\psi : r \to m, \delta \, ; \quad F(H_\phi^0) \subseteq H_\psi : \emptyset \to r, m, \delta \, ;$$
$$F(H_\phi) \subseteq H_\psi^0 : r, m \to \delta \, ; \quad F(H_\phi^0) \subseteq H_\psi^0 : m \to r, \delta \, .$$

Let us make some remarks on Theorems 7.1 and 7.2. First of all, none of the acting conditions for F implies the continuity of the generating function f; we shall give a corresponding example in the next section. This fact is very surprising: one should actually expect that, whenever F acts between two Hölder spaces then F acts also in the space C of continuous functions, since "half of the norm" in H_ϕ is made up of the norm in C, see (7.6). Observe, moreover, that condition (7.15) shows that, if $F(H_\phi^0) \subseteq H_\psi$ and $\phi \not< \psi$, then the function f does not depend on u, i.e. F is constant! In particular, this holds if $F(H_\alpha^0) \subseteq H_\beta$ for $\alpha < \beta$.

7.3 Boundedness conditions

Now we shall give boundedness conditions for F between Hölder spaces. Fortunately, these conditions are in some sense easier to verify than just the acting conditions given in the preceding section.

Theorem 7.3   *The superposition operator F generated by f acts from H_ϕ (or H_ϕ^0) into H_ψ and is bounded if and only if for each $r > 0$ one can* find an $m(r) > 0$ such that

$$|f(s,u) - f(t,v)| \le m(r)\{\psi(|s-t|) + \psi[\phi^{-1}(\frac{|u-v|}{r})]\}. \qquad (7.22)$$

Moreover, in this case the growth function (2.24) of F satisfies the bilateral estimate

$$\frac{1}{2k+1}m(r) \le \mu_F(r) \le m(r), \qquad (7.23)$$

where $k = 2\phi(\frac{1}{2})^{-1}$ and $m(r)$ is the minimal constant occurring in (7.22).

$\Rightarrow$ The sufficiency of (7.22) for the boundedness of F between H_ϕ and H_ψ is obvious. To prove the necessity, we observe that, if F is bounded on $B_r(H_\phi)$ by $\mu_F(r)$ (see (2.24)), then all functions of the form

$$x_{s,t;u,v}(\tau) = \begin{cases} u & \text{if } 0 \le \tau \le s, \\ u + \frac{v-u}{t-s}(\tau - s) & \text{if } s \le \tau \le t, \\ v & \text{if } t \le \tau \le 1, \end{cases}$$

where $0 \le s \le t \le 1$ and $|u-v| \le r\phi(|s-t|)$, are uniformly bounded by $\mu_F(r)$ for $|u|, |v| \le r$. Applying F to the functions $x_{s,t;u,v}$ yields, in particular, at $\tau = s$ and $\tau = t$, the estimate $|f(s,u) - f(t,v)| \le \mu_F(r)\psi(|s-t|)$. We claim that this estimate implies (7.22). To see this, we first observe that $|f(s,u) - f(t,u)| \le \mu_F(r)\psi(|s-t|)$ for $|u| \le r$. If, on the one hand, $|u-v| \le r\phi(\frac{1}{2})$, then for each $s \in [0,1]$ we can find $\tilde{s} \in [0,1]$ such that $|u-v| = r\phi(|s-\tilde{s}|)$, hence

$$|f(s,u) - f(s,v)| \le |f(s,u) - f(\tilde{s},u)| + |f(\tilde{s},u) - f(s,v)|$$
$$\le 2\mu_F(r)\psi(|s-\tilde{s}|) = 2\mu_F(r)\psi[\phi^{-1}(\frac{|u-v|}{r})]. \qquad (7.24)$$

If, on the other hand, $|u-v| > r\phi(\frac{1}{2})$, and we put $k = 2\phi(\frac{1}{2})^{-1}$ then, by (7.24),

$$|f(s,u) - f(s,v)| \le \sum_{j=1}^{k}|f(s,u+\frac{j}{k}(v-u)) - f(s,u+\frac{j-1}{k}(v-u))|$$
$$\le 2k\mu_F(r)\psi[\phi^{-1}(\frac{|u-v|}{k})] \le 2k\mu_F(r)\psi[\phi^{-1}(|u-v|)].$$

Thus (7.22) holds with $m(r) = (2k+1)\mu_F(r)$.
The estimate (7.23) for the growth function of F follows easily. $\Leftarrow$

The estimate (7.23) can be made more precise by taking into account the two different components $\|x\|_C$ and $h_\phi(x)$ of the norm (7.6) in the space H_ϕ. For $r > 0$, let

$$W(r, \tau, \rho) = \sup |f(s, u) - f(t, v)|, \qquad (7.25)$$

where the supremum is taken over $(s, u), (t, v) \in [0, 1] \times \mathbb{R}$ with $|s - t| \leq \tau$, $|u|, |v| \leq r$, and $|u - v| \leq \rho$.

Theorem 7.4   *If the superposition operator F generated by f is bounded from H_ϕ into H_ψ, its growth function (2.24) may be estimated by $\mu_F(r) \leq \max\{\nu_f(r), \lambda_f(r)\}$ ($r > 0$), where*

$$\nu_f(r) = \max_{\substack{0 \leq s \leq 1 \\ |u| \leq r}} |f(s, u)| \qquad (7.26)$$

and

$$\lambda_f(r) = \sup_{0 < \tau \leq 1} \frac{W(r, \tau, r\phi(\tau))}{\psi(\tau)}.$$

$\Rightarrow$ It is evident that $\nu_f(r)$ is just the growth function of F, considered as an operator in the space C (see (6.8)). Now, if $|s - t| \leq \tau$, then $|x(s) - x(t)| \leq h_\phi(x)\phi(|s - t|) \leq r\phi(\tau)$ (see (7.4)), hence $|Fx(s) - Fx(t)| \leq W(r, \tau, r\phi(\tau))$. The statement follows now from the definition of the norm (7.3). $\Leftarrow$

Condition (7.22) shows that, if F is bounded between H_ϕ and H_ψ, the generating function f is necessarily continuous on $[0, 1] \times \mathbb{R}$; just this fact makes it possible to consider the growth function (7.26). It is surprising that without the boundedness requirement for F this is false! Let us give an example. Let $\phi(t) = t^\alpha$ ($0 < \alpha < 1$), and let

$$f(s, u) = \begin{cases} u^{-2/\alpha} - su^{-4/\alpha} & \text{if } u > s^{\alpha/2}, \\ 0 & \text{if } u \leq s^{\alpha/2}. \end{cases} \qquad (7.27)$$

To see that the corresponding operator F maps the space H_α into itself we first observe that every function $x \in H_\alpha$ is bounded away from zero on the set $\Delta_x = \{s : x(s) > s^{\alpha/2}\}$, say $|x(s)| \geq \gamma_x > 0$ ($s \in \Delta_x$).

Let $y = Fx$, and keep $s, t \in [0, 1]$ fixed. To show that $y \in H_\alpha$, we distinguish three cases.

Case 1 $s \in \Delta_x$, $t \in \Delta_x$; in this case we get from the trivial estimate $|x(s)^{2/\alpha} - x(t)^{2/\alpha}| \leq |x(s) - x(t)| \leq \|x\|_\alpha |s - t|^\alpha$ and the definition (7.27) of f that

$$|y(s) - y(t)| \leq x(s)^{-2/\alpha} x(t)^{-2/\alpha} [|x(s)^{2/\alpha} - x(t)^{2/\alpha}|$$
$$+ t|x(s)^{2/\alpha} + x(t)^{2/\alpha}| \; |x(s)^{2/\alpha} - x(t)^{2/\alpha}| + x(t)^{4/\alpha}|s - t|]$$
$$\leq \gamma_x^{-4/\alpha} \|x\|_\alpha |s - t|^\alpha + 2\gamma_x^{-8/\alpha} \|x\|_\alpha^{2/\alpha} |s - t|^\alpha + \gamma_x^{-8/\alpha} \|x\|_\alpha^{4/\alpha} |s - t|.$$

<u>Case 2</u> $s \in \Delta_x$, $t \notin \Delta_x$; here we get $|y(s) - y(t)| = x(s)^{-4/\alpha}|x(s)^{2/\alpha} - s|$
$\leq \gamma_x^{-4/\alpha} \|x\|_\alpha \; |s - t|^\alpha + \gamma_x^{-4/\alpha} |s - t|$.
<u>Case 3</u> $s \notin \Delta_x$, $t \notin \Delta_x$; here we have simply $|y(s) - y(t)| = 0$.

In any case, $|y(s) - y(t)| = O(|s - t|^\alpha)$, hence $y \in H_\alpha$. Consider now the constant functions $x_k(s) \equiv k^{-\alpha/2}$ which converge to θ as $k \to \infty$. Since $\|Fx_k\|_\alpha \geq |f(0, x_k(0))| = k$, F is unbounded on every ball $B_r(H_\alpha)$ (hence even locally unbounded at θ!). Moreover, the same sequence shows that the function f is discontinuous at $(0, 0)$, and hence the operator F does not act in the space C of continuous functions!

The fact that the function f constructed in this counterexample depends on both s and u is not accidental. In fact, in the autonomous case $f = f(u)$ the situation is much simpler:

Theorem 7.5 *Suppose that the function $f = f(u)$ does not depend on s, and the superposition operator F generated by f maps H_ϕ into H_ψ. Then F is bounded on every ball $B_r(H_\phi)$, and f is continuous on $\mathbb{R}$.*

$\Rightarrow$ By Theorem 7.1, the acting condition $F(H_\phi) \subseteq H_\psi$ implies that for each $r > 0$ there exists $m(r) > 0$ such that

$$|f(u) - f(v)| \leq m(r)\psi[\phi^{-1}\left(\frac{|u - v|}{r}\right)] \qquad (7.28)$$

for $|u|, |v| \leq r$. For $v = 0$ this gives, on the one hand, $\|Fx\|_C \leq m(r)\psi(\phi^{-1}(1)) + |f(0)|$ if $\|x\|_C \leq r$, $u = x(s)$ and $v = x(t)$ and, on the other, $h_\psi(Fx) \leq m(r)$ if $h_\phi(x) \leq r$, which shows that F is bounded.

The continuity of the function f follows trivially from (7.28). $\Leftarrow$

7.4 Continuity conditions

As we have seen, the fact that the superposition operator F acts between two Hölder spaces H_ϕ and H_ψ does not imply its boundedness, except for the autonomous case; see Theorem 7.5. The problem of finding continuity conditions is still more delicate. First of all we observe that, even if the function $f = f(u)$ does not depend on s, the corresponding superposition operator F need not be continuous in H_ϕ. Consider, for example, the function

$$f(u) = \min\{u, 1\}. \tag{7.29}$$

Since f satisfies (7.28) (for $\psi = \phi$ and $m(r) = r$), the corresponding operator F maps H_ϕ into itself and is bounded. However, F is not continuous at $x_0(s) = \phi(s)$. To see this, let $x_n(s) = \phi(s) + \varepsilon_n$ where $0 < \varepsilon_n < 1$, $\varepsilon_n \to 0$. Then $\|x_n - x_0\|_\phi \to 0$ as $n \to \infty$, but $\sup_{0 < \sigma \le 1} \omega(Fx_n - Fx_0, \sigma)/\phi(\sigma) = 1$.

The fact that the operator F generated by the function (7.29) maps the space H_ϕ into itself, but not into the smaller space H_ϕ^0, is again not accidental. Indeed, the following remarkable result is true:

Theorem 7.6   *If the superposition operator F generated by $f = f(u)$ maps the space H_ϕ into the space H_ψ^0, then F is continuous.*

$\Rightarrow$ Fix $x_0 \in H_\phi$, and let x_n be a sequence in H_ϕ which converges in the norm (7.6) to x_0. By Theorem 7.2, we have $|f(u) - f(v)| \le m\psi\big[\phi^{-1}\big(\frac{|u-v|}{r}\big)\big]$ for $|u - v| \le r$ $(m, r > 0)$. Since $f = f(u)$ does not depend on s, f is continuous (Theorem 7.5); hence $\|Fx_n - Fx_0\|_C \to 0$ $(n \to \infty)$, by Theorem 6.5; it remains to show that $h_\psi(Fx_n - Fx_0) \to 0$ $(n \to \infty)$. For $|s - t| \le \delta$ and $|x_n(s) - x_n(t)| \le r\phi(|s - t|)$ we have $|Fx_n(s) - Fx_n(t)| \le m\psi\big[\phi^{-1}\big(\frac{|x_n(s)-x_n(t)|}{r}\big)\big] \le m\psi(|s - t|)$; consequently,

$$\lim_{\sigma \to 0} \frac{\omega(Fx_n - Fx_0, \sigma)}{\psi(\sigma)} = 0$$

uniformly in $n \in \mathbb{N}$, since $m > 0$ is arbitrary. This implies that, given $\varepsilon > 0$, we have $\omega(Fx_n - Fx_0, \sigma) \le \varepsilon\psi(\sigma)$ for $\sigma \le \sigma(\varepsilon)$ and all $n \in \mathbb{N}$. On the other hand, we have $\omega(Fx_n - Fx_0, \sigma) \le \varepsilon\psi(\sigma)$ for $\sigma \ge \sigma(\varepsilon)$ and $n \ge n(\varepsilon)$, since f is continuous. We conclude that $\sup_{\sigma > 0} \omega(Fx_n - Fx_0) \le \varepsilon\psi(\sigma)$ for $n \ge n(\varepsilon)$, and hence we are done. $\Leftarrow$

The following theorem contains three general continuity results on the superposition operator in Hölder spaces.

Theorem 7.7   *Suppose that the superposition operator F generated by f acts between two Hölder spaces H_ϕ and H_ψ. Then F is continuous at $x_0 \in H_\phi$ if and only if for each $\varepsilon > 0$ there exists a $\delta > 0$ such that*

$$|f(s, x_0(s) + h) - f(s, x_0(s)) - f(t, x_0(t) + k) + f(t, x_0(t))|$$
$$\le \varepsilon\Big\{\psi(|s - t|) + \psi\big[\phi^{-1}\big(\frac{|h - k|}{\delta}\big)\big]\Big\} \tag{7.30}$$

for $|h|, |k| \le \delta$. Moreover, F is continuous on H_ϕ (respectively H_ϕ^0) if for each $(s_0, u_0) \in [0,1] \times \mathbb{R}$ and each $\varepsilon > 0$, $r > 0$, there exists a $\delta > 0$ (respectively for each $\varepsilon > 0$ there exist $r > 0$, $\delta > 0$) such that

$$|f(s, u+h) - f(s, u) - f(t, v+k) + f(t, v)|$$
$$\le \varepsilon \{\psi(|s-t|) + \psi[\phi^{-1}(\frac{|u-v|}{r})] + \psi[\phi^{-1}(\frac{|h-k|}{\delta})]\} \tag{7.31}$$

for $(s, u), (s, u+h), (t, v), (t, v+k) \in W_\phi(s_0, u_0, r, \delta)$ and $|h|, |k| \le \delta$. Finally, F is uniformly continuous (on bounded sets) in H_ϕ if and only if for each $\varepsilon > 0$, $r > 0$ there exists a $\delta > 0$ such that (7.31) holds for $|u|, |v| \le r$ and $|h|, |k| \le \delta$.

$\Rightarrow$ Let $x_0 \in H_\phi$, $\|x_0\|_\phi < r$, and $\varepsilon > 0$. First we observe that (7.31) implies the continuity of the function f on $[0,1] \times \mathbb{R}$. By assumption, for each $s_0 \in [0,1]$ we find a $\delta > 0$ such that

$$|f(s, x_0(s) + h) - f(s, x_0(s)) - f(t, x_0(t) + k) + f(t, x_0(t))|$$
$$\le \varepsilon \{\psi(|s-t|) + \psi[\phi^{-1}(\frac{|x_0(s) - x_0(t)|}{r})] + \psi[\phi^{-1}(\frac{|h-k|}{\delta})]\}$$

for $|s - s_0|, |t - s_0| \le \delta$ and $|h|, |k| \le \delta$. By the compactness of $[0,1]$, we may choose finitely many points $s_1, \cdots, s_m$ and some universal $\delta > 0$ such that

$$|f(s, x_0(s) + h) - f(s, x_0(s)) - f(t, x_0(t) + k) + f(t, x_0(t))|$$
$$\le \varepsilon \{2\psi(|s-t|) + \psi[\phi^{-1}(\frac{|h-k|}{\delta})]\} \tag{7.32}$$

for $|s - s_j|, |t - s_j| \le \delta$ ($j = 1, \cdots, m$) and $|h|, |k| \le \delta$. Let now $x_n \in H_\phi$ be a sequence which converges to x_0; for large n we have then $\|x_n\|_\phi < r$ and $\|x_n - x_0\|_\phi \le \delta$, hence, by (7.32), $|f(s, x_n(s)) - f(s, x_0(s)) - f(t, x_n(t)) + f(t, x_0(t))| \le \varepsilon \{2\psi(|s-t|) + \psi[\phi^{-1}(\delta^{-1}|x_n(s) - x_0(s) - x_n(t) + x_0(t)|)]\} \le 3\varepsilon\psi(|s-t|)$. Now, for large n and $|s-t| \le \delta$ we get the estimate $|f(s, x_n(s)) - f(s, x_0(s)) - f(t, x_n(t)) + f(t, x_0(t))| \le 3\varepsilon\psi(|s-t|)$, while for $|s-t| \ge \delta$ we get, by the continuity of f, the estimate $\psi(|s-t|)^{-1} |f(s, x_n(s)) - f(s, x_0(s)) - f(t, x_n(t)) + f(t, x_0(t))| \le 2\psi(\delta)^{-1}\|Fx_n - Fx_0\|_C \le 3\varepsilon$. This shows that $h_\psi(Fx_n - Fx_0) \to 0$ as $n \to \infty$, and hence F is continuous at x_0.

Suppose now that F is continuous at $x_0 \in H_\phi$. This means that, given $\varepsilon > 0$, one can find a $\delta > 0$ such that the superposition operator $\tilde{F}$ generated by the function (1.11) maps the ball $B_\delta(H_\phi)$ into the ball $B_\varepsilon(H_\psi)$. By Theorem 7.3, condition (7.30) follows.

So far we have proved the necessity and sufficiency of (7.30) for the continuity of F at $x_0 \in H_\phi$. The proof of the sufficiency of (7.31) for the continuity (respectively uniform continuity) of F on H_ϕ follows the same reasoning. Finally, the necessity of (7.31) for the uniform continuity of F follows from the fact that in this case the estimates for the operator $\tilde{F}h = F(x_0 + h) - Fx_0$ generated by the function (7.32) do not depend on x_0. $\Leftarrow$

Observe that the condition for the continuity of F on H_ϕ is only sufficient, in contrast to that for the continuity of F at a single point, or for the uniform continuity on balls. Conditions for continuity on the whole space H_ϕ which are both necessary and sufficient are not known.

7.5 Lipschitz and Darbo conditions

In this section we shall be concerned with the Lipschitz condition

$$\|Fx_1 - Fx_2\|_\psi \leq k(r)\|x_1 - x_2\|_\phi \quad (x_1, x_2 \in B_r(H_\phi)) \tag{7.33}$$

and the Darbo condition

$$\alpha_\psi(FN) \leq k(r)\alpha_\phi(N) \quad (N \subseteq B_r(H_\phi)) \tag{7.34}$$

for the superposition operator F between two Hölder spaces H_ϕ and H_ψ, where α_ϕ denotes the Hausdorff measure of noncompactness (2.32) in the space H_ϕ.

In all spaces considered so far, these two conditions turned out to be equivalent (see Theorem 2.13 for general ideal spaces, Theorem 3.11 for Lebesgue spaces, Theorem 4.10 for Orlicz spaces, Theorem 5.7 for Lorentz and Marcinkiewicz spaces, and Theorem 6.6 for the space C). Hölder spaces give the first example for which there is a large "gap" between these two conditions. In fact, the Darbo condition (7.34) holds for a reasonably large class of function f, while the Lipschitz condition (7.33) leads to a strong degeneracy for the function f, provided the Lipschitz constant k in (7.33) is actually independent of r.

By the remark at the end of Section 7.2, only the case $\phi < \psi$ is interesting.

Theorem 7.8 *Suppose that the superposition operator F generated by f acts between two Hölder spaces H_ϕ and H_ψ over $[0,1]$, where $\phi < \psi$.*

Then F satisfies a global Lipschitz condition

$$\|Fx_1 - Fx_2\|_\psi \leq k\|x_1 - x_2\|_\phi \quad (x_1, x_2 \in H_\phi) \qquad (7.35)$$

if and only if the function f has the form

$$f(s, u) = a(s) + b(s)u \qquad (7.36)$$

for some $a, b \in H_\psi$.

$\Rightarrow$ For the proof we use the norm (7.13). The sufficiency of (7.36) for (7.35) is a simple consequence of the formula (7.12).

To prove the necessity, suppose that F satisfies the condition (7.35), and fix $\sigma, \tau \in [0, 1]$ ($\sigma < \tau$) and $u_1, u_2, v_1, v_2 \in \mathbb{R}$. Define two functions $x_1, x_2 \in H_\phi$ by

$$x_i(s) = \begin{cases} u_i & \text{if } 0 \leq s \leq \sigma, \\ \frac{v_i - u_i}{\tau - \sigma}(s - \sigma) + u_i & \text{if } \sigma < s < \tau, \\ v_i & \text{if } \tau \leq s \leq 1 \end{cases}$$

($i = 1, 2$). Obviously, $\|x_1 - x_2\|_\phi = |u_1 - u_2| + \phi(\tau - \sigma)^{-1}|v_1 - v_2 - u_1 + u_2|$. By (7.35) we have

$$|f(0, u_1) - f(0, u_2)| + \frac{|f(\tau, v_1) - f(\tau, v_2) - f(\sigma, u_1) + f(\sigma, u_2)|}{\psi(\tau - \sigma)}$$

$$\leq k|u_1 - u_2| + k\frac{|v_1 - v_2 - u_1 + u_2|}{\psi(\tau - \sigma)}.$$

Setting $v_1 = \xi + \eta$, $v_2 = \xi$, $u_1 = \eta$, and $u_2 = 0$ gives

$$|f(0, u_1) - f(0, u_2)| + \frac{|f(\tau, \xi + \eta) - f(\tau, \xi) - f(\sigma, \eta) + f(\sigma, 0)|}{\psi(\tau - \sigma)} \leq k|\eta|.$$

Multiplying both sides of this inequality by $\psi(\tau - \sigma)$ and letting τ tend to σ yields $f(\sigma, \xi + \eta) - f(\sigma, \xi) - f(\sigma, \eta) + f(\sigma, 0) = 0$. This shows that, for fixed $\sigma \in [0, 1]$, the function $\ell_\sigma(t) = f(\sigma, t) - f(\sigma, 0)$ ($t \in \mathbb{R}$) is additive and continuous, hence linear. Consequently, we have $\ell_\sigma(t) = b(\sigma)t$ for some function b. Finally, setting $a(s) = f(s, 0)$ gives the representation (7.36), where $b = f(\cdot, 1) - f(\cdot, 0) \in H_\psi$ and $a = f(\cdot, 0) \in H_\psi$ as claimed. $\Leftarrow$

We remark that an analogous result holds for the superposition operator F between the spaces H_ϕ^0 and H_ψ^0; the proof is exactly the same. Note that Theorem 7.8 is completely analogous to Theorem 6.15 for the space NBV.

Theorem 7.8 shows that the global Lipschitz condition (7.35) is very restrictive in the space H_ϕ. If we replace (7.35) by the local condition (7.33), however, we get a less restrictive condition of f, at least in the autonomous case.

Theorem 7.9   *Suppose that the superposition operator F generated by $f = f(u)$ acts between two Hölder spaces H_ϕ and H_ψ over $[0,1]$, where $\phi < \psi$. Then the local Lipschitz condition (7.33) holds if and only if f is continuously differentiable on $\mathbb{R}$ with*

$$|f'(u) - f'(v)| \leq k(r)\psi\left[\phi^{-1}\left(\frac{|u - v|}{r}\right)\right] \tag{7.37}$$

for $|u|, |v| \leq r$.

$\Rightarrow$ The sufficiency of (7.37) for (7.33) is a simple consequence of the Newton–Leibniz formula. To prove the necessity, observe first that the local Lipschitz condition (7.33) for F implies that f is locally Lipschitz on $\mathbb{R}$, and hence f' exists almost everywhere on $\mathbb{R}$. Fix $x \in H_\phi$ and $\lambda \in \mathbb{R}$ such that $\|x\|_\phi < r$ and $\|x + \lambda\|_\phi < r$. Let $g(s, \lambda) = \lambda^{-1}[f(x(s) + \lambda) - f(x(s))]$. By assumption, $\|F(x + \lambda) - Fx\|_\psi \leq k(r)\lambda$, hence $|g(s, \lambda)| \leq k(r)$ and $|g(s, \lambda) - g(t, \lambda)| \leq k\psi(|s - t|)$.

By the Arzelà–Ascoli compactness criterion, we may find a sequence λ_n in $(0, \infty)$ such that $\lambda_n \to 0$ and the sequence of functions $g_n = g(\cdot, \lambda_n)$ converges uniformly on $[0, 1]$ to some continuous functions g. But $f' = g$ almost everywhere on $\mathbb{R}$, and hence f is continuously differentiable. Passing to the limit $\lambda \to 0$, we conclude that the superposition operator G generated by g maps the space H_ϕ into the space H_ψ. The assertion follows now from Theorem 7.5 (see (7.28)). $\Leftarrow$

Let us now analyze the Darbo condition (7.34). We shall discuss (7.34) in the subspace H_ϕ^0, rather than in the whole space H_ϕ. The reason for this is that simple compactness criteria in the space H_ϕ do not exist, while compact sets in H_ϕ^0 are easily described:

Lemma 7.3   *The Hausdorff measure of noncompactness (2.32) in the space H_ϕ^0 is given by the formula*

$$\alpha_\phi(N) = \lim_{\sigma \to 0} \sup_{x \in N} \frac{\omega(x, \sigma)}{\phi(\sigma)}. \tag{7.38}$$

$\Rightarrow$ Denote the right-hand side of (7.38) by $\tilde{\alpha}_\phi(N)$. Let $\eta > \alpha_\phi(N)$, and let $\{z_1, \cdots, z_m\}$ be a finite η-net for N in H_ϕ^0. Fix $x \in N$ and choose z_j such that $h_\phi(x - z_j) \leq \eta$. Since $z_j \in H_\phi^0$, we have $\omega(z_j, \sigma) \leq \varepsilon\phi(\sigma)$ $(\sigma \leq \delta)$, hence $\omega(x, \sigma) \leq \omega(x - z_j, \sigma) + \omega(z_j, \sigma) \leq (\eta + \varepsilon)\phi(\sigma)$. This shows that $\alpha_\phi(N) \geq \tilde{\alpha}_\phi(N)$.

Conversely, given $\varepsilon > 0$, we can find a $\delta \in (0, 1)$ such that $\omega(x, \sigma) \leq (\tilde{\alpha}_\phi(N) + \varepsilon)\phi(\sigma)$ for $\sigma \leq \delta$, uniformly in $x \in N$. As a bounded set in H_ϕ, N is precompact in the space C; thus we may find a finite $\varepsilon\phi(\delta)$-net $\{z_1, \cdots, z_m\}$ in $N \subset C$. Given $x \in N$, we choose z_j such that $\|x - z_j\|_C \leq \varepsilon\phi(\delta)$ and obtain $\omega(x - z_j, \sigma) \leq 2\|x - z_j\|_C \leq 2\varepsilon\phi(\delta) \leq 2\varepsilon\phi(\sigma)$ for $\sigma \geq \delta$, and $\omega(x - z_j, \sigma) \leq \omega(x, \sigma) + \omega(z_j, \sigma) \leq (\tilde{\alpha}_\phi(N) + 2\varepsilon)\phi(\sigma)$ for $\sigma \leq \delta$. This shows that $\|x - z_j\|_\phi \leq \tilde{\alpha}_\phi(N) + 2\varepsilon$, hence $\alpha_\phi(N) \leq \tilde{\alpha}_\phi(N)$. $\Leftarrow$

Unfortunately, we do not know conditions for (7.34) in terms of f which are both necessary and sufficient. Nevertheless, Lemma 7.3 allows us to formulate the following sufficient condition:

Theorem 7.10 *Suppose that the superposition operator F generated by f acts between two Hölder spaces H_ϕ^0 and H_ψ^0. Then F satisfies the Darbo condition (7.34) if*

$$k(r) = \sup_{0 < \rho \leq r} \lim_{\tau \to 0} \frac{W(r, \tau, \rho\phi(\tau))}{\rho\psi(\tau)} < \infty,$$

with W given by (7.25).

$\Rightarrow$ We use Lemma 7.3 to verify (7.34). Let $N \subseteq B_r(H_\phi^0)$ and $\eta > \alpha_\phi(N)$. Given $\varepsilon > 0$, choose $\delta > 0$ such that $\omega(x, \tau) \leq \eta\phi(\tau)$ and $\psi(\tau)^{-1}W(r, \tau, \rho\phi(\tau)) \leq (k(r) + \varepsilon)\rho$ for $\tau \leq \delta$. For $x \in N$ we have then

$$\frac{\omega(Fx, \tau)}{\psi(\tau)} \leq \frac{W(r, \tau, \omega(x, \tau))}{\psi(\tau)} \leq (k(r) + \varepsilon)\eta \quad (\tau \leq \delta);$$

hence, by (7.38), $\alpha_\psi(FN) \leq (k(r) + \varepsilon)\eta$, which proves (7.34). $\Leftarrow$

7.6 Differentiability conditions

Suppose that the superposition operator F acts between two Hölder spaces H_ϕ and H_ψ and is differentiable at some point $x_0 \in H_\phi$. Consider the auxiliary function (1.11). The differentiability of F at x_0 means that $\mu_{\tilde{F}}(r) = o(r)$ $(r \to 0)$, where $\mu_{\tilde{F}}(r)$ is the growth function (2.24) of the superposition operator $\tilde{F}$ generated by the function (1.11). This relation is in turn equivalent, by (7.23), to $m(r) = o(r)$ $(r \to 0)$, where $m(r)$ is the constant occurring in (7.22), with f replaced by $\tilde{f}$. By this reasoning, we arrive at the following differentiability criterion:

Theorem 7.11   *Suppose that the superposition operator F generated by f acts between two Hölder spaces H_ϕ and H_ψ. Then F is differentiable at $x \in H_\phi$ if and only if the limit (6.18) exists as a continuous function, and for each $\varepsilon > 0$ there exists a $\delta > 0$ such that*

$$|f(s, x(s) + u) - f(s, x(s)) - a(s)u - f(t, x(t) + v) + f(t, x(t)) + a(t)v|$$
$$\leq \varepsilon r\{\psi(|s - t|) + \psi[\phi^{-1}(\frac{|u - v|}{r})]\}$$

for $r \leq \delta$ and $|u|, |v| \leq r$. In this case, the derivative of F at x has the form (6.16).

Similarly, the following holds:

Theorem 7.12   *Suppose that the superposition operator F generated by f acts between two Hölder spaces H_ϕ and H_ψ. Then F is asymptotically linear if and only if the limit (6.19) exists as a continuous function, and for each $\varepsilon > 0$ there exists an $\omega > 0$ such that*

$$|f(s, u) - a_\infty(s)u - f(t, v) + a_\infty(t)v|$$
$$\leq \varepsilon r\{\psi(|s - t|) + \psi[\phi^{-1}(\frac{|u - v|}{r})]\}$$

for $r \geq \omega$ and $|u|, |v| \leq r$. In this case the asymptotic derivative of F has the form (6.17).

As a corollary to these differentiability criteria, we get the following degeneracy result:

Theorem 7.13   *Suppose that the superposition operator F generated by f is differentiable in the space H_ϕ, and the derivative $F'(x)$ is uniformly bounded in x. Then the function f has the form (7.36), i.e. is affine in u.*

We remark that, by considering conditions of the form $m(r) = o(r^k)$ ($r \to 0$; $k = 2, 3, \cdots$) one gets criteria for the existence of higher derivatives of F which are analogous to that given in Theorem 7.11.

7.7 The superposition operator in the space $\mathcal{J}_{\phi,p}$

In the preceding sections we have given a rather complete discussion of the properties of the superposition operator in Hölder spaces. Much less is known for the Hölder-type spaces $\mathcal{J}_{\phi,p}$ introduced in Section 7.1. First of all, we remark that the Hausdorff measure of noncompactness (2.32) in the space $\mathcal{J}_{\phi,p}$ is given by

$$\alpha_{\phi,p}(N) = \lim_{\sigma \to 0} \sup_{x \in N} \left\{ \int_0^\sigma \frac{\omega(x,t)^p}{\phi(t)^p} \frac{dt}{t} \right\}^{1/p} ; \qquad (7.39)$$

this may be proved as the similar statement of Lemma 7.3.

We summarize some properties of F with the following:

Theorem 7.14   *Let ϕ and ψ be two Hölder functions, and let $1 \le p < \infty$. Suppose that the function f is continuous on $\Omega \times \mathbb{R}$, and that for any $r > 0$ the estimate*

$$\frac{W(r,\tau,\rho\phi(t))^p}{\phi(\tau)^p} \le \tau c(r,\tau,\lambda) + \frac{\lambda \rho^p}{\phi(\tau)^p} \quad (\rho \le r, \ \lambda \in \Lambda(r)) \qquad (7.40)$$

holds for some set $\Lambda(r) \subseteq (0,\infty)$, where W is given by (7.25), and $c(r,\cdot,\lambda)$ is some integrable function. Then the superposition operator F generated by f maps $\mathcal{J}_{\phi,p}$ into $\mathcal{J}_{\psi,p}$ and is bounded and continuous. Its growth function (2.24) in the norm (7.9) may be estimated by $\mu_F(r) \le \max\{\nu_f(r), \lambda_f(r)\}$ $(r > 0)$, where $\nu_f(r)$ is given by (7.26) and

$$\lambda_f(r) = \inf_{\lambda \in \Lambda(r)} \left\{ \int_0^1 c(r,t,\lambda)dt + \lambda r^p \right\}^{1/p} .$$

Moreover, the Darbo condition

$$\alpha_{\psi,p}(FN) \le k(r)\alpha_{\phi,p}(N) \quad (N \subseteq B_r(\mathcal{J}_{\phi,p})) \qquad (7.41)$$

holds with $k(r) = [\inf \Lambda(r)]^{1/p}$, where $\alpha_{\phi,p}$ is given by (7.39).

$\Rightarrow$ Let $x \in \mathcal{J}_{\phi,p}$ with $\|x\|_{\phi,p} \le r$. By definition of W and (7.40) we get for every $\lambda \in \Lambda(r)$

$$\int_0^1 \frac{\omega(Fx,t)^p}{\psi(t)^p} \frac{dt}{t} \le \int_0^1 \frac{W(r,t,\omega(x,t))^p}{\psi(t)^p} \frac{dt}{t}$$

$$\le \int_0^1 c(r,t,\lambda)dt + \lambda \int_0^1 \frac{\omega(x,t)^p}{\phi(t)^p} \frac{dt}{t}$$

$$\le \int_0^1 c(r,t,\lambda)dt + \lambda j_{\phi,p}(x) < \infty .$$

This shows that $Fx \in \mathcal{J}_{\psi,p}$, $j_{\psi,p}(Fx) \leq \int_0^1 c(r,t,\lambda)dt + \lambda j_{\phi,p}(x)$, and F is bounded and continuous. The estimate for $\mu_F(r)$ follows by an easy computation. It remains to show that F satisfies the Darbo condition (7.41). To see this, let $\eta > \alpha_{\phi,p}(N)$ and fix $x \in N$ and $\lambda \in \Lambda(r)$. For small $\sigma > 0$ we get, by (7.39), uniformly in $x \in N$,

$$\int_0^\sigma \frac{\omega(Fx,t)^p}{\psi(t)^p} \frac{dt}{t} \leq \int_0^\sigma c(r,t,\lambda)dt + \lambda \int_0^\sigma \frac{\omega(x,t)^p}{\phi(t)^p} \frac{dt}{t} \leq \int_0^\sigma c(r,t,\lambda)dt + \lambda\eta^p \,.$$

This means that

$$\lim_{\sigma \to 0} \sup_{x \in N} \int_0^\sigma \frac{\omega(Fx,t)^p}{\psi(t)^p} \frac{dt}{t} \leq \lambda\eta^p \,;$$

hence, again by (7.39), $\alpha_{\psi,p}(FN) \leq \lambda^{1/p}\alpha_{\phi,p}(N)$, as claimed. $\Leftarrow$

7.8 Notes, remarks and references

1. All the material presented on Hölder spaces may be found in many textbooks on function spaces, partial differential equations, or singular integral equations. The fact that we assumed that diam $\Omega = 1$ and $\phi(1) = 1$ is of course only technical. The compactness of Ω, however, is important, although some results may also be stated for noncompact Ω. A study of what we call Hölder functions may be found in the book [138]. The spaces $\mathcal{J}_{\phi,p}$ are also studied in detail in [138], in particular, compactness properties and imbedding theorems between them. We remark that the spaces $\mathcal{J}_{\phi,p}$ are special cases of so-called *Besov spaces* $B_{\infty,p}^\phi$ which have been studied, for instance, in [49], [106], and [121].

We point out that there is a formal analogy between the spaces $\mathcal{J}_{\alpha,\beta}$, H_α and H_α^0, on the one hand, and the Lorentz and Marcinkiewicz spaces $\Lambda_{p,q}$, M_p and M_p^0, considered in Section 5.2, on the other. Moreover, if the construction which leads to these spaces is considered from an axiomatic point of view, one gets various classes of known function spaces, such as Besov spaces, Gusejnov spaces, Morrey–Campanato spaces, and others (see [20], [50], [244] for details).

2. The superposition operator in Hölder spaces was studied by many authors, first of all by M.Z. Berkolajko, V.A. Bondarenko, Ja.B. Rutitskij, and P.P. Zabrejko [46–48], [51–56]; a very concise and self-contained presentation of the main properties of F in Hölder spaces is, in particular, [55]; for proofs see [56].

The first acting conditions for F in the Hölder spaces H_α and H_ϕ (however, combined with boundedness of F) seem to be due to A.A. Babajev [40], M.Z. Berkolajko [46] and H.Sh. Muhtarov [233] in the autonomous case $f = f(u)$. The general Theorems 7.1 and 7.2, as well as Lemma 7.2, are taken from [56].

3. As pointed out in Theorem 7.5, in the case $f = f(u)$ the acting condition $F(H_\phi) \subseteq H_\psi$ implies the boundedness of F, and is equivalent to the (local) condition (7.28). This was observed by many authors independently (see [40], [137] and [327] for $H_\phi = H_\alpha$, [46] for general H_ϕ, and [98] for vector-valued functions in H_α). The fact that the acting conditions $F(H_\phi) \subseteq H_\psi$ or $F(H_\phi^0) \subseteq H_\psi$ in case $\phi \not\ll \psi$ implies that $f = f(s)$ (i.e. F is constant) was first observed in [54].

If the function f depends on both s and u, the situation becomes more complicated. Theorem 7.3, which gives a criterion for both acting and boundedness of F, is taken from [55], for the proof see [56], where the two-sided estimate (7.23) for the growth function μ_F between H_ϕ and H_ψ can also be found. Analogous results are contained in [51] and [52]. The counterexample (7.27) after Theorem 7.4 is given in the following more general form in [52]. Consider two concave differentiable Hölder functions ϕ and ψ on $[0, \infty)$ such that $\phi \ll \psi$ and $|\psi(s) - \psi(t)| \geq c|s - t|$ for some $c > 0$. Let

$$f(s, u) = \begin{cases} \frac{1}{\psi^{-1}(u)}\left[1 - \frac{s}{\psi^{-1}(u)}\right] & \text{if } u > \psi(s), \\ 0 & \text{if } u \leq \psi(s). \end{cases} \tag{7.42}$$

Then the superposition operator F generated by f maps the space H_ϕ into itself, but is locally unbounded at θ. To prove the latter assertion, it suffices to consider the constant functions $x_k(s) \equiv \xi_k$, where $0 \leq \xi_k \leq 1/k$ and $\psi(\xi_{k+1}) \leq \frac{1}{2}\psi(\xi_k)$; in fact, although $\|x_k\|_\phi = \xi_k \to 0$, we have $\|Fx_k\|_\phi \geq |f(0, x_k(0))| = \xi_k^{-1} \to \infty$. The same reasoning shows that the function (7.42) is discontinuous at $(0, 0)$. Of course, (7.27) is contained in (7.42) by the special choice $\phi(t) = t^\alpha$ and $\psi(t) = t^{\alpha/2}$.

We point out again that, as is shown by the counterexample, the operator F may act in a Hölder space without acting in the space C of continuous functions. Conversely, the crucial condition (7.22) implies the somewhat strange fact that the *boundedness* of the operator F (in some space H_ϕ) guarantees the *continuity* of the corresponding function f (on $[0, 1] \times \mathbb{R}$).

4. The counterexample (7.29), which is taken from [46], shows that even in the autonomous case $f = f(u)$ the operator F need not be continuous in H_ϕ. The paper [46] contains also the proof of Theorem 7.6 and related results; for example, if F maps H_ϕ into H_ϕ then F is continuous on H_ϕ^0

(compare this again with the counterexample (7.29), where the discontinuity point $x_0 = \phi$ belongs, of course, to $H_\phi \setminus H_\phi^0$).

In case $f = f(s, u)$ one can find all continuity results of Section 7.4 in [55]. Some continuity and boundedness conditions of this type in the space H_ϕ^0 are contained in [47]. In [327–329] it is shown that, in the autonomous case $f = f(u)$, the operator F is continuous from H_α into H_β if and only if

$$\lim_{|h| \to 0} \frac{|f(u + h) - f(u)|}{|h|^{\beta/\alpha}} = 0,$$

uniformly on bounded intervals; for $\alpha = \beta$, this amounts to differentiability of the function f, see also [98].

In [55] also a compactness criterion for F between H_ϕ^0 and H_ψ^0 is mentioned; this criterion implies, in particular, that F is necessarily *constant* if F is compact in H_ϕ (see also [47]); this is of course parallel to the spaces L_p (Theorem 3.11) and C (Theorem 6.6).

The above-mentioned analogy between the spaces $\mathcal{J}_{\alpha,\beta}$ and H_α, on the one hand, and the Lorentz and Marcinkiewicz spaces $\Lambda_{p,q}$ and M_p, on the other, has more than formal character. We recall that, if X is an ideal space of measurable functions, the subspace X^0 was defined as a set of all functions in X with absolutely continuous norms (see (2.1)). Likewise, one may show [20] that H_ϕ^0 consists of all functions in H_ϕ whose norm tends to 0 as their domain of definition is "shrinking". From this point of view, Theorem 7.6 reminds very much of Theorem 2.7, although ideal spaces and Hölder spaces are of quite different nature.

Some applications of the "little" Hölder spaces H_α^0 to interpolation theory may be found in [322], to differential equations in [201], and to integral equations in [25].

In rather the same way as special Young functions generate "nice" Orlicz spaces (see Section 4.1) and superposition operators between them, the superposition operator also behaves "nicer" between H_ϕ and H_ψ if the Hölder functions ϕ and ψ have additional properties. We say that ψ satisfies a Δ_2 *condition with respect to* ϕ if $\psi \circ \phi^{-1} \in \Delta_2$, i.e.

$$\psi(\phi^{-1}(2t)) \leq C\psi(\phi^{-1}(t))$$

for some $C > 0$. In this case, it can be shown that the acting condition $F(H_\phi^0) \subseteq H_\psi$ implies the acting condition $F(H_\phi) \subseteq H_\psi$ [55]; moreover, local boundedness implies global boundedness, as condition (7.22) shows.

The above-mentioned Δ_2 condition holds of course for the classical case $\phi(t) = t^\alpha$ and $\psi(t) = t^\beta$ ($0 < \alpha, \beta \leq 1$). For the space $H_1 = \text{Lip}$ the fact that the local boundedness implies the global boundedness of F is proved in

[53]. Interestingly, it is shown there that the acting condition $F(C^1) \subseteq$ Lip implies the acting condition $F(\text{Lip}) \subseteq$ Lip.

5. As already mentioned in Section 7.1, the "little" Hölder space H_ϕ^0 is, in some sense, much more "tractable" than the "big" Hölder space H_ψ. This is illustrated, for instance, by the fact that precompact sets are easily described in H_ϕ^0 (see Lemma 7.3, which is taken from [25]), while no simple compactness criterion in the space H_ϕ is known (some criteria mentioned in the literature, e.g. [110] or [138], are false). The space H_ϕ^0 may be obtained as closure of the space C^1 of all continuously differentiable functions with respect to the norm (7.6); this is true, in particular, for the space H_α^0 with $0 < \alpha < 1$. Observe, however, that the space H_1^0 consists only of constant functions.

The local Lipschitz condition (7.33) was studied in [327] in case $f = f(u)$ (see below). The fact that the global Lipschitz condition (7.35) is much more restrictive is illustrated by Theorem 7.8. This theorem was proved for $H_1 = $ Lip in [219], for H_α $(0 < \alpha < 1)$ in [218] and, independently, in [25], and for H_α^0 $(0 < \alpha < 1)$ also in [25]; for vector-valued functions see [222], [223].

Theorem 7.8 shows that, roughly speaking, it does not make sense to apply the Banach contraction mapping principle (at least globally) to nonlinear problems involving superposition operators in Hölder spaces. Theorem 7.9 which is much less restrictive is proved in [328]; Theorem 7.10 may be found in the paper [25].

6. The necessary and sufficient conditions for differentiablity and asymptotic linearity given in Theorems 7.11 and 7.12 are taken from [16]; they are, of course, immediate consequences of the estimate (7.23). The fact that every asymptotically linear superposition operator in H_ϕ degenerates (Theorem 7.13) is proved in [48]. The sufficient conditions contained in [245] follow from the general Theorems 7.11 and 7.12. On the other hand, [245] deals also with the *vector-valued case*, mainly in view of applications to nonlinear elliptic boundary value problems. It is shown in [120] that, in case $f = f(u)$, every C^1 function f generates a continuous superposition operator F in H_α, and every C^2 function f generates a continuously Fréchet differentiable superposition operator F in H_α. This shows that one must not drop the boundedness assumption on $F'(x)$ in Theorem 7.13.

In view of applications to *nonlinear integral equations* of Hammerstein type, Theorem 7.13 is very bad, since linear integral operators may be considered in general only from a Hölder space into itself. Thus, such equations may not be studied in H_ϕ by methods which build on smoothness assumptions.

One may consider this problem also from the "reversed" viewpoint; the higher are the smoothness assumptions on the superposition operator between two Hölder spaces H_ϕ and H_ψ, the more different should be the generating Hölder functions ϕ and ψ. In the extreme case, namely that of *analytic* superposition operators, the functions ϕ and ψ should be "completely different"; this is the same phenomenon as that discussed in Section 4.5 for Orlicz spaces.

7.   As already pointed out, no attention has been given so far to the spaces $\mathcal{J}_{\phi,p}$ from the viewpoint of nonlinear superposition operators. Theorem 7.14, which is by no means exhaustive, is proved in [25]. Some application of the space $\mathcal{J}_{\phi,p}$ in the theory of singular integral equations may be found in the book [138].

We repeat here our general remark on the underlying domain Ω. As can be seen from the proofs, all statements given in Sections 7.2 – 7.7 are equally valid for general connected compact subsets of Euclidean space, as far as only *sufficient* conditions are involved (for acting, boundedness, continuity etc.). On the other hand, the "necessity parts" of our proofs mostly build on the construction of special functions requiring the specific structure of intervals; see e.g. the proofs of Lemma 7.2 and Theorem 7.1. This is the reason why we constricted ourselves to the case $\Omega = [0,1]$ throughout this chapter.

Chapter 8

The superposition operator
in spaces of smooth functions

In this chapter we study the superposition operator in various spaces of functions which are characterized by certain smoothness properties. We begin with a necessary and sufficient acting and continuity condition for F in the space C^k of k-times continuously differentiable functions. Surprisingly, without the continuity requirement for F the generating function f need not even be continuous. Afterwards, we show that a (global) Lipschitz condition for F is "never" satisfied, while a (local) Darbo condition holds "always". This is in sharp contrast to the situation in spaces of measurable functions dealt with in Chapters 2 – 5, and also in the space C.

In the second part we try to develop a parallel theory in the spaces H_ϕ^k of all functions from C^k whose k-th derivatives belong to the Hölder space H_ϕ. In particular, we give a sufficient acting and boundedness condition.

The last part is concerned with the superposition operator in various classes of smooth (i.e. C^∞) functions, including Roumieu spaces, Beurling spaces, Gevrey spaces, and their projective and inductive limits. It turns out that an acting condition for the operator F in such classes, together with suitable additional growth conditions on the derivatives of the function f, guarantees not only the boundedness and continuity, but also the compactness of F.

8.1 The spaces C^k and H_ϕ^k

Let Ω be a compact domain in Euclidean space $\mathbb{R}^N$ without isolated points. By $C^k = C^k(\Omega)$ we denote the space of all k-times continuously differentiable functions x on Ω, equipped with the natural algebraic operations and the norm

$$\|x\|_k = \sum_{|\alpha| \le k} \|D^\alpha x\|_C, \qquad (8.1)$$

where $\|z\|_C$ is the norm defined in (6.1), and $|\alpha| = \alpha_1 + \cdots + \alpha_N$, as usual. Obviously, C^k is a Banach space, and convergence with respect to the norm (8.1) means uniform convergence in all derivatives up to order k.

If Ω is not compact, one may define a locally convex topology on $C^k(\Omega)$ by means of the seminorms

$$p_K(x) = \sum_{|\alpha| \le k} \max_{s \in K} |D^\alpha x(s)|,$$

where K runs over all compact subsets of Ω; this topology is metrizable if Ω is *countably compact*, i.e. may be represented as a countable union of compact sets; see the analogous statements for $C(\Omega)$ in Section 6.1.

Let Ω again be compact, and let ϕ be a Hölder function (see Section 7.1). By $H_\phi^k = H_\phi^k(\Omega)$ we denote the space of all functions $x \in C^k(\Omega)$ for which the norm

$$\|x\|_{k,\phi} = \|x\|_k + \sum_{|\alpha|=k} h_\phi(D^\alpha x) \tag{8.2}$$

is finite, where $h_\phi(z)$ is defined as in (7.4). Thus, a k-times continuously differentiable function x belongs to H_ϕ^k if and only if $\omega(D^\alpha x, \sigma) = O(\phi(\sigma))$ for $|\alpha| = k$, as $\sigma \to 0$. Likewise, by $H_\phi^{k,0} = H_\phi^{k,0}(\Omega)$ we denote the subspace of all functions $x \in H_\phi^k$ satisfying $\omega(D^\alpha x, \sigma) = o(\phi(\sigma))$ for $|\alpha| = k$, as $\sigma \to 0$ (see (7.7)). The spaces H_ϕ^k and $H_\phi^{k,0}$ are sometimes called *Schauder spaces* in the literature.

In the following two sections, we shall consider the superposition operator in the spaces C^k and H_ϕ^k (or $H_\phi^{k,0}$). As in Chapter 7, sufficient conditions may easily be obtained for arbitrary (compact) domains Ω, while proving their necessity requires that Ω is a more specific set, say $\Omega = [0,1]$. In this case the norms (8.1) and (8.2) may be replaced, of course, by the more tractable (equivalent) norms

$$\|x\|_k = \sum_{j=0}^{k-1} |x^{(j)}(0)| + \|x^{(k)}\|_C \tag{8.3}$$

and

$$\|x\|_{k,\phi} = \sum_{j=0}^{k} |x^{(j)}(0)| + h_\phi(x^{(k)}), \tag{8.4}$$

respectively; compare this with (7.13). We remark that the spaces C^k and H_ϕ^k are also algebras; this follows from the elementary formula for the derivative of products. For $x, y \in C^k$ we have, for instance

$$\|xy\|_k \leq C(y)\|x\|_k ,\tag{8.5}$$

where the constant $C(y)$ depends on the norms $\|y\|_C, \cdots, \|y^{(k)}\|_C$; likewise, for $x, y \in H_\phi^k$ we have

$$\|xy\|_{k,\phi} \leq C_\phi(y)\|x\|_{k,\phi} ,\tag{8.6}$$

where the constant $C_\phi(y)$ depends on the norms $\|y\|_\phi, \cdots, \|y^{(k)}\|_\phi$.

As usual, by

$$C^\infty = C^\infty(\Omega) = \bigcap_{k=1}^{\infty} C^k(\Omega)$$

we denote the linear space of all infinitely differentiable functions on Ω. The space $C^\infty(\Omega)$ may be equipped, if Ω is compact, with the seminorms

$$p_m(x) = \sum_{|\alpha| \leq m} \max_{s \in \Omega} |D^\alpha x(s)| ,\tag{8.7}$$

or, if Ω is not compact, with the double family of seminorms

$$p_{K,m}(x) = \sum_{|\alpha| \leq m} \max_{s \in K} |D^\alpha x(s)| ,\tag{8.8}$$

where K runs over all compact subsets of Ω.

8.2 The superposition operator in the space C^k

In order to study the superposition operator between two spaces of continuously differentiable functions, we recall some elementary facts about derivatives of composite functions. For reasons similar to those in Chapter 7, we shall assume in this section that Ω is the unit interval $[0, 1]$; all *sufficient* acting, boundedness, and continuity conditions easily carry over to arbitrary (compact) sets $\Omega \subset \mathbb{R}^N$.

To simplify the notation, we shall use the following abbreviations in the sequel. For a multi-index $\alpha = (\alpha_0, \alpha_1, \cdots, \alpha_N) \in \mathbb{N}^{N+1}$ we write

$$\begin{aligned} |\alpha| &= \alpha_0 + \alpha_1 + \alpha_2 + \cdots + \alpha_N , \\ \|\alpha\| &= \alpha_0 + 1\alpha_1 + 2\alpha_2 + \cdots + N\alpha_N , \\ C_k^\alpha &= \frac{k!}{\alpha_0!\alpha_1!\cdots\alpha_k!(1!)^{\alpha_1}(2!)^{\alpha_2}\cdots(k!)^{\alpha_k}} . \end{aligned}\tag{8.9}$$

For instance, in this notation the formula for the derivatives of the composite function $y(s) = f(s, x(s))$ becomes

$$y^{(k)}(s) = \sum_{\|\alpha\|=k} C_k^\alpha D_s^{\alpha_0} D_u^{|\alpha|-\alpha_0} f[s, x(s)] x'(s)^{\alpha_1} \cdots x^{(k)}(s)^{\alpha_k}, \qquad (8.10)$$

where D_s and D_u denote the corresponding partial derivatives with respect to s and u.

For some purpose it is convenient to write the formula (8.10) in a more explicit form. Given a real function f on $[0,1] \times \mathbb{R}$, we define inductively functions $f_0, f_1, f_2, \cdots, f_k, \cdots$, on $[0,1] \times \mathbb{R}$, $[0,1] \times \mathbb{R}^2$, $[0,1] \times \mathbb{R}^3$, $\cdots$, $[0,1] \times \mathbb{R}^{k+1}$, $\cdots$ by

$$
\begin{aligned}
f_0(s, u_0) &= f(s, u_0), \\
f_1(s, u_0, u_1) &= D_s f_0(s, u_0) + D_{u_0} f_0(s, u_0) u_1, \\
f_2(s, u_0, u_1, u_2) &= D_s f_1(s, u_0, u_1) + D_{u_0} f_1(s, u_0, u_1) u_1 \\
&\quad + D_{u_1} f_1(s, u_0, u_1) u_2 \\
&\ \ \vdots \\
f_k(s, u_0, \cdots, u_k) &= D_s f_{k-1}(s, u_0, \cdots, u_{k-1}) \\
&\quad + \sum_{j=0}^{k-1} D_{u_j} f_{k-1}(s, u_0, \cdots, u_{k-1}) u_{j+1}
\end{aligned}
\qquad (8.11)
$$

$(s \in [0,1]; u_0, u_1, \cdots, u_k \in \mathbb{R})$.

Lemma 8.1 *If the function f is of class C^m on $[0,1] \times \mathbb{R}$, then $f_k(\cdot, \cdot, u_1, \cdots, u_k)$ is of class C^{m-k} on $[0,1] \times \mathbb{R}$ for each $u_1, \cdots, u_k \in \mathbb{R}$, and $f_k(s, u_0, \cdot, \cdots, \cdot)$ is of class C^∞ on $\mathbb{R}^k$ for each $s \in [0,1]$ and $u_0 \in \mathbb{R}$. Moreover, there exist functions g_k on $[0,1] \times \mathbb{R}^k$ $(k = 1, \cdots, m)$ such that*

$$f_k(s, u_0, \cdots, u_k) = g_k(s, u_0, \cdots, u_{k-1}) + D_{u_0} f(s, u_0) u_k. \qquad (8.12)$$

Finally, whenever f is of class C^m on $[0,1] \times \mathbb{R}$, then, for each $x \in C^m$, the function $y = Fx$ has derivatives

$$y^{(k)}(s) = f_k(s, x(s), x'(s), \cdots, x^{(k)}(s)) \quad (k = 0, 1, \cdots, m). \qquad (8.13)$$

⇒ The claimed smoothness properties of the functions f_k are easily proved by induction. A comparison of (8.11) and (8.12) shows that one has to choose $g_1(s, u_0) = D_s f_0(s, u_0)$, $g_2(s, u_0, u_1) = D_s f_1(s, u_0, u_1) + D_{u_0} f_1(s, u_0,$ $u_1) u_1$, $\cdots$, $g_k(s, u_0, \cdots, u_k) = D_s f_{k-1}(s, u_0, \cdots, u_{k-1}) + \sum_{j=0}^{k-2} D_{u_j} f_{k-1}(s,$ $u_0, \cdots, u_{k-1}) u_{j+1}$. Finally, (8.13) may be obtained by combining the formula (8.10) and the definition (8.11). $\Leftarrow$

The chain rule (8.13) implies that, if the function f is of class C^m, then the corresponding superposition operator F maps the space C^m into itself; this is, of course, well known to every first-year calculus student. The question arises whether or not the converse is also true. Surprisingly enough, the answer is negative!

Consider, for example, the function

$$f(s, u) = \begin{cases} 0 & \text{if } u \leq 0, \\ -2u^3 s^{-3/2} + 3u^2 s^{-1} & \text{if } 0 < u \leq s^{1/2}, \\ 1 & \text{if } s^{1/2} < u. \end{cases} \qquad (8.14)$$

We show that this function generates a superposition operator F which maps the space C^1 into itself. To this end, it suffices to show that, for any $x \in C^1$ with $x(0) = 0$, the function $y = Fx$ is continuously differentiable at $s = 0$. For small $\delta > 0$ we have $x(s) < \sqrt{s}$ for $0 < s < \delta$, hence either $x(s) \leq 0$ (hence $y(s) = 0$) or

$$\lim_{s \to 0} \frac{y(s)}{s} = \lim_{s \to 0} \frac{-2s^{-3/2} x(s)^3 + 3s^{-1} x(s)^2}{s} = 3x'(0)^2,$$

which shows that y is differentiable at 0 with $y'(0) = 3x'(0)^2$. A straightforward calculation shows that, for $0 < x(s) < \sqrt{s}$, we have

$$\lim_{s \to 0} y'(s) = \lim_{s \to 0} [D_s f(s, x(s)) + D_u f(s, x(s)) x'(s)]$$
$$= \lim_{s \to 0} [3s^{-5/2} x(s)^3 - 3s^{-2} x(s)^2 - 6s^{-3/2} x(s)^2 x'(s)$$
$$+ 6s^{-1} x(s) x'(s)] = 3x'(0)^2 = y'(0),$$

and hence y' is in fact continuous at $s = 0$.

It is obvious, however, that the generating function (8.14) is even *discontinuous* at $(0,0)$, and thus F does not act in the space C! Observe that this counterexample is very similar to the function (7.27) which generates a superposition operator in the Hölder space H_α, despite being discontinuous at $(0,0)$. Recall also that the counterexample (7.27) was due to the *lack of boundedness* of the corresponding operator F in the space H_α (see Theorem 7.3). Here it turns out that the crucial point in the counterexample (8.14) is the *lack of continuity* of the operator F in the space C^1; this follows from the following:

Theorem 8.1   *The superposition operator F generated by f maps the space C^1 into itself and is continuous if and only if the function f is continuously differentiable on $[0, 1] \times \mathbb{R}$.*

$\Rightarrow$ The "if" part is an immediate consequence of the chain rule (8.13) and the definition (8.3) of the norm in the space C^1.

To prove the "only if" part, denote by x_u the function with constant value u ($u \in \mathbb{R}$). We have $y_u(s) = Fx_u(s) = f(s, u)$, hence $y'_u(s) = D_s f(s, u) = g_1(s, u)$ with g_1 as in Lemma 8.1. Moreover, $|g_1(s, u) - g_1(t, v)| = |y'_u(s) - y'_v(t)| \leq |y'_u(s) - y'_u(t)| + |y'_u(t) - y'_v(t)| \leq \varepsilon + \|y_u - y_v\|_1 \leq 2\varepsilon$, if $|s - t| \leq \delta$ and $|u - v| = \|x_u - x_v\|_1 \leq \delta$. This shows that the partial derivative g_1 exists and is continuous on $[0, 1] \times \mathbb{R}$.

To prove the existence and continuity of the partial derivative $h_1(s, u) = D_u f(s, u)$ is somewhat harder. For $t \in [0, 1]$ and $v \in \mathbb{R}$ let $x_{t,v}(s) = v + s - t$, such that $y_{t,v}(s) = Fx_{t,v}(s) = f(s, v + s - t)$. Given $s_0 \in [0, 1]$, $u_0 \in \mathbb{R}$, and $h \in \mathbb{R}$, we have

$$f(s_0, u_0 + h) - f(s_0, u_0)$$
$$= f(s_0 + h, u_0 + h) - f(s_0, u_0) - f(s_0 + h, u_0 + h) + f(s_0, u_0 + h)$$
$$= y_{s_0, u_0}(s_0 + h) - y_{s_0, u_0}(s_0) - \int_{s_0}^{s_0 + h} g_1(t, u_0 + h) dt$$
$$= y_{s_0, u_0}(s_0 + h) - y_{s_0, u_0}(s_0) - g_1(s_0, u_0) h$$
$$- \int_{s_0}^{s_0 + h} [g_1(t, u_0 + h) - g_1(t, u_0)] dt .$$

Since g_1 is continuous, as shown above, we conclude that $h_1(s_0, u_0) = \lim\limits_{h \to 0} \frac{1}{h}$ $[f(s_0, u_0 + h) - f(s_0, u_0)] = y'_{s_0, u_0}(s_0) - g_1(s_0, u_0)$. Furthermore,

$$|h_1(s, u) - h_1(t, v)| = |y'_{s,u}(s) - g_1(s, u) - y'_{t,v}(t) + g_1(t, v)|$$
$$\leq |y'_{s,u}(s) - y'_{s,u}(t)| + |y'_{s,u}(t) - y'_{t,v}(t)| + |g_1(s, u) - g_1(t, v)|$$
$$\leq \varepsilon + \|y_{s,u} - y_{t,v}\|_1 + \varepsilon \leq 3\varepsilon$$

for $|s - t| \leq \delta$ and $|u - v| \leq \delta$, since $\|x_{s,u} - x_{t,v}\|_1 = |u - v - s + t| \leq |s - t| + |u - v|$, and F is continuous on C^1, by assumption. This shows that the partial derivative h_1 exists and is continuous on $[0, 1] \times \mathbb{R}$ as well. The proof is complete. $\Leftarrow$

Let us make some remarks on Theorem 8.1. First of all, this theorem is in contrast to Theorem 6.3, which states that merely the acting condition $F(C) \subseteq C$ is necessary and sufficient for the continuity of the generating function f. This is due to the remarkable fact that, in case of a compact domain Ω without isolated points, the acting condition $F(C) \subseteq C$ implies both the continuity and boundedness of F in C (Theorems 6.4 and 6.5). On the other hand, it was shown in Theorem 7.5 that in the autonomous case $f = f(u)$ the acting condition $F(H_\phi) \subseteq H_\phi$ implies at least the *boundedness* of F in H_ϕ (but not its continuity, see (7.29)). Obviously, in the case $f = f(u)$ the acting conditon $F(C^1) \subseteq C^1$ implies that f is C^1 on $\mathbb{R}$. Moreover, it follows from the definition of the norm (8.1) that, even in the non-autonomous case $f = f(s, u)$, the continuous differentiability of f implies the boundedness of F in C^1. We make this more precise in the following:

Theorem 8.2   *Suppose that the partial derivatives*

$$g(s, u) = D_s f(s, u), \quad h(s, u) = D_u f(s, u) \qquad (8.15)$$

of the function f exist and are continuous. Then the superposition operator F generated by f is bounded in the space C^1, and its growth function (2.24) may be estimated by $\mu_F(r) \leq \max\{\nu_f(r), \lambda_f(r)\}$ $(r > 0)$, where $\nu_f(r)$ is given by (7.26) and

$$\lambda_f(r) = \nu_g(r) + r\nu_h(r). \qquad (8.16)$$

$\Rightarrow$ The proof follows directly from the definition of the norm (8.1) and the fact that the derivative of $y = Fx$ is $y'(s) = g(s, x(s)) + h(s, x(s))x'(s)$, by (8.13). $\Leftarrow$

Observe that the estimate for the growth function $\mu_F(r)$ given in Theorem 8.2 is in some sense parallel to that given in Theorem 7.4 for H_ϕ and Theorem 7.14 for $\mathcal{J}_{\phi,p}$.

So far we have dealt essentially with the case when the operator F maps the space C^1 into itself. Similar results hold when F maps the space C^k over $[0, 1]$, with norm (8.1) or (8.3), into itself. However, if F maps C^m into the "smaller" space C^n (i.e. $n > m$), a strong degeneration occurs:

Theorem 8.3   *The superposition operator F generated by f maps the space C^m into the space C^n $(m < n)$ if and only if F is constant, i.e. the function f does not depend on u.*

$\Rightarrow$ Suppose that F is not constant, but maps C^m into C^n for $1 \leq m < n$; then there is a point $(s_0, u_0) \in [0, 1] \times \mathbb{R}$ such that $D_u f(s_0, u_0) \neq 0$. Choose a function $x_0 \in C^m$ such that $x_0(s_0) = u_0$ and $x_0^{(m+1)}(s_0)$ does not exist. By assumption, we have that $y_0 = Fx_0 \in C^n$; moreover, we have, with g_m

as in Lemma 8.1,

$$x_0^{(m)}(s) = \frac{y_0^{(m)}(s) - g_m(s, x_0(s), \cdots, x_0^{(m-1)}(s))}{D_u f(s, x_0(s))}$$

for $|s - s_0| \leq \delta$. But this means that $x_0^{(m)}$ is differentiable at s_0, a contradiction. $\Leftarrow$

Now we turn to the problem of characterizing the (global) Lipschitz condition

$$\|Fx_1 - Fx_2\|_n \leq k\|x_1 - x_2\|_m \quad (x_1, x_2 \in C^m) \tag{8.17}$$

for the superposition operator F in terms of the generating function f. By Theorem 8.3, only the case $m \geq n$ is interesting. It turns out that (8.17) leads to the same degeneracy as in the spaces H_ϕ (see Theorem 7.8). Thus, from this point of view there is an essential difference between the spaces C^k and C, as a comparison with Theorem 6.6 shows.

Theorem 8.4 *Suppose that the superposition operator F generated by f maps the space C^m into the space C^n, where $m \geq n$. Then F satisfies the Lipschitz condition (8.17) if and only if the function f has the form (7.36) for some $a, b \in C^n$.*

$\Rightarrow$ For the proof we use the norm (8.3). First, let f be of the form (7.36), and let $x_1, x_2 \in C^m$. By (8.5), we get then $\|Fx_1 - Fx_2\|_n \leq C(b)\|x_1 - x_2\|_m$, where the constant $C(b)$ depends on $\|b\|_C, \cdots, \|b^{(n)}\|_C$, and hence (8.17) holds.

Conversely, suppose that the operator F satisfies the Lipschitz condition (8.17). Fix $\sigma, \tau \in [0, 1]$ ($\sigma < \tau$) and $u_1, u_2 \in \mathbb{R}$, and define two continuous functions z_1, z_2 by

$$z_i(s) = \begin{cases} 0 & \text{if } 0 \leq s \leq \sigma, \\ u_i(\tau - \sigma)^{-2}(s - \sigma) & \text{if } \sigma \leq s \leq \tau, \\ u_i(\tau - \sigma)^{-1} & \text{if } \tau \leq s \leq 1. \end{cases}$$

Further, define two functions x_1, x_2 by

$$x_i(s) = v_i + \frac{1}{(m-1)!} \int_0^s (s-t)^{m-1} z_i(t) dt \tag{8.18}$$

($i = 1, 2$), where $v_1, v_2 \in \mathbb{R}$ are arbitrary. Obviously, $x_1, x_2 \in C^m$ and the relations $x_i(s) = v_i$, $x_i'(s) = \cdots = x_i^{(m)}(s) = 0$ ($0 \leq s \leq \sigma$) hold; moreover,

$x_i^{(k)}(\tau) = \gamma_k u_i (\tau - \sigma)^{n-k-1}$ where we put $\gamma_k = 1/(n-k+1)!$ for $k = 1, \cdots, n$.
The inequality (8.17) gives for the special functions (8.18)

$$\sum_{j=0}^{n-1} |f_j(0, v_1, 0, \cdots, 0) - f_j(0, v_2, 0, \cdots, 0)|$$

$$+ \max_{0 \le s \le 1} |g_n(s, x_1(s), \cdots, x_1^{(n-1)}(s)) - g_n(s, x_2(s), \cdots, x_2^{(n-1)}(s))$$

$$+ h_1(s, x_1(s)) x_1^{(n)}(s) - h_1(s, x_2(s)) x_2^{(n)}(s)|$$

$$\le k|v_1 - v_2| + k\frac{|u_1 - u_2|}{\tau - \sigma},$$

where g_n and h_1 are defined as in Lemma 8.1 and Theorem 8.1, respectively.
Omitting the first sum and the maximum sign, and putting $s = \tau$ yields

$$|g_n(\tau, x_1(\tau), \cdots, x_1^{(n-1)}(\tau)) - g_n(\tau, x_2(\tau), \cdots, x_2^{(n-1)}(\tau))$$

$$+ \gamma_n(\tau - \sigma)^{m-1-n}[h_1(\tau, x_1(\tau))u_1 - h_1(\tau, x_2(\tau))u_2]|$$

$$\le k|v_1 - v_2| + k\frac{|u_1 - u_2|}{\tau - \sigma}.$$

Next, multiplying both sides of this inequality by $\tau - \sigma$ and putting $u_1 = u_2 = (\tau - \sigma)^{n-m}$ we get

$$|(\tau - \sigma)[g_n(\tau, x_1(\tau), \cdots, x_1^{(n-1)}(\tau)) - g_n(\tau, x_2(\tau), \cdots, x_2^{(n-1)}(\tau))]$$
$$+ \gamma_n[h_1(\tau, x_1(\tau)) - h_1(\tau, x_2(\tau))] \le k(\tau - \sigma)|v_1 - v_2|. \tag{8.19}$$

By the definition of γ_k and the choice of u_i, (8.19) implies that

$$\lim_{\tau \to \sigma} x_i^{(k)}(\tau) = \begin{cases} v_i & \text{if } k = 0, \\ 0 & \text{if } k = 1, \cdots, n-2, \\ \gamma_n & \text{if } k = n-1, \end{cases}$$

$(i = 1, 2)$. Consequently, letting τ tend to σ in (8.19), we conclude that
$h_1(\sigma, v_1) = h_1(\sigma, v_2)$. Since $\sigma \in [0, 1]$ and $v_1, v_2 \in \mathbb{R}$ are arbitrary, this
shows that the function $h_1(s, u) = D_u f(s, u)$ is constant with respect to u,
and thus the representation (7.36) holds. $\Leftarrow$

Now we shall study the (local) Darbo condition

$$\alpha_k(FN) \le k(r)\alpha_k(N) \quad (N \subseteq B_r(C^k)) \tag{8.20}$$

for F, where α_k denotes the Hausdorff measure of noncompactness (2.32) in
the space C^k. As a consequence of the two-sided estimate (6.14), one may

show that

$$\frac{1}{2}\alpha_k(N) \le \tilde{\alpha}_k(N) \le 2\alpha_k(N)\,, \tag{8.21}$$

where

$$\tilde{\alpha}_k(N) = \lim_{\sigma \to 0} \sup_{x \in N} \omega(x^{(k)}, \sigma) \tag{8.22}$$

with $\omega(z, \sigma)$ denoting the modulus of continuity (7.5). The fact that only the highest derivative $x^{(k)}$ appears in (8.22) is explained by the formula (8.3) for the norm in the space C^k over $[0, 1]$; in fact, if N is a bounded subset of C^k, then all points $(x(0), x'(0), \cdots, x^{(k-1)}(0))$ with $x \in N$ form a precompact subset of $\mathbb{R}^k$.

By means of the estimate (8.21) we show now that, whenever F acts in the space C^k, then F *always* satisfies the Darbo condition (8.20). This is of course in sharp contrast to, say, Theorems 2.13, 3.11, 4.10, and 6.6.

Theorem 8.5   *Suppose that the superposition operator F maps the space C^k into itself and is continuous. Then the Darbo condition (8.20) holds.*

$\Rightarrow$ In order to estimate the measure of noncompactness $\alpha_k(N)$ of a bounded set $N \subset C^k$, we use formula (8.22). Given $N \subseteq B_r(C^k)$ and $x \in N$, for $y = Fx$ we have, by (8.13),

$$
\begin{aligned}
y^{(k)}(s) - y^{(k)}(t) &= f_k(s, x(s), \cdots, x^{(k)}(s)) - f_k(t, x(t), \cdots, x^{(k)}(t)) \\
&= g_k(s, x(s), \cdots, x^{(k-1)}(s)) - g_k(t, x(t), \cdots, x^{(k-1)}(t)) \\
&\quad + h_1(s, x(s))x^{(k)}(s) - h_1(t, x(t))x^{(k)}(t)\,,
\end{aligned}
$$

where g_k is defined as in Lemma 8.1 and h_1 as in Theorem 8.1. Now, the proof of Lemma 8.1 shows that g_k contains only partial derivatives of f_{k-1}, multiplied by derivatives of x up to order $k-1$; the corresponding superposition operator therefore maps $B_r(C^k)$ into precompact subsets of C^m. In addition, for the remainder term we have $|h_1(s, x(s))x^{(k)}(s) - h_1(t, x(t))x^{(k)}(t)|$ $\le |h_1(s, x(s))| \, |x^{(k)}(s) - x^{(k)}(t)| + |x^{(k)}(t)| \, |h_1(s, x(s)) - h_1(t, x(t))|$. Since $|x^{(k)}(t)|$ is bounded (by r) and $|h_1(s, x(s)) - h_1(t, x(t))|$ tends to zero (uniformly in $x \in N$) as $|s - t| \to 0$, we get (8.20) with

$$k(r) = \max_{\substack{s \in \Omega \\ |u| \le r}} |h_1(s, u)|\,,$$

as claimed. $\Leftarrow$

Observe that the "Darbo constant" $k(r) = \nu_{h_1}(r)$ (see (7.26) and (8.16)) in (8.20) depends only on the derivative of f with respect to u.

8.3 The superposition operator in the space H_ϕ^k

The purpose of this section is to develop some results on the superposition operator in the Hölder space H_ϕ^k which are parallel to those in C^k given in Section 8.2. As before, we restrict ourselves to functions over the interval $\Omega = [0, 1]$. Unfortunately, an acting condition for F in the space H_ϕ^1, say, which is both necessary and sufficient is not known. However, an elementary sufficient condition is contained in the following:

Theorem 8.6   *Suppose that the partial derivatives (8.15) exist, and the superposition operators G and H generated by g and h, respectively, map the space H_ϕ^1 into the space H_ϕ and are bounded and continuous. Then the superposition operator F generated by f maps the space H_ϕ^1 into itself and is bounded and continuous. Moreover, its growth function (2.24) may be estimated by $\mu_F(r) \leq \max\{\nu_f(r), \lambda_f(r)\}$ $(r > 0)$, where ν_f is given by (7.26) and*

$$\lambda_f(r) = \nu_g(r) + r\nu_f(r) + \mu_G(r) + 6r\mu_H(r).$$

$\Rightarrow$ The boundedness and continuity of F follows from the definition of the norm (8.4) in the space H_ϕ^1. To prove the estimate for $\mu_F(r)$, we observe that $\|x\|_{1,\phi} \leq r$ implies that $|x(0)| \leq r$, $|x'(0)| \leq r$, and $|x'(s) - x'(t)| \leq r\phi(|s - t|)$, hence $|x'(s)| \leq 2r$, $|x(s) - x(t)| \leq 2r|s - t|$, and $|x(s)| \leq 3r$. Putting these estimates together and taking into account the chain rule (8.13) proves the assertion. $\Leftarrow$

Let us now discuss the (global) Lipschitz condition

$$\|Fx_1 - Fx_2\|_{1,\phi} \leq k\|x_1 - x_2\|_{1,\phi} \quad (x_1, x_2 \in H_\phi^1) \tag{8.23}$$

in the space H_ϕ^1. The following is completely analogous to Theorems 7.8 and 8.4:

Theorem 8.7   *Suppose that the superposition operator F generated by f maps the space H_ϕ^1 into itself. Then F satisfies the Lipschitz condition (8.23) if and only if the function f has the form (7.36) for some $a, b \in H_\phi^1$.*

$\Rightarrow$ For the proof we use the norm (8.4). If the function f is of the form (7.36), the condition (8.23) follows again from (8.6) with $k = C_\phi(b)$.

Conversely, suppose that the operator F satisfies the Lipschitz condition (8.23). Fix $\sigma, \tau \in [0, 1]$ $(\sigma < \tau)$ and $0 < u_i < v_i < \infty$ $(i = 1, 2)$, and define two functions $x_1, x_2 \in H_\phi^1$ by

$$x_i(s) = \begin{cases} u_i s & \text{if } 0 \leq s \leq \sigma, \\ u_i s + \frac{1}{2}\frac{v_i - u_i}{\tau - \sigma}(s - \sigma)^2 & \text{if } \sigma \leq s \leq \tau, \\ v_i s + \frac{1}{2}(u_i - v_i)(\sigma + \tau) & \text{if } \tau \leq s \leq 1. \end{cases}$$

It is easy to see that $\|x_1 - x_2\|_{1,\phi} = |u_1 - u_2| + \frac{|v_1 - v_2 - u_1 + u_2|}{\phi(\tau - \sigma)}$. Writing $\varphi(s, u, v) = g(s, u) + h(s, u)v$ with g and h as in (8.15), we get by assumption that

$$\frac{|\varphi(\sigma, u_1\sigma, u_1) - \varphi(\sigma, u_2\sigma, u_2) - \varphi(\tau, v_1\tau, v_1) + \varphi(\tau, v_2\tau, v_2)|}{\phi(\tau - \sigma)}$$

$$\leq k(r)|u_1 - u_2| + k(r)\frac{|v_1 - v_2 - u_1 + u_2|}{\phi(\tau - \sigma)}.$$

Putting $u_1 = \xi/\sigma$, $u_2 = 0$, $v_1 = (\xi + \eta)/\tau$, $v_2 = \eta/\tau$, multiplying both sides by $\phi(\tau - \sigma)$ and letting τ tend to σ yields $|\varphi(\sigma, \xi, \frac{\xi}{\sigma}) - \varphi(\sigma, 0, 0) - \varphi(\sigma, \xi + \eta, \frac{\xi + \eta}{\sigma}) + \varphi(\sigma, \eta, \frac{\eta}{\sigma})| = 0$. This means that, for fixed σ, the function $\ell_\sigma(t) = \varphi(\sigma, t, t/\sigma) - \varphi(\sigma, 0, 0)$ is additive, hence $sg(s, u) + uh(s, u) = \alpha(s) + \beta(s)u$. But the only solutions of the linear differential equation

$$s\frac{\partial z}{\partial s} + u\frac{\partial z}{\partial u} = \alpha(s) + \beta(s)u$$

are of the form $z = f(s, u) = a(s) + b(s)u$, by Euler's theorem on homogeneous functions (of degree 0). $\Leftarrow$

We remark that an analogous result holds if F satisfies the Lipschitz condition (8.23) in the little Hölder space $H_\phi^{1,0}$ (see Section 8.1); here the functions a and b in (7.36) belong, of course, to $H_\phi^{1,0}$.

The little Hölder space $H_\phi^{k,0}$ has again the advantage that precompact sets are easily described. We drop the proof of the following lemma, because it is literally the same as that of Lemma 7.3.

Lemma 8.2   *The Hausdorff measure of noncompactness (2.32) in the space $H_\phi^{k,0}$ is given by the formula*

$$\alpha_{k,\phi}(N) = \lim_{\sigma \to 0} \sup_{x \in N} \frac{\omega(x^{(k)}, \sigma)}{\phi(\sigma)}. \tag{8.24}$$

By means of the formula (8.24) for the Hausdorff measure of noncompactness in the space $H_\phi^{k,0}$ one can prove the following result which is completely analogous to Theorem 8.5:

Theorem 8.8   *Suppose that the superposition operator F maps the space $H_\phi^{k,0}$ into itself and is continuous. Then the Darbo condition*

$$\alpha_{k,\phi}(FN) \leq k(r)\alpha_{k,\phi}(N) \quad (N \subseteq B_r(H_\phi^{k,0}))$$

holds.

8.4 The superposition operator in the space $R_\mu(L)$

It follows from the formula (8.10) for the derivatives of a composite function $y(s) = f(s, x(s))$ that, if f is a smooth (i.e. C^∞) function on $[0,1] \times \mathbb{R}$, then the superposition operator F generated by f maps the space $C^\infty = C^\infty([0,1])$ of all smooth functions on $[0,1]$ into itself. It is also known that the composition of two analytic functions is again analytic. The question arises if the same is true for certain intermediate classes between smooth and analytic functions. For instance, the answer is positive for the class of so-called *quasi-analytic functions*; recall that a smooth function x is called quasi-analytic if the fact that x vanishes on some interval implies that x vanishes everywhere.

This problem amounts to studying the superposition operator F generated by f between spaces of smooth functions. As in the first part of this chapter, we confine ourselves again to functions over the interval $\Omega = [0,1]$, although most (sufficient) conditions may be formulated for general (compact) Ω. Thus, $D_s^i D_u^j f(s, u)$ will denote throughout the partial derivative $\partial^{i+j} f(s, u)/\partial s^i \partial u^j$.

Let $\mu = (M_0, M_1, M_2, \cdots)$ be an increasing sequence of positive numbers ($M_0 = 1$), and let $L > 0$. By $R_\mu(L)$ we denote the set of all smooth functions x on $[0,1]$ such that $\|x^{(k)}\| \leq C L^k M_k$ for some $C > 0$, where $\|z\|$ denotes throughout this section the norm (6.1) of the space $C = C([0,1])$. Equipped with the norm

$$\|x\|_{\mu,L} = \sup_{k \geq 0} \frac{\|x^{(k)}\|}{L^k M_k}, \tag{8.25}$$

the set $R_\mu(L)$ becomes a Banach space, sometimes called the *Roumieu space* generated by the sequence μ. It is not hard to see that the imbedding $R_\mu(L) \subseteq R_\mu(L')$ is bounded for $L \leq L'$, and compact for $L < L'$.

Some examples are in order. A particularly important example is

$$M_n = (n!)^p \quad (0 < p < \infty), \tag{8.26}$$

which leads to the so-called *Gevrey spaces* $G^p = G^p(L)$. For $0 < p < 1$, the class G^p consists of *entire analytic functions* on $[0,1]$ of order at most $(1-p)^{-1}$. For $p = 1$, G^p is just the set of all analytic functions on $[0,1]$, while G^p contains also nonanalytic functions in case $p > 1$.

Another interesting example is

$$M_n = H^{n^p} \quad (1 < p < \infty), \tag{8.27}$$

where $H > 1$ is some fixed number. We shall return to the examples (8.26) and (8.27) later.

We are now going to study acting conditions for the superposition operator in the spaces $R_\mu(L)$. For $n = 0, 1, \cdots$ and $u > 0$, $v > 0$, $\tau > 0$, consider the sequence

$$\psi_n(u, v, \tau) = \sum_{\|\alpha\|=n} C_n^\alpha M_{|\alpha|} M_n^{-1} M_1^{\alpha_1} \cdots M_n^{\alpha_n} u^{\alpha_0} v^{|\alpha|-\alpha_0} \tau^{\|\alpha\|-|\alpha|+\alpha_0},$$

(8.28)

(see (8.9)), and denote by W_μ the set of all (u, v, τ) for which the sequence (8.28) is bounded, i.e.

$$\psi(u, v, \tau) = \sup_{n=0,1,\cdots} \psi_n(u, v, \tau) < \infty.$$

(8.29)

The proplem arises to characterize those sequences μ for which the set W_μ is nonempty. Two classes of such sequences will be given in the following:

Lemma 8.3   *Suppose that the sequence $\mu = (M_0, M_1, \cdots)$ satisfies the estimate*

$$C_{i+j}^i M_i M_j \le (i + j) M_{i+j-1} \quad (i, j = 1, 2, \cdots).$$

(8.30)

Then the set $Q_0 = \{(u, v, \tau) : u, v, \tau > 0, u\tau + v + \tau \le 1\}$ is contained in W_μ. Similarly, suppose that μ satisfies the estimate

$$C_{i+j}^i M_i M_j \le (i + j) H^{i+j-1} M_{i+j-1} \quad (i, j = 1, 2, \cdots)$$

(8.31)

for some $H > 1$. Then the set $Q_H = \{(u, v, \tau) : u, v, \tau > 0, (u\tau + v + \tau)H \le 1\}$ is contained in W_μ.

$\Rightarrow$ Given $i_1, i_2, \cdots, i_k \in \mathbb{N}$, by a repeated application of (8.30) we obtain

$$M_{i_1} M_{i_2} \cdots M_{i_k} \le \frac{i_1! i_2! \cdots i_k!}{(i_1 + \cdots + i_k - k + 1)!} M_{i_1+\cdots+i_k-k+1};$$

(8.32)

hence $M_{|\alpha|} M_1^{\alpha_1} \cdots M_n^{\alpha_n} \le (n!)^{-1} M_n |\alpha| (1!)^{\alpha_1} \cdots (n!)^{\alpha_n}$.

By the well known combinatorial identity

$$\sum_{\substack{|\beta|=k \\ \|\beta\|=n}} \frac{k!}{\beta_1! \cdots \beta_n!} = C_{n-1}^{k-1}$$

and (8.32) we get for $\beta_1 = \alpha_0 + \alpha_1$, $\beta_2 = \alpha_2, \cdots, \beta_n = \alpha_n$

$$\psi_n(u, v, \tau) \leq \sum_{\|\alpha\|=n} \frac{|\alpha|!}{\alpha_0! \alpha_1! \cdots \alpha_n!} u^{\alpha_0} v^{|\alpha|-\alpha_0} \tau^{\|\alpha\|-|\alpha|+\alpha_0}$$

$$= \sum_{\|\alpha\|=n} \sum_{\beta_1=0}^{n} \sum_{\gamma=0}^{\beta_1} \frac{(\gamma+\alpha_1)!}{\gamma! \alpha_1!} \frac{|\alpha|!}{(\gamma+\alpha_1)! \alpha_2! \cdots \alpha_n!} \cdot u^\gamma \left(\frac{v}{\tau}\right)^{\alpha_1} \left(\frac{v}{\tau}\right)^{|\alpha|-\gamma-\alpha_1} \tau^n$$

$$= \sum_{\|\beta\|=n} \frac{|\beta|!}{\beta_1! \cdots \beta_n!} \left(u + \frac{v}{\tau}\right)^{\beta_1} \left(\frac{v}{\tau}\right)^{|\beta|-\beta_1} \tau^n$$

$$= \sum_{\|\beta\|=1}^{n} C_{n-1}^{|\beta|-1} \left(u + \frac{v}{\tau}\right)^{|\beta|} \tau^n \leq (u\tau + v + \tau)^n,$$

which implies that $Q_0 \subseteq W_\mu$.

The inclusion $Q_H \subseteq W_\mu$ is a consequence of the inclusion $Q_0 \subseteq W_\mu$. $\Leftarrow$

We remark that the sequence (8.26) satisfies the estimate (8.30) for $p \geq 1$, and the sequence (8.27) satisfies the estimate (8.31). Moreover, the sequence (8.26) satisfies (8.31) for $0 < p < 1$, provided we take $H \geq 2^{1-p}$.

We shall now assume that f is a smooth function satisfying

$$|D_s^i D_u^j f(s, u)| \leq \gamma A^i B^j M_{i+j} \tag{8.33}$$

for $i, j = 0, 1, \cdots, 0 \leq s \leq 1$, and $|u| \leq r$, where the constants A, B and γ depend, in general, on $r > 0$.

Theorem 8.9   Suppose that the sequence μ is such that $(A/L, Br\tau, \tau) \in W_\mu$ for some fixed constants $A, B, r, \tau, L > 0$. Assume that the function f satisfies (8.33), and let $\rho = \gamma \psi(A/L, Br\tau, \tau)$ with ψ as in (8.29). Then the superposition operator F generated by f maps the ball $B_r(R_\mu(L))$ into the ball $B_\rho(R_\mu(L/\tau))$.

$\Rightarrow$ From the formula (8.10) for the derivatives of $y = Fx$ we conclude that, for $0 \leq s \leq 1$ and $|x(s)| \leq r$,

$$
\begin{aligned}
|y^{(n)}(s)| &\leq \sum_{\|\alpha\|=n} C_n^{\alpha} \gamma A^{\alpha_0} B^{\alpha_1 + \cdots + \alpha_n} M_{|\alpha|} (rLM_1)^{\alpha_1} \cdots (rLM_n)^{\alpha_n} \\
&\leq \sum_{\|\alpha\|=n} C_n^{\alpha} \gamma A^{\alpha_0} B^{|\alpha|-\alpha_0} r^{|\alpha|-\alpha_0} L^{n-\alpha_0} M_{|\alpha|} M_1^{\alpha_1} \cdots M_n^{\alpha_n} \\
&= \gamma \sum_{\|\alpha\|=n} C_n^{\alpha} (\frac{A}{L})^{\alpha_0} (Br\tau)^{|\alpha|-\alpha_0} \tau^{n-|\alpha|+\alpha_0} M_{|\alpha|} M_1^{\alpha_1} \cdots M_n^{\alpha_n} (\frac{L}{\tau})^n \\
&\leq \gamma \psi(\frac{A}{L}, Br\tau, \tau)(\frac{L}{\tau})^n M_n = \rho L^n \tau^{-n} M_n .
\end{aligned}
$$

The statement follows from the definition of the norm (8.25). $\Leftarrow$

The crucial point in Theorem 8.9 is, of course, to find elements in W_μ; this can be done by means of Lemma 8.3. For instance, combining Theorem 8.9 and Lemma 8.3 gives the following result: if μ satisfies (8.31) and f satisfies (8.33), then the operator F maps the ball $B_r(R_\mu(L))$ into the ball $B_\rho(R_\mu(KL))$, where $K = H(Br+1+A/L)$ and $\rho = \gamma\psi(A/L, Br/K, 1/K)$.

As the preceding theorem shows, sufficient conditions for the operator F to act between two spaces $R_\mu(L_1)$ and $R_\mu(L_2)$ may be obtained by combining a boundedness condition (like (8.29)) on the sequence μ with a growth condition (like (8.33)) on the function f. Now we shall prove a certain converse of this. First of all, we need a technical auxiliary result.

Lemma 8.4   Let f be a smooth function on $[0,1] \times \mathbb{R}$, and suppose that the superposition operator F generated by f maps the set of all linear functions

$$
x_{k,n}(s) = \frac{k+1}{n+1}(s - s_0) + u_0 \quad (k = 0, 1, \cdots, n) \tag{8.34}
$$

($s_0 \in [0,1]$, $u_0 \in \mathbb{R}$ fixed with $|u_0| < r$) into the ball $B_\gamma(R_\mu(L))$. Then f satisfies (8.33) for $i, j = 0, 1, \cdots$, $0 \leq s \leq 1$, and $|u| \leq r$, where $A = B = 2eL$.

$\Rightarrow$ By hypothesis, $\|y_{k,n}\|_{\mu,L} \leq \gamma$ for $y_{k,n} = Fx_{k,n}$; by (8.10) we have

$$
\begin{aligned}
y_{k,n}^{(n)}(s_0) &= \sum_{\|\alpha\|=n} C_n^{\alpha} D_s^{\alpha_0} D_u^{|\alpha|-\alpha_0} f(s_0, u_0) x_{k,n}'(s_0)^{\alpha_1} \\
&= \sum_{i=0}^{n} C_n^i D_s^{n-i} D_u^i f(s_0, u_0)[\frac{k+1}{n+1}]^i \quad (k = 0, \cdots, n).
\end{aligned}
$$

This system of linear equations for $D_s^n f(s_0, u_0)$, $D_s^{n-1} D_u f(s_0, u_0)$, $\cdots$,

$D_u^n f(s_0, u_0)$ has a unique solution for each right-hand side, since the determinant of its coefficients is

$$\Delta_n = C_n^0 C_n^1 \cdots C_n^n \bar{\Delta}_n,$$

where

$$\bar{\Delta}_n = \begin{vmatrix} 1 & \theta_1 & \cdots & \theta_1^{n-1} & \theta_1^n \\ 1 & \theta_2 & \cdots & \theta_2^{n-1} & \theta_2^n \\ \vdots & \vdots & & \vdots & \vdots \\ 1 & \theta_n & \cdots & \theta_n^{n-1} & \theta_n^n \\ 1 & 1 & \cdots & 1 & 1 \end{vmatrix}$$

$(\theta_i = \frac{i}{n+1})$ is a Vandermonde determinant. Thus, $D_s^{n-i} D_u^i f(s_0, u_0) = \Delta_n^{-1} \Delta_{n,i}$, where $\Delta_{n,i} = C_n^0 C_n^1 \cdots C_n^{i-1} C_n^{i+1} \cdots C_n^n \bar{\Delta}_{n,i}$ with

$$\bar{\Delta}_{n,i} = \begin{vmatrix} 1 & \theta_1 & \cdots & \theta_1^{i-1} & y_{0,n}^{(n)}(s_0) & \theta_1^{i+1} & \cdots & \theta_1^n \\ 1 & \theta_2 & \cdots & \theta_2^{i-1} & y_{1,n}^{(n)}(s_0) & \theta_2^{i+1} & \cdots & \theta_2^n \\ \vdots & \vdots & & \vdots & \vdots & \vdots & & \vdots \\ \vdots & \vdots & & \vdots & \vdots & \vdots & & \vdots \\ 1 & 1 & \cdots & 1 & y_{n,n}^{(n)}(s_0) & 1 & \cdots & 1 \end{vmatrix}.$$

By evaluating this determinant with the usual product rule, a straightforward but cumbersome calculation leads to the estimate

$$|D_s^{n-i} D_u^i f(s_0, u_0)| \leq \sum_{k=0}^{n} C_n^k \frac{(n+1)^{i+1}}{(i+1)!} \gamma L^n M_n$$

$$= \frac{(n+1)^{i+1}}{(i+1)!} \gamma (2L)^n M_n \leq \frac{(n+1)^{n+1}}{(n+1)!} \gamma (2L)^n M_n$$

$$\leq \frac{(n+1)^{n+1} e^{n+1}}{(n+1)^{n+1} \sqrt{2\pi(n+1)}} \gamma (2L)^n M_n \leq \gamma (2eL)^n M_n,$$

where we used the fact that $\|y_{k,n}^{(n)}\| \leq L^n M_n \|y_{k,n}\|_{\mu,L} \leq L^n M_n \gamma$. This proves the assertion. $\Leftarrow$

Observe that Theorem 8.9 gives only a *sufficient* acting condition for the superposition operator, while Lemma 8.4 gives a *necessary* acting condition on a very restrictive subclass of $R_\mu(L)$. To combine these results to both necessary and sufficient acting conditions it turns out to be useful to pass from the space $R_\mu(L)$ to certain "limiting" spaces $R_\mu(0)$ and $R_\mu(\infty)$. In this connection, one may even prove boundedness, continuity, and compactness results for F; this will be carried out in the next section. Before doing so, however, we still show that the usual degeneration of Lipschitz continuous superposition operators occurs also in the space $R_\mu(L)$:

Theorem 8.10   *Suppose that the superposition operator F generated by f maps the space $R_\mu(L)$ into itself. Then F satisfies the Lipschitz condition*

$$\|Fx_1 - Fx_2\|_{\mu,L} \leq k\|x_1 - x_2\|_{\mu,L} \quad (x_1, x_2 \in R_\mu(L)) \tag{8.35}$$

if and only if the function f has the form (7.29) for some $a, b \in R_\mu(L)$.

$\Rightarrow$ It is again easy to see that an affine function (7.29) generates a Lipschitz continuous superposition operator in the space $R_\mu(L)$. Conversely, if F satisfies (8.35), for the linear functions $x_1(s) = \tau_1 s$ and $x_2(s) = \tau_2 s$ ($\tau_1, \tau_2 \in \mathbb{R}$) we have $\|x_1 - x_2\|_{\mu,L} = \gamma|\tau_1 - \tau_2|$ with $\gamma = \max\{1, 1/LM_1\}$. Taking into account only the second derivative of $y_i = Fx_i$ ($i = 1, 2$) in the left-hand side of (8.35), we get

$$|D_{ss}(s, \tau_1 s) + 2D_{su}f(s, \tau_1 s)\tau_1 + D_{uu}(s, \tau_1 s)\tau_1^2$$
$$- D_{ss}(s, \tau_2 s) - 2D_{su}f(s, \tau_2 s)\tau_2 - D_{uu}(s, \tau_2 s)\tau_2^2|$$
$$\leq L^2 M_2 k\gamma|\tau_1 - \tau_2|,$$

where $D_{ss}f$, $D_{su}f$, and $D_{uu}f$ are the corresponding second derivatives of f. Putting $u_i = \tau_i s$ ($i = 1, 2$), multiplying by s^2 and letting s tend to 0 yields $|D_{uu}f(s, u_1)u_1^2 - D_{uu}f(s, u_2)u_2^2| = 0$; from this the assertion follows easily. $\Leftarrow$

8.5 The superposition operator in Roumieu classes

Let $R_\mu(L)$ be the Banach space defined in the previous section (see (8.25)). The set

$$R_\mu(0) = \bigcap_{0 < L < \infty} R_\mu(L), \tag{8.36}$$

equipped with the topology of the projective limit, is called the *projective Roumieu class* generated by μ; likewise, the set

$$R_\mu(\infty) = \bigcup_{0 < L < \infty} R_\mu(L), \tag{8.37}$$

equipped with the topology of the inductive limit, is called the *inductive Roumieu class* generated by μ. The sets $R_\mu(0)$ and $R_\mu(\infty)$ are not normed spaces, but locally convex Fréchet spaces; moreover, $R_\mu(\infty)$ is a Montel space, i.e. every bounded set in $R_\mu(\infty)$ is precompact.

The following two theorems show that the behaviour of the superposition operator in the classes $R_\mu(0)$ and $R_\mu(\infty)$ is much nicer than in other spaces of differentiable functions.

Theorem 8.11   Let $W_\mu \neq \emptyset$. The superposition operator F generated by f acts in the inductive Roumieu class $R_\mu(\infty)$ and is bounded if and only if (8.33) holds for some $A > 0$, $B > 0$, and $\gamma > 0$. Moreover, in this case F is always continuous and compact.

$\Rightarrow$ Let first $x \in R_\mu(\infty)$; by definition (8.37), we then have $\|x\|_{\mu,L} = r < \infty$ for some $L > 0$. Applying (8.33) yields $|D_s^i D_u^j f(s, x(s))| \leq \gamma A^i B^j M_{i+j}$ for some $\gamma > 0$. By assumption, we may choose $(u, v, \tau) \in W_\mu$; for $L' \geq \max\{L, A/u\}$ and $\tau' \leq \min\{\tau, v/Br\}$ we get then, by Theorem 8.9, that $\|Fx\|_{\mu,L'/\tau'} \leq \rho$ with $\rho = \gamma\psi(u, v, \tau)$ which shows that F maps $R_\mu(\infty)$ into itself. Now, if $Q \subset R_\mu(\infty)$ is bounded, then $Q \subset R_\mu(L)$ for some $L > 0$, since $R_\mu(\infty)$ is a regular inductive limit. By what has been proved, the set $F(Q)$ is then bounded in $R_\mu(L')$ for some $L' \geq L$, and hence also in $R_\mu(\infty)$.

Conversely, suppose that F acts in $R_\mu(\infty)$ and is bounded. Since the set Q of all affine functions $x(s) = \alpha(s - s_0) + u_0$ ($0 \leq \alpha \leq 1$, $s_0 \in [0, 1]$, $|u_0| \leq r$) contains all functions (8.34) and is bounded in each space $R_\mu(L)$, we get (8.33) by Lemma 8.4.

We show now that F is continuous and compact. Again by the regularity of the inductive limit $R_\mu(\infty)$, continuity is equivalent to sequential continuity; moreover, it suffices to prove the continuity of F at θ, under the hypothesis that $f(s, 0) \equiv 0$ (otherwise one may pass from the function f to the function $f(s, u) - f(s, 0)$). Now, if x_n is a sequence in $R_\mu(\infty)$ converging to zero, by the regularity of $R_\mu(\infty)$ and the boundedness of $\{x_n\}$ we have $x_n \in R_\mu(L)$ for some $L > 0$. By what has been proved before, the set $\{Fx_n\}$ is bounded in $R_\mu(L')$ for some $L' \geq L$. Moreover, $\{Fx_n\}$ is precompact in $R_\mu(L'')$ for $L'' > L'$, and hence has an accumulation point y^*. Since the function f is continuous on $[0, 1] \times \mathbb{R}$, the operator F is continuous in the space C (see Theorem 6.5), and hence Fx_n converges to θ in C. But this implies that $y^* = \theta$, and therefore F is continuous at θ, considered as an operator in $R_\mu(\infty)$.

The compactness of F follows from the fact that the bounded sets and precompact sets in $R_\mu(\infty)$ coincide. $\Leftarrow$

Similarly, for the class $R_\mu(0)$ we get the following:

Theorem 8.12   Let $W_\mu \neq \emptyset$. The superposition operator F generated by f acts in the projective Roumieu class $R_\mu(0)$ and is bounded if and only *if, for each $A > 0$ and $B > 0$, (8.33) holds for some $\gamma > 0$. Moreover, in this case F is always continuous and compact.*

$\Rightarrow$ Since the parameter τ' in the proof of Theorem 8.11 does not depend on L', the operator F actually maps the space $R_\mu(0)$ into itself. Moreover, since any bounded set in $R_\mu(0)$ is trivially bounded in $R_\mu(L)$ for each $L > 0$, the hypothesis on (8.33) implies that F is also bounded in $R_\mu(0)$. This proves the "if" part of Theorem 8.12. To prove the "only if" part, we remark that the set Q mentioned in the proof of Theorem 8.11 is course bounded in $R_\mu(0)$, and thus it suffices to repeat the reasoning of the "only if" part of Theorem 8.11. The rest of the proof proceeds along the lines of that of Theorem 8.11. $\Leftarrow$

8.6 Notes, remarks and references

1. All material presented about the spaces C^k and H_ϕ^k may be found in standard textbooks on functional analysis, differential equations, or integral equations. A concise recent reference is Section 2.3 of [64]. The structure of the underlying set Ω (an interval, a compact set in $\mathbb{R}^N$, a countably compact set, etc.) determines the topology of the spaces considered in this section, but is not really essential for proving the basic properties of the superposition operator F by means of *sufficient* conditions on f. On the other hand, as in Chapter 7, *necessary* conditions require the more specific structure of the interval $\Omega = [0, 1]$.

2. The formula (8.10) for the derivatives of a composite function may be found in many papers and books; see e.g. [114], [180], [274] or [286]. The elementary Lemma 8.1 is taken from [220]. The counterexample after Lemma 8.1 is due to J. Matkowski [221]; see also [189]. The proof of Theorem 8.1 is implicitly contained in [53], but our proof [70] has the advantage of carrying over to the space C^k.

We remark that the paper [189] is also concerned with the following *representation problem*. An operator F in the space C (over $\Omega = [0, 1]$) is called *locally determined* if, whenever two functions x_1 and x_2 coincide on a subinterval J of $[0, 1]$, then the functions Fx_1 and Fx_2 also coincide on J; this is of course analogous to the property (c) stated in Lemma 1.1 for operators in the space S. It is proved in [189] that every locally determined operator F from C^m into C has the form $Fx(s) = f(s, x(s), x'(s), \cdots, x^{(m)}(s))$ for some (unique) function f on $[0, 1] \times \mathbb{R}^{m+1}$, and every locally determined operator F from C^n into C^1 has the form $Fx(s) = f(s, x(s), x'(s), \cdots, x^{(n-1)}(s))$ for some (unique) function f on $[0, 1] \times \mathbb{R}^n$. In particular, this holds for $m = 0$ and $n = 1$, i.e. for superposition operators in the spaces C and C^1.

Theorem 8.3 is contained in [220] and shows that there are no nontrivial superposition operators which increase the degree of smoothness. The degeneracy of Lipschitz continuous superposition operators in C^k is also proved in [220]; the proof is of course similar to that of the corresponding result in Hölder spaces (Theorem 7.8). The estimate (8.21) for the Hausdorff measure of noncompactness in C^k is a simple consequence of the corresponding estimate (6.14) in the space C. Theorem 8.5 is completely elementary and contained in [24].

3. Very little attention has been given to the superposition operator in the "higher-order" Hölder spaces H_ϕ^k or $H_\phi^{k,0}$. Some elementary sufficient conditions are contained in [245], [352], and [356]. The sufficient condition given in Theorem 8.6 is, of course, straightforward. We remark that similar sufficient conditions for the *differential operator* $\Phi x(s) = \phi(s, x(s), x'(s))$ between H_ϕ^1 and H_ϕ are given in [329].

Theorem 8.7 is of course analogous to Theorems 7.8 and 8.4; see also [24]. The proofs of Lemma 8.2 and Theorem 8.7 can also be found in [24].

A rather detailed description of the continuity and smoothness properties of superposition operators in the Schauder spaces $H_\alpha^k(\Omega)$ (i.e. $H_\phi^k(\Omega)$ in case $\phi(t) = t^\alpha$, $0 < \alpha \leq 1$) may be found in Chapter 2 of [356]. For instance, it is shown there that, if f is of class $C^{1+\max\{1,k\}}$ (respectively of class $C^{2+\max\{1,k\}}$) on $\Omega \times \mathbb{R}$, then F is continuous (respectively continuously Fréchet differentiable) on H_α^k; moreover, a sufficient condition for analyticity of F on H_α^k is also given in [356].

4. The Banach spaces $R_\mu(L)$ considered in Section 8.4 are important in the theory of *ultradistributions* and *partial differential equations*. The first results on the composition of functions from the Gevrey spaces G^p seem to be due to M. Gevrey [118]; see also [107], [108]. The general spaces $R_\mu(L)$ have been introduced in [275]; the first systematic investigation of the superposition operator in these spaces, as well as in the projective and inductive classes $R_\mu(0)$ and $R_\mu(\infty)$, is due to V.I. Nazarov [236]; for applications see [237]. Various special classes of such spaces, such as those generated by the sequences (8.26) and (8.27), are dealt with in Chapter 6 of [115] or Chapter 7 of [190]. We remark that the space $R_\mu(L)$ consists only of *quasi-analytic functions* if and only if

$$\sum_{n=1}^{\infty} \frac{M_n}{M_{n+1}} = \infty \quad \text{or} \quad \sum_{n=1}^{\infty} \frac{1}{\sqrt[n]{M_n}} = \infty. \tag{8.38}$$

For example, (8.38) *fails* for the sequences μ given in (8.26) and (8.27); thus, the corresponding spaces $R_\mu(L)$ contain also functions which are *not* quasi-analytic.

All results proved in Section 8.4 may be found in [28], where even Banach
space valued functions are considered.

5. As already observed at the beginning of Section 8.5, the classes $R_\mu(0)$
and $R_\mu(\infty)$ are not normed spaces, but only locally convex. Nevertheless,
these classes have such a "nice" structure that, whenever the superposition
operator F acts in them, F is always bounded and continuous, as Theorems
8.11 and 8.12 show; this is somewhat similar to the situation in the normed
spaces L_p (Theorems 3.3 and 3.7) and C (Theorems 6.4 and 6.5). In contrast
to those spaces, however, F is also *compact*; compare this, for instance, with
the remark after Theorem 6.2.

We remark that the projective Roumieu classes $R_\mu(0)$ are also called
Beurling classes in the literature (see e.g. [190]).

All results proved in Section 8.5 may also be found in [28], except for
Theorem 8.10 which is proved in [24]. Moreover, the paper [28] contains
some remarks on the "little" Roumieu spaces $R_\mu^0(L)$ of all functions $x \in$
$R_\mu(L)$ satisfying

$$\lim_{k \to 0} \frac{\|x^{(k)}\|}{L^k M_k} = 0. \tag{8.39}$$

For example, the Hausdorff measure of noncompactness (2.32) in the space
$R_\mu^0(L)$ satisfies the equality

$$\alpha_{\mu,L}(N) = \lim_{k \to \infty} \sup_{x \in N} \frac{\|x^{(k)}\|}{L^k M^k} \tag{8.40}$$

(see [24]); compare this with the corresponding formulas (7.38) and (8.24).
In particular, (8.40) shows that a subset $N \subset R_\mu^0(L)$ is precompact if and
only if (8.39) holds uniformly in $x \in N$.

Chapter 9

The superposition operator in Sobolev spaces

In spite of the importance of Sobolev spaces in many fields of mathematical analysis, the theory of the superposition operator in these spaces is not yet sufficiently complete. In particular, acting conditions for the superposition operator between the spaces W_p^k and W_q^k which are both necessary and sufficient are known only in case $f = f(u)$ and $k = 1$. One may show, however, that these acting conditions imply the boundedness and continuity of the operator F.

For higher–order Sobolev spaces, we give some sufficient acting, boundedness, and continuity conditions for F which build essentially on the well known Sobolev imbedding theorems. Most of these results refer to bounded domains $\Omega \subset \mathbb{R}^N$ with sufficiently regular boundary. Interestingly, the requirement that F maps the space $W_p^k(\mathbb{R}^N)$ into itself is reasonable, roughly speaking, only in case $k = 1$ and $p \geq 1$, or $k \geq 2$ and $kp \geq N$, and leads to a strong degeneracy of f in the other cases.

In the final section, we discuss some acting, boundedness, and continuity properties of F between Sobolev–Orlicz spaces.

9.1 Sobolev spaces

In this section we recall some facts about Sobolev spaces and the relations between them. We shall consider throughout functions over special domains Ω in the Euclidean space $\mathbb{R}^N$. The two types of bounded domains Ω we shall deal with are *convex polygons*, i.e. $\Omega = \{s : s \in \mathbb{R}^N, \ell_j s > c_j \ (j = 1, \cdots, m)\}$ for some linear functionals $\ell_1, \cdots, \ell_m$ and scalars $c_1, \cdots, c_m$, or *smooth domains*, which means, loosely speaking, that the last coordinate s_N of every point $s = (s_1, \cdots, s_N)$ on the boundary $\partial\Omega$ behaves locally like a smooth function of the other coordinates $s_1, \cdots, s_{N-1}$. Moreover, as examples of unbounded domains we shall consider the *whole space* $\Omega = \mathbb{R}^N$, or the *half space* $\Omega = \mathbb{R}_+^N = \{s : s \in \mathbb{R}^N, s_N > 0\}$. Thus, "$\Omega$ bounded" will mean in the sequel that Ω is either a convex polygon or a bounded smooth domain, while "any Ω" will mean that Ω is one of the four types of domains described above.

For $k \in \mathbb{N}$ and $1 \le p \le \infty$ we define the *Sobolev space* $W_p^k = W_p^k(\Omega)$ as set of all functions $x \in L_p(\Omega)$ for which all distributional derivatives $D^\alpha x$ ($|\alpha| \le k$) belong to $L_p(\Omega)$, equipped with the norm

$$\|x\|_{k,p} = \begin{cases} \displaystyle\sum_{|\alpha| \le k} \|D^\alpha x\|_p & \text{if } 1 \le p < \infty, \\ \displaystyle\max_{|\alpha| \le k} \|D^\alpha x\|_\infty & \text{if } p = \infty, \end{cases} \tag{9.1}$$

where $\|z\|_p$ denotes, of course, the L_p norm (3.3) of z. In particular, W_p^0 coincides with the Lebesgue space L_p. Moreover, for Ω as above and $k \ge 1$, the space $W_\infty^k(\Omega)$ coincides with the Hölder space $H_1^k(\bar{\Omega})$ (see Section 8.1), and the norms (9.1) and (8.2) are equivalent.

It is clear that, for bounded Ω, $k \ge 0$, and $1 \le p \le \infty$, the space $C^k(\bar{\Omega})$ (see Section 8.1) is imbedded in $W_p^k(\Omega)$. In general, one can show that $C^\infty(\bar{\Omega}) \cap W_p^k(\Omega)$ is dense in $W_p^k(\Omega)$ for any Ω, and $C_0^\infty(\Omega)$ (the space of all C^∞ functions with *compact support* in Ω) is dense in $W_p^k(\Omega)$ if $\Omega = \mathbb{R}^N$ or $k = 0$. On the other hand, if $\Omega \ne \mathbb{R}^N$ and $k > 0$, then $C_0^\infty(\Omega)$ is not dense in $W_p^k(\Omega)$. The closure of $C_0^\infty(\Omega)$ with respect to the norm (9.1) (for $1 \le p < \infty$) is usually denoted by $W_p^{k,0}(\Omega)$. Thus $W_p^{k,0}(\Omega) = W_p^k(\Omega)$ if and only if $k = 0$ or $\Omega = \mathbb{R}^N$.

One may also define Sobolev spaces of negative order. Given $k \in \mathbb{N}$ and $1 \le p \le \infty$, the space $W_p^{-k} = W_p^{-k}(\Omega)$ consists, by definition, of all distributions y of the form $y(s) = \sum_{|\alpha| \le k} D^\alpha x_\alpha(s)$, with $x_\alpha \in L_p$ for $|\alpha| \le k$. It turns out that $W_{\tilde{p}}^{-k}$ ($p^{-1} + \tilde{p}^{-1} = 1$) is nothing else but the *associate space* of $W_p^{k,0}$ (see (2.6)); moreover, for $1 \le p < \infty$, the space $W_{\tilde{p}}^{-k}$ is also the *dual space* of $W_p^{k,0}$. In particular, in case $k = 0$ we get again the corresponding formulas $L_{\tilde{p}} = \tilde{L}_p = L_p^*$ ($1 \le p < \infty$) and $L_1 = \tilde{L}_\infty \ne L_\infty^*$ for Lebesgue spaces.

Now we recall the definition of the Sobolev space W_p^s for arbitrary real s. For $1 \le p < \infty$ and $0 < \sigma < 1$, the space $W_p^\sigma = W_p^\sigma(\Omega)$ consists, by definition, of all functions $x \in L_p(\Omega)$ for which the norm

$$\|x\|_{\sigma,p} = \left\{ \int_\Omega |x(s)|^p ds + \int_\Omega \int_\Omega \frac{|x(s) - x(t)|^p}{|s - t|^{N+\sigma p}} ds\, dt \right\}^{1/p}$$

is finite. Similarly, the space $W_\infty^\sigma(\Omega)$ is defined; this space coincides with the

Hölder space $H_\sigma(\bar\Omega)$ defined by the Hölder function $\phi(t) = t^\sigma$ (see Section 7.1). Finally, for $1 \le p < \infty$ and any real $s \ge 0$ we write $s = \lfloor s \rfloor + \sigma$ (where $\lfloor \vartheta \rfloor$ denotes the largest integer $\le \vartheta$) and define $W_p^s(\Omega)$ by means of the norm

$$\|x\|_{s,p} = \begin{cases} \displaystyle\sum_{|\alpha| \le \lfloor s \rfloor} \|D^\alpha x\|_{\sigma,p} & \text{if } 1 \le p < \infty, \\ \displaystyle\max_{|\alpha| \le \lfloor s \rfloor} \|D^\alpha x\|_{\sigma,\infty} & \text{if } p = \infty. \end{cases} \tag{9.2}$$

As a consequence, $W_\infty^s(\Omega)$ coincides with the Hölder space $H_\sigma^k(\bar\Omega)$ for $k = \lfloor s \rfloor$ (see Section 8.1).

At this point, we can make the same remarks as for the "entire" Sobolev spaces W_p^k above. For instance, $C^\infty(\bar\Omega) \cap W_p^s(\Omega)$ is dense in $W_p^s(\Omega)$ for any Ω, and $C_0^\infty(\Omega)$ is dense in $W_p^s(\Omega)$ if and only if $\Omega = \mathbb{R}^N$ or $0 \le s \le \frac{1}{p}$, $(s,p) \ne (1,1)$. As above, $W_p^{s,0}(\Omega)$ is the closure of $C_0^\infty(\Omega)$ with respect to the norm (9.2) for $1 \le p < \infty$ and $s \ge 0$.

Finally, for $s \in \mathbb{R}$, $s \ge 0$, and $1 \le p < \infty$, the space $W_{\tilde p}^{-s}(\Omega)$ is defined as the dual space of $W_p^{s,0}(\Omega)$, $p^{-1} + \tilde p^{-1} = 1$.

There are some fundamental relations between the spaces introduced above which are basic in many fields of mathematical analysis, and are usually referred to as *Sobolev imbedding theorems*. For later convenience, we summarize these results in the following:

Lemma 9.1 Let $k \in \mathbb{N}$, $0 < \sigma < 1$, $r, s \in \mathbb{R}$ with $0 \le r \le s$, and $p, q \in \mathbb{R}$ with $1 \le p \le q \le \infty$.

(a) If $1 < p \le q < \infty$ and $r - \frac{N}{q} \le s - \frac{N}{p}$, then $W_p^s(\Omega)$ is continuously imbedded in $W_q^r(\Omega)$.

(b) If $r - \frac{N}{q} < s - \frac{N}{p}$ and Ω is bounded, the imbedding $W_p^s(\Omega) \subseteq W_q^r(\Omega)$ is even compact.

(c) If $k + \sigma \le s - \frac{N}{p}$, then $W_p^s(\Omega)$ is continuously imbedded in $H_\sigma^k(\bar\Omega)$.

(d) If $k + \sigma < s - \frac{N}{p}$ and Ω is bounded, the imbedding $W_p^s(\Omega) \subseteq H_\sigma^k(\bar\Omega)$ is even compact.

All statements hold also with W^0 in the place of W.

We remark that the case $p = 1$ plays a particular role in the theory of the Sobolev spaces W_p^s. For example, in case $p = 1$ one may take $q = \infty$ in Lemma 9.1 (a), and $\sigma = 0$ or $\sigma = 1$ in Lemma 9.1(c). As we shall see, for studying the superposition operator in Sobolev spaces W_p^k, one has to distinguish sometimes the cases $k > 1$ and $k = 1$, and also the cases $p > 1$ and $p = 1$. In this connection, we mention a characterization of functions in W_p^1 which is closely related to the space AC of (locally) *absolutely continuous functions* on Ω. Recall that the space $AC = AC(\Omega)$ consists of all measurable functions x on Ω such that, for every line τ parallel to one of the s_i-coordinate axes $\{(s_1, \cdots, s_N) : s_1 = \cdots = s_{i-1} = s_{i+1} = \cdots = s_N = 0\}$, x is locally absolutely continuous on $\Omega \cap \tau$. The following characterization which is due to E. Gagliardo will be used several times in the sequel:

Lemma 9.2   *A function x belongs to $W_p^1(\Omega)$ $(1 \le p < \infty)$ if and only if there exists a function $\bar{x} \in AC(\Omega)$ such that $x(s) = \bar{x}(s)$ almost everywhere in Ω, $D_i \bar{x} \in L_p(\Omega)$ for $i = 1, \cdots, N$, and $\bar{x} \in L_p(\Omega)$. If Ω is bounded, the requirement that $\bar{x} \in L_p(\Omega)$ is superfluous.*

9.2 Sufficient acting conditions in W_p^1

We begin now the study of the superposition operator between the Sobolev spaces $W_p^1(\Omega)$ and $W_q^1(\Omega)$, where $\Omega \subset \mathbb{R}^N$ is bounded. In most applications, e.g. to nonlinear partial differential equations, it is assumed that the function $f = f(s, u)$ is smooth on $\Omega \times \mathbb{R}$; under this hypothesis, the classical chain rule (8.10) applies, and $y = Fx$ has the first derivatives

$$D_i y(s) = D_i f(s, x(s)) + D_u f(s, x(s)) D_i x(s) , \qquad (9.3)$$

where $D_i f(s, u)$ $(i = 1, \cdots, N)$ and $D_u f(s, u)$ denote the partial derivatives of f with respect to s_i and u, respectively. Moreover, the imbedding results given in Lemma 9.1 imply the following sufficient acting condition for F: if the partial derivatives of f satisfy the growth conditions

$$|D_i f(s, u)| \le a(s) + b|u|^{p(N-q)/q(N-p)} \qquad (9.4)$$

$(i = 1, \cdots, N)$ and

$$|D_u f(s, u)| \le c(s) + d|u|^{N(p-q)/q(N-p)} \qquad (9.5)$$

for some $a \in L_q$ and $c \in L_{pq/(p-q)}$ $(1 < q \le p < N)$, then the superposition operator F generated by f maps the space W_p^1 into the space W_q^1.

As a matter of fact, the hypothesis on the smoothness of f may be essentially weakened. The investigation of this phenomenon, which requires a very delicate analysis of the chain rule and of regularity properties of real (non-smooth) functions, was carried out by M. Marcus and V.J. Mizel in a series of papers in a much more general setting than that presented here.

Given a measurable function f on $\Omega \times \mathbb{R}$, consider the set of all points $(s, u) \in \Omega \times \mathbb{R}$, where f does not have a (total) derivative. There are essentially two classes of functions f for which one can establish reasonable acting conditions for F. First, if f is locally Lipschitz continuous on $\Omega \times \mathbb{R}$, then the above mentioned set has $(N+1)$-dimensional Lebesgue measure zero, and one may verify (9.3) in certain cases, as well as the sufficiency of (9.4) and (9.5) for $F(W_p^1) \subseteq W_q^1$. Second, if f is a locally absolutely continuous Carathéodory function (for the definition see below), then the chain rule (9.3) is generally not valid, but one may obtain acting conditions by means of the characterization of W_p^1 given in Lemma 9.2.

Theorem 9.1   *Suppose that f is locally Lipschitz continuous on $\Omega \times \mathbb{R}$, where $\Omega \subset \mathbb{R}^N$ is bounded. Let $1 < q \le p < N$. Assume that the estimates (9.4) and (9.5) hold for some $a \in L_q$, $c \in L_{pq/(p-q)}$, $b \ge 0$, and $d \ge 0$, where the partial derivatives exist. Then the superposition operator F generated by f maps $W_p^1(\Omega)$ into $W_q^1(\Omega)$.*

$\Rightarrow$ Since f is locally Lipschitz on $\Omega \times \mathbb{R}$, we get from (9.5) that

$$\begin{aligned} |f(s,u) - f(s,0)| &\le \left| \int_0^u D_u f(s,v) dv \right| \\ &\le c(s)|u| + d|u|^{p(N-q)/q(N-p)} . \end{aligned} \tag{9.6}$$

Again by the local Lipschitz continuity of f, the function $y_u(s) = f(s,u)$ belongs to $AC(\Omega)$ for each $u \in \mathbb{R}$, and hence to $W_p^1(\Omega)$, by Lemma 9.2 and (9.4). Now, by Lemma 9.1 (a), applied to $q = Np/(N-p)$, $r = 0$, and $s = 1$, we see that W_p^1 is imbedded in $L_{Np/(N-p)}$; moreover, we use the fact that $L_{pq/(p-q)}$ coincides with the multiplicator space L_q/L_p (see (3.8)). Given $x \in W_p^1$, we get from Lemma 9.2 and the chain rule (9.3) that $y = Fx$ coincides almost everywhere in Ω with $\bar{y} = F\bar{x} \in AC$, where $\bar{x}$ in turn coincides almost everywhere in Ω with x. Applying the estimates (9.4) and (9.5) to formula (9.3) yields

$$\begin{aligned} \|D_i \bar{y}\|_q &\le \|a + b|\bar{x}|^{p(N-q)/q(N-p)}\|_q + \|[c + d|\bar{x}|^{N(p-q)/q(N-p)}]D_i\bar{x}\|_q \\ &\le \|a\|_q + b\|\bar{x}\|_{p(N-q)/(N-p)}^{p(N-q)/q(N-p)} + \|c\|_{pq/(p-q)}\|D_i\bar{x}\|_p + \\ d\|\bar{x}\|_{Np/(N-p)}^{(p-q)/p}\|D_i\bar{x}\|_p^{q/p} &\le C(1 + \|\bar{x}\|_{1,p}^{p(N-q)/q(N-p)}) . \end{aligned} \tag{9.7}$$

Again by Lemma 9.2, this shows that $y \in W_q^1$. $\Leftarrow$

To give a parallel result to Theorem 9.1 in the case of an absolutely continuous function f, some definitions are in order. Given a space X of measurable functions over Ω, we denote by $\bar{X}$ the space of all measurable functions x such that there exists $\bar{x} \in X$ with $x(s) = \bar{x}(s)$ almost everywhere in Ω. In particular, for $x \in \overline{AC}$ we denote by $\bar{D}_i x$ the derivative $D_i \bar{x}$, where $\bar{x} \in AC$

is equivalent to x. Note that Lemma 9.2 states that $x \in W_p^1$ if and only if $x \in \overline{AC}$ and $\bar{D}_i\, x \in L_p$. Moreover, it is easy to see that if $x \in L_1 \cap \overline{AC}$ and $\bar{D}_i\, x \in L_1$, then $\bar{D}_i\, x(s) = D_i x(s)$ almost everywhere in Ω.

Following Marcus and Mizel, we call a function f on $\Omega \times \mathbb{R}$ a *locally absolutely continuous Carathéodory function* if $f(\cdot, u)$ belongs to $\overline{AC}$ for every $u \in \mathbb{R}$, and $f(s, \cdot)$ is continuous and locally absolutely continuous for almost all $s \in \Omega$. It is clear that every such function is a Carathéodory function in the sense of Section 1.4.

Given a point $s \in \Omega$ and $h \neq 0$, we set

$$\Delta_i^h z(s) = \frac{z(s_h) - z(s)}{h}, \qquad (9.8)$$

where $s_h = s + he_i$, e_i being the i-th unit vector $(i = 1, \cdots, N)$ in $\mathbb{R}^N$. There is a well known relation between the increment (9.8) and the partial derivative $D_i z(s)$ which we state without proof in the following:

Lemma 9.3 *Let Ω' be an open subdomain of Ω with $\bar{\Omega}' \subset \Omega$, and let $h' = \text{dist}\,(\Omega', \partial\Omega)$. If $z \in L_q(\Omega)$ with $1 < q < \infty$ is such that $\|\Delta_i^h z\|' \leq C$ for $0 < h < h'$, where $\|\cdot\|'$ denotes the function norm over Ω', then also $D_i z \in L_q(\Omega)$ and $\|D_i z\|_q \leq C$. Conversely, if $z \in W_q^1(\Omega)$ then always*

$$\|\Delta_i^h z\|_q' \leq \|D_i z\|_q.$$

We are now in a position to state the second sufficient acting condition, which uses much weaker hypotheses than Theorem 9.1.

Theorem 9.2 *Suppose that f is a locally absolutely continuous Carathéodory function on $\Omega \times \mathbb{R}$, where $\Omega \subset \mathbb{R}^N$ is bounded. Let $1 < q \leq p < \infty$. Assume that the estimates*

$$|\bar{D}_i\, f(s, u)| \leq a(s) + b(u) \qquad (9.9)$$

and

$$|D_u f(s, u)| \leq c(s) + d(u) \qquad (9.10)$$

hold for almost all $s \in \Omega$ and all $u \in \mathbb{R}$, where $a \in L_q$, $b \in C(\mathbb{R})$, $c \in L_{pq/(p-q)}$, and $d \in L_1(\mathbb{R})$ are nonnegative functions. Given $x \in W_p^1$, suppose that $b \circ x \in L_q$, $d \circ x \in L_{pq/(p-q)}$, and $(d \circ x) D_i x \in L_q$ $(i = 1, \cdots, N)$. Then $y = Fx \in W_q^1$.

$\Rightarrow$ Without loss of generality we may suppose that $f(\cdot, u) \in AC$ for all $u \in \mathbb{Q}$; this implies that, for $u \in \mathbb{Q}$, we have $\bar{D}_i\, f(s, u) = D_i f(s, u)$ almost everywhere in Ω, and hence we may replace $\bar{D}_i$ by D_i in (9.9). For any fixed $u_0 \in \mathbb{R}$, the function $y_0 = Fx_0$, with $x_0(s) \equiv u_0$, belongs to $\overline{AC}$, hence $\bar{D}_i y_0 \in L_q$; by Lemma 9.2, we conclude that $y_0 \in W_q^1$. Let

$$\beta(t) = \int_0^t d(\tau) d\tau, \tag{9.11}$$

with d given in (9.9). Since $\beta \in AC(\mathbb{R})$ and, by hypothesis, $(d \circ x) D_i x \in L_q$ for $i = 1, \cdots, N$, we know that $\beta \circ x \in W_q^1$. Now, (9.10) implies that for almost all $s \in \Omega$ and all $u \in \mathbb{R}$

$$|f(s, u) - f(s, u_0)| \le |\int_0^u D_u f(s, v) dv| \tag{9.12}$$
$$\le c(s)|u - u_0| + \beta(u) - \beta(u_0).$$

Consequently, $y = Fx$ and y_0 as above satisfy $\|y - y_0\|_q \le \|c\|_{pq/(p-q)}\|x - x_0\|_p + \|\beta \circ x - \beta \circ x_0\|_q$, which shows that $y \in L_q$. On the other hand, if s' and s'' are two points in Ω which lie on some straight line parallel to the s_i-axis, then

$$|f(s', u) - f(s'', u)| \le |\int_{s_i'}^{s_i''} a(s) ds_i| + b(u)|s' - s''|, \tag{9.13}$$

by (9.9). Combining (9.12) and (9.13) and using the notation (9.8) we get

$$|\Delta_i^h y(s)| \le \frac{1}{|h|}|f(s_h, x(s_h)) - f(s, x(s_h))| + \frac{1}{|h|}|f(s, x(s_h)) - f(s, x(s))|$$
$$\le \frac{1}{|h|}\int_0^{|h|} a(s + \tau e_i) d\tau + b(x(s_h)) + c(s)|\Delta_i^h x(s)| + |\Delta_i^h \beta(x(s))|. \tag{9.14}$$

Since $x \in W_p^1$ and $\beta \circ x \in W_q^1$, it follows from Lemma 9.3 that

$$\|\Delta_i^h x\|_p' \le \|D_i x\|_p, \quad \|\Delta_i^h(\beta \circ x)\|_q' \le \|D_i(\beta \circ x)\|_q, \tag{9.15}$$

where the prime denotes the norm over Ω', $\bar{\Omega}' \subset \Omega$. Putting (9.15) into (9.14) yields

$$\|\Delta_i^h y\|_q' \le \|a\|_q + \|b \circ x\|_q + \|c\|_{pq/(p-q)}'\|D_i x\|_p + \|D_i(\beta \circ x)\|_q. \tag{9.16}$$

234 CHAPTER 9

Since $y \in L_q$, again from Lemma 9.3 we conclude that $D_i y \in L_q$, hence $y \in W_q^1$ as claimed. $\Leftarrow$

Observe that the only place where we have used the fact that $q > 1$ is the application of Lemma 9.3. In fact, by a somewhat different technique one may show that the statement of Theorem 9.2 is also true in case $q = 1$.

9.3 Necessary acting conditions in W_p^1

In the previous section we have discussed two sufficient conditions on the function f under which the corresponding superposition operator F maps the space W_p^1 into the space W_q^1. In this connection, the growth conditions (9.4) and (9.5) played a crucial role. The question arises whether these conditions are also *necessary* for F to act from W_p^1 into W_q^1. A complete (affirmative) answer has been given by M. Marcus and V.J. Mizel in the autonomous case $f = f(u)$. Interestingly, they obtained also a necessary and sufficient acting condition for the nonlinear *first-order differential operator*

$$\Phi x(s) = \phi(x(s), Dx(s)) \qquad (9.17)$$

between W_p^1 and L_q, where $Dx = \{D_1 x, \cdots, D_N x\}$. The necessity of this condition then implies rather easily the necessity of the corresponding condition for F.

Lemma 9.4 Let $\Omega = (0,1)^N$ be the open unit cube in $\mathbb{R}^N$, and let ϕ be a nonnegative Borel function on $\mathbb{R} \times \mathbb{R}^N$. Suppose that the operator (9.17) maps $W_p^1(\Omega)$ into $L_1(\Omega)$ $(1 \leq p < N)$. Then the function ϕ satisfies the growth condition

$$|\phi(u, \eta)| \leq C(1 + |u|^{Np/(N-p)} + |\eta|^p) \qquad (9.18)$$

for almost all $u \in \mathbb{R}$ and all $\eta \in \mathbb{R}^N$.

$\Rightarrow$ Let $\psi(u, \eta) = 1 + |u|^{Np/(N-p)} + |\eta|^p$, and suppose that there exists a sequence $(u_n, \eta_n) \in \mathbb{R}^{N+1}$ such that

$$\alpha_n = \frac{\phi(u_n, \eta_n)}{\psi(u_n, \eta_n)} \to \infty \quad (n \to \infty). \qquad (9.19)$$

To prove (9.18), we distinguish two cases.

<u>Case 1</u> The sequence $|u_n|^{Np/(N-p)}(1 + |\eta_n|^p)^{-1}$ is bounded away from 0, i.e.

$$0 < \beta \leq \frac{|u_n|^{Np/(N-p)}}{1 + |\eta_n|^p}. \qquad (9.20)$$

We may then assume that

$$\sum_{n=1}^{\infty} \alpha_n^{-1/N} \leq \frac{1}{2} \beta^{1/N}, \quad \sum_{n=1}^{\infty} \alpha_n^{-(N-p)/N} < \infty, \tag{9.21}$$

with α_n from (9.19). Let $\delta_n = \alpha_n^{-1/N} |u_n|^{-p/(N-p)}$; then

$$\delta = \sum_{n=1}^{\infty} \delta_n \leq \sum_{n=1}^{\infty} \alpha_n^{-1/N} \beta^{-1/N} \leq \frac{1}{2}. \tag{9.22}$$

Now we use the following geometric fact: given $\eta \in \mathbb{R}^N \setminus \{0\}$ and $\sigma > 0$, one can find a piecewise linear function $w : [-\frac{\sigma}{2}, \frac{\sigma}{2}]^N \to \mathbb{R}$, which vanishes on the boundary of $[-\frac{\sigma}{2}, \frac{\sigma}{2}]^N$, and is such that $Dw(s) \equiv \eta$ in a cube $F \subset [-\frac{\sigma}{2}, \frac{\sigma}{2}]^N$ with N-dimensional Lebesgue measure

$$\mu(F) \geq a\sigma^N > 0. \tag{9.23}$$

Further, for every cube B with dist $(B, 0) \leq \sqrt{N} a^{-1/N} \mu(B)^{1/N}$ (a from (9.23)) we may choose $\rho_n > 0$ such that $\mu(B) \leq \rho_n^N$ and

$$\frac{1}{\mu(B)} \int_B \phi((\eta, \eta_n) + u_n, \eta_n) d\eta \geq \frac{\phi(u_n, \eta_n)}{2}, \tag{9.24}$$

where (η, ζ) denotes the scalar product in $\mathbb{R}^N$. Let σ_n be the largest number of the form $\delta_n/2j$ ($j = 1, 2, \cdots$) such that $\sigma_n \leq \rho_n$. Denote by w_n ($n = 1, 2, \cdots$) the piecewise linear function on $E_n^* = [-\frac{\sigma_n}{2}, \frac{\sigma_n}{2}]^N$ corresponding to $\eta = \eta_n$ as indicated above, and denote by F_n^* the corresponding cube satisfying (9.23). One can show that w_n has the form $w_n(s) = \sum_{i=1}^{n} \eta_{n,i} s_i + b_n$ ($s \in E_n^*$), where $|b_n| \leq 2\sqrt{N} \sigma_n |\eta_n|$. Setting $\tilde{b}_n = u_n - b_n$ and $\tilde{w}_n(s) = w_n(s) + \tilde{b}_n$, we have that

$$\tilde{w}_n(s) = \begin{cases} \sum_{i=1}^{n} \eta_{n,i} s_i + \tilde{b}_n & \text{if } s \in F_n^*, \\ \tilde{b}_n & \text{if } s \in \partial E_n^*. \end{cases} \tag{9.25}$$

Now let $E_n = [-\frac{\delta_n}{2}, \frac{\delta_n}{2}]^N$ (recall that $\delta_n = \alpha_n^{-1/N} |u_n|^{-p/(N-p)}$), and let $\{E_{n,1}, \cdots, E_{n,k}\}$ be a family of cubes with no common interior points, such that each of these cubes is a translate of E_n^*; for $q = 1, \cdots, k$, denote by $s^{(q)}$ the centre of $E_{n,q}$. Define functions x_n on $E_n = E_{n,1} \cup \cdots \cup E_{n,k}$ by

$x_n(s) = \tilde{w}_n(s - s^{(q)})$, and let $F_n = (s^{(1)} + F_1^*) \cup \cdots \cup (s^{(k)} + F_k^*)$. We then have, by (9.24) and (9.25),

$$
\int_{F_n} \phi(x_n(s), Dx_n(s)) ds = k \int_{F_n^*} \phi(\tilde{w}_n(s), D\,\tilde{w}_n(s)) ds
$$

$$
\geq k\mu(F_n^*) \frac{\phi(u_n, \eta_n)}{2} \geq ka\mu(E_n^*) \frac{\phi(u_n, \eta_n)}{2} \tag{9.26}
$$

$$
= \frac{a}{2} \delta_n^N \phi(u_n, \eta_n) = \frac{a}{2\alpha_n} |u_n|^{-Np/(N-p)} \phi(u_n, \eta_n),
$$

since dist $(F_n^*, 0) \leq \sqrt{N} \mu(F_n^*)^{1/N} a^{-1/N}$. Finally, set $D_n = = [-\delta_n, \delta_n]^N$ and $D_{n,i} = \{s : \frac{\delta_n}{2} \leq |s_i| \leq \delta_n,\ |s_i| \geq \max_{1 \leq k \leq N} |s_k|\}$; consequently, $D_n = D_{n,1} \cup \cdots \cup D_{n,N} \cup E_n$. Define $\tilde{x}_n$ on D_n by

$$
\tilde{x}_n(s) = \begin{cases} x_n(s) & \text{if } s \in E_n, \\ 2\frac{b_n}{s_n}(\delta_n - |s_i|) & \text{if } s \in D_{n,i} \end{cases}
$$

$(i = 1, \cdots, N)$. Then $\tilde{x}_n$ is piecewise linear and continuous on D_n, vanishes on ∂D_n, and satisfies

$$
|D\tilde{x}_n(s)| \leq \begin{cases} |\eta_n| & \text{if } s \in E_n, \\ 2\frac{|\bar{b}_n|}{\delta_n} & \text{if } s \in D_n \setminus E_n. \end{cases} \tag{9.27}
$$

With $\tau_n = 2(\delta_1 + \cdots + \delta_{n-1}) + \delta_n$, denote by $\tilde{s}_n$ the point $(\tau_n, \frac{1}{2}, \cdots, \frac{1}{2}) \in \mathbb{R}^N$, and let $\tilde{D}_n = D_n + \tilde{s}_n$. (Observe that $\tilde{D}_n \subseteq \Omega$, since $2\delta \leq 1$, δ from (9.22).) Finally, define a function x on $\bar{\Omega}$ by

$$
x(s) = \begin{cases} \tilde{x}_n(s - \tilde{s}_n) & \text{if } s \in \tilde{D}_n, \\ 0 & \text{if } s \in \bar{\Omega} \setminus \bigcup_{n=1}^{\infty} \tilde{D}_n. \end{cases} \tag{9.28}
$$

We claim that $x \in W_p^1(\Omega)$, but $\Phi x \notin L_1(\Omega)$. By construction, the function x is continuous on $\bar{\Omega}$, except at the point $(2\delta, \frac{1}{2}, \cdots, \frac{1}{2})$, and vanishes on $\partial\Omega$. By (9.27) and the definition of x we have

$$
\int_{\tilde{D}_n} |Dx(s)|^p ds = \int_{D_n \setminus E_n} |D\tilde{x}_n(s)|^p ds + \int_{E_n} |D\tilde{x}_n(s)|^p ds
$$

$$
\leq (2^N - 1)\delta_n^N (2\sqrt{N})^p (|u_n|\delta_n^{-1} + |\eta_n|)^p + \delta_n^N |\eta_n|^p \tag{9.29}
$$

$$
\leq C_1 (\delta_n^{N-p} |u_n|^p + \delta_n^N |\eta_n|^p).
$$

But from the definition of δ_n and (9.20) we know that $\delta_n^N |\eta_n|^p \leq \delta_n^N$ $|u_n|^{Np/(N-p)}\beta^{-1} = (\alpha_n \beta)^{-1}$ and $\delta_n^{N-p}|u_n|^p = \alpha_n^{-(N-p)/N}$; hence, by (9.21) and (9.29),

$$\int_\Omega |Dx(s)|^p ds \leq C_2 \{\sum_{n=1}^\infty \alpha_n^{-(N-p)/N} + \frac{1}{\beta} \sum_{n=1}^\infty \alpha_n^{-1}\} < \infty;$$

this shows that $D_i x \in L_p$ for $i = 1, \cdots, N$, and therefore $x \in W_p^1$. On the other hand, from (9.26) we know that

$$\int_{\tilde{D}_n} \phi(x(s), Dx(s)) ds = \int_{D_n} \phi(\tilde{x}_n(s), D\tilde{x}_n(s)) ds$$
$$\geq \int_{F_n} \phi(x_n(s), Dx_n(s)) ds \geq \frac{a}{2\alpha_n} |u_n|^{-Np/(N-p)} \phi(u_n, \eta_n),$$

hence, by (9.19),

$$\int_\Omega \Phi x(s) ds \geq \frac{a}{2} \sum_{n=1}^\infty \frac{1}{\alpha_n} |u_n|^{-Np/(N-p)} \phi(u_n, \eta_n)$$
$$= \frac{a}{2} \sum_{n=1}^\infty |u_n|^{-Np/(N-p)} \psi(u_n, \eta_n)$$
$$\geq \frac{a}{2} \sum_{n=1}^\infty |u_n|^{-Np/(N-p)} (1 + |u_n|^{Np/(N-p)}) = \infty;$$

this shows that $\Phi x \notin L_1$, contradicting the hypothesis $\Phi(W_p^1) \subseteq L_1$.

<u>Case 2</u>   The sequence $|u_n|^{Np/(N-p)}(1 + |\eta_n|^p)^{-1}$ is bounded from above, and $|\eta_n|^p$ is bounded away from 0, i.e.

$$\frac{|u_n|^{Np/(N-p)}}{1 + |\eta_n|^p} \leq \gamma < \infty, \quad 0 < \beta \leq |\eta_n|^p. \tag{9.30}$$

In this case we set $\delta_n = \alpha_n^{-1/N} |\eta_n|^{-p/N}$, such that again $\delta = \sum_{n=1}^\infty \delta_n \leq \frac{1}{2}$. Now we have $\delta_n^{N-p}|u_n|^p = (\delta_n^N |u_n|^{Np/(N-p)})^{(N-p)/N} \leq (\gamma \delta_n^N [1 + |\eta_n|^p])^{(N-p)/N} \leq C\alpha_n^{-(N-p)/N}$. Let x be again defined by (9.28). We then get, on the one hand,

$$\int_\Omega |Dx(s)|^p ds \leq C\{\sum_{n=1}^\infty \alpha_n^{-(N-p)/N} + \sum_{n=1}^\infty \alpha_n^{-1}\} < \infty,$$

and, on the other, as before,

$$\int_\Omega \Phi x(s)ds \geq \frac{a}{2}\sum_{n=1}^{\infty}\frac{1}{\alpha_n}|u_n|^{-Np/(N-p)}\phi(u_n,\eta_n)$$

$$= \frac{a}{2}\sum_{n=1}^{\infty}|u_n|^{-Np/(N-p)}\psi(u_n,\eta_n)$$

$$\geq \frac{a}{2}\sum_{n=1}^{\infty}|u_n|^{-Np/(N-p)}(1+|u_n|^{Np/(N-p)}) = \infty,$$

again contradicting the hypothesis $\Phi(W_p^1) \subseteq L_1$.

The fact that it suffices to consider the Cases 1 and 2 may be seen as follows. By the contradiction reached in the first case, (9.19) implies that $(1+|\eta_n|^p)|u_n|^{-Np/(N-p)} \to \infty$ as $n \to \infty$. Now there are two possibilities: either we find a subsequence of η_n which is bounded away from 0; in this case (9.30) is satisfied and we arrive at the contradicition of the second case; or η_n tends to 0; in this case both sequences u_n and η_n are bounded contradicting (9.19).

This completes the proof. $\Leftarrow$

Lemma 9.4 allows us to obtain in a fairly straightforward manner the following necessary acting condition:

Theorem 9.3   *Let $\Omega \subset \mathbb{R}^N$ be bounded, and let $f = f(u)$ be a Borel function on $\mathbb{R}$. Suppose that the superposition operator F generated by f maps $W_p^1(\Omega)$ into $W_q^1(\Omega)$ ($1 \leq q \leq p < N$). Then the function f is locally Lipschitz, and its derivative satisfies the growth condition*

$$|f'(u)| \leq a + b|u|^{N(p-q)/q(N-p)} \tag{9.31}$$

for almost all $u \in \mathbb{R}$.

$\Rightarrow$ Without loss of generality we may assume that Ω is the unit cube as in Lemma 9.4. By choosing special linear functions x and using that $F(W_p^1) \subseteq W_q^1$ one sees that f is locally absolutely continuous. As in the previous section, we may then conclude that f is locally Lipschitz, and f' is a Borel function (up to "corrections" on null sets). Now, if we define a map ϕ on $\mathbb{R} \times \mathbb{R}^N$ by $\phi(u,\eta) = |f'(u)\eta|^q$, then the corresponding differential operator (9.17) maps, by hypothesis, the space $W_p^1(\Omega)$ into $L_1(\Omega)$. By Lemma 9.4, the growth condition $|f'(u)\eta|^q \leq C(1+|u|^{Np/(N-p)} + |\eta|^p)$ holds for almost all $u \in \mathbb{R}$ and all $\eta \in \mathbb{R}^N$. Writing

$$|f'(u)|^q \leq C(1+|u|^{Np/(N-p)} + |\eta|^p)|\eta|^{-q}, \tag{9.32}$$

a trivial calculation shows that the minimum of the right-hand side of (9.32) with respect to $|\eta|$ is achieved for $|\eta| \sim |u|^{N/(N-p)}$, and (9.31) follows easily. $\Leftarrow$

We remark that there is a parallel result in the case $p > N$; for sake of completeness, we state this without proof.

Theorem 9.4   *Suppose that the hypotheses of Theorem 9.3 are satisfied, where $1 \leq q \leq p$ and $p > N$. Then the function f is locally Lipschitz, and its derivative satisfies*

$$|f'(u)| \leq m(r) \quad (|u| \leq r) \tag{9.33}$$

for almost all $u \in \mathbb{R}$.

The Theorems 9.1, 9.3 and 9.4 together give a complete characterization of superposition operators which act from W_p^1 into W_q^1 and are generated by locally Lipschitz continuous maps $f = f(u)$. Unfortunately, the problem of finding a necessary and sufficient acting condition in the non-autonomous case $f = f(s, u)$ is open. There is an interesting counterexample, again due to Marcus and Mizel, which shows that the condition $F(W_p^1) \subseteq L_q$ does not imply a growth condition of Krasnosel'skij type (3.19) with $a \in L_q$ and $b \geq 0$, and hence the condition $F(W_p^1) \subseteq L_q$ does *not* imply the condition $F(L_p) \subseteq L_q$. In fact, for $\Omega = (0, 1)$, $\varepsilon > 0$, and $p > 1$ let

$$f(s, u) = \begin{cases} (\frac{3}{2} - \alpha)(\alpha - \frac{1}{2})s^{-1} & \text{if } u = \alpha s^\varepsilon \text{ for } \frac{1}{2} < \alpha < \frac{3}{2}, \\ 0 & \text{otherwise.} \end{cases} \tag{9.34}$$

Given $x \in W_p^1$, the function $y = Fx$ is obviously in $L_q(\delta, 1)$ for every $\delta \in (0, 1)$; on the other hand, every function $x \in W_p^1$ (with $x(0) = 0$) fulfills $|x(s)| = O(s^{1-1/p})$, and hence the set of all $s \in (0, \delta)$ with $y(s) = f(s, x(s)) \neq 0$ becomes small enough, for ε sufficiently close to 0, to guarantee that $y \in L_q(0, \delta)$ as well. This shows that F maps W_p^1 into L_q; on the other hand, the function (9.34) may not be bounded by a sum of an L_q function and a function of u.

9.4 Boundedness and continuity conditions in W_p^1

In the preceding section we have obtained an acting condition for the superposition operator F between W_p^1 and W_q^1 which is both necessary and sufficient in the autonomous case $f = f(u)$. The explicit form of this condition (see (9.31) and (9.33)) implies that, whenever F acts between W_p^1 and W_q^1 in the autonomous case, then F is bounded and continuous.

Unfortunately, since we are also interested in the non-autonomous case $f = f(s, u)$, we cannot use the conditions of Theorems 9.3 and 9.4 directly. Nevertheless, it turns out that appropriate growth conditions on the partial derivatives $D_i f$ and $D_u f$ ensure, by the Sobolev imbedding theorems, not only the acting $F(W_p^1) \subseteq W_q^1$, but also the boundedness and continuity of F.

In deriving the main result of Section 9.3, the first-order differential operator (9.17) was useful. For the ease of the reader, we state a sufficient acting condition for the *non-autonomous* differential operator

$$\Phi x(s) = \phi(s, x(s), Dx(s)) \tag{9.35}$$

which is a rather straightforward consequence of Lemma 9.1.

Lemma 9.5   *Let $1 \le p, q < \infty$, and let ϕ be a Carathéodory function on $\Omega \times \mathbb{R} \times \mathbb{R}^N$ which satisfies the following growth conditions. For $p < N$, suppose that*

$$|\phi(s, u, \eta)| \le a(s) + b|u|^{Np/q(N-p)} + c|\eta|^{p/q} \tag{9.36}$$

for some $a \in L_q$ and $b, c \ge 0$; for $p = N$, suppose that

$$|\phi(s, u, \eta)| \le a(s) + b|u|^{r/q} + c|\eta|^{p/q} \tag{9.37}$$

for some $a \in L_q$ and $b, c \ge 0$, where $r \ge 1$ is arbitrary; for $p > N$, suppose that

$$|\phi(s, u, \eta)| \le b(u)[a(s) + c|\eta|^{p/q}] \tag{9.38}$$

for some $a \in L_q$, $b \in C(\mathbb{R})$, and $c \ge 0$. Then the differential operator (9.35) generated by ϕ maps the space W_p^1 into the space L_q and is bounded and continuous.

Now, if f is some Carathéodory function on $\Omega \times \mathbb{R}$ such that the chain rule (9.3) holds for $i = 1, \cdots, N$, then Lemma 9.5 may be applied to the functions $\phi_i(s, u, \eta) = D_i f(s, u) + D_u f(s, u)\eta_i$ $(i = 1, \cdots, N)$; however, this procedure is not very explicit. It turns out that one can give rather natural growth conditions on $D_i f$ and $D_u f$ separately which imply the boundedness and continuity of F and, moreover, yield an explicit estimate for the growth function (2.24) of F between W_p^1 and W_q^1. This will be carried out in the following:

Theorem 9.5   *Let $\Omega \subset \mathbb{R}^N$ be bounded, let $1 < q \le p < \infty$, and let f be a locally absolutely continuous Carathéodory function on $\Omega \times \mathbb{R}$. Suppose that* the partial derivatives $\bar{D}_i f$ $(i = 1, \cdots, N)$ generate superposition operators from L_r into L_q, and the partial derivative $D_u f$ a superposition operator from L_r into $L_{pq/(p-q)}$, where $r = Np/(N-p)$ for $p < N$, $r \geq 1$ arbitrary for $p = N$, and $r = \infty$ for $p > N$. Then the superposition operator F generated by f maps W_p^1 into W_q^1 and is bounded and continuous.

$\Rightarrow$ We restrict ourselves to the case $p < N$; the other cases are treated similarly by means of the corresponding imbeddings. By a reasoning similar to that used in Section 9.2, one may show that $y = Fx$ satisfies the generalized chain rule $D_i y(s) = \bar{D}_i f(s, x(s)) + D_u f(s, x(s)) D_i x(s)$, where $\bar{D}_i z(s) = D_i \bar{z}(s)$, with $\bar{z}$ being a locally absolutely continuous function which coincides almost everywhere in Ω with z.

To prove the boundedness of F, fix $x \in W_p^1$ with $\|x\|_{1,p} \leq r$. The assumptions on $D_i f$ and $D_u f$ imply that

$$|\bar{D}_i f(s, u)| \leq a(s) + b|u|^{Np/q(N-p)}$$

and

$$|D_u f(s, u)| \leq c(s) + d|u|^{N(p-q)/q(N-p)},$$

where $a \in L_q$, $c \in L_{pq/(p-q)}$, and $b, d \geq 0$. As in Theorem 9.1 (see (9.7)), we have $\|D_i y\|_q \leq C(1 + r^{p(N-q)/q(N-p)})$; similarly (9.6) shows that

$$\|y - F\theta\|_q \leq \|c\|_{pq/(p-q)} \|x\|_p + d\|x\|_{p(N-q)/(N-p)}^{p(N-q)/q(N-p)}. \tag{9.39}$$

By Lemma 9.1 (a), W_p^1 is continuously imbedded in $L_{p(N-q)/(N-q)}$; thus (9.39) implies that $\|y\|_q \leq \|F\theta\|_q + \|c\|_{pq/(p-q)} r + dr^{p(N-q)/q(N-p)}$. Altogether, the growth function (2.24) of F between $X = W_p^1$ and $Y = W_q^1$ satisfies

$$\mu_F(r) \leq C(1 + r^{p(N-q)/q(N-p)}). \tag{9.40}$$

Note that (9.40) gives a *linear* estimate if and only if $p = q$; this is similar to the Lebesgue space situation discussed in detail in Chapter 3.

The proof of the continuity of F is a consequence of the assumptions on $D_i f$ and $D_u f$, and the fact that superposition operators between Lebesgue spaces are always continuous (see Theorem 3.7). $\Leftarrow$

We point out that Theorem 9.5 is true in the case $q = 1$ as well, with some technical modifications.

When dealing with applications of the superposition operator F between Sobolev spaces, it is sometimes not necessary to require the continuity of F, but only its *demi-continuity*, which means that the convergence of a sequence x_n in W_p^1 to some $x_0 \in W_p^1$ implies the *weak* convergence of Fx_n to Fx_0 in W_q^1. It turns out that the general growth conditions given in Theorem 9.2, together with certain acting conditions for the functions b and d in (9.9) and (9.10), respectively, are sufficient to ensure the demi-continuity of F.

Theorem 9.6 *Suppose that f is a locally absolutely continuous Carathéodory function on $\Omega \times \mathbb{R}$, where $\Omega \subset \mathbb{R}^N$ is bounded. Let $1 < q \le p < \infty$, and assume that the estimates (9.9) and (9.10) hold for almost all $s \in \Omega$ and all $u \in \mathbb{R}$, where $a \in L_q$, $c \in L_{pq/(p-q)}$, and b and d are continuous nonnegative functions which generate superposition operators from L_r into L_q and from L_r into $L_{pq/(p-q)}$, respectively, where $r = Np/(N-p)$ for $p < N$ and $r \ge p$ arbitrary for $p \ge N$. Then the superposition operator F generated by f is demi-continuous between W_p^1 and W_q^1.*

$\Rightarrow$ Assume, without loss of generality, that $x_n \in W_p^1$ converges to θ in W_p^1, $f(s,0) \equiv 0$ for almost all $s \in \Omega$, and $p < N$. The latter assumption means, by Theorem 3.1, that

$$|b(u)| \le C(1 + |u|^{Np/q(N-p)}) \tag{9.41}$$

and

$$|d(u)| \le C(1 + |u|^{N(p-q)/q(N-p)}). \tag{9.42}$$

We have to show that $y_n = Fx_n$ converges weakly to θ in W_q^1. For β defined as in (9.11) we have $\|y_n\|_q \le \|c\|_{pq/(p-q)}\|x_n\|_p + \|\beta \circ x_n - \beta \circ \theta\|_q$. Since, by (9.42),

$$\|\beta \circ x_n - \beta \circ \theta\|_q \le C[\|x_n\|_q + \|x_n\|_{p(N-q)/(N-p)}^{p(N-q)/q(N-p)}],$$

the sequence y_n converges in L_q to θ. Since the space L_q is reflexive for $q > 1$, it suffices to show the boundedness of the sequence $D_i y_n$ $(i = 1, \cdots, N)$ in L_q. For it then follows that $D_i y_n \to \theta$ weakly in L_q, and hence $y_n \to \theta$ weakly in W_q^1. By the chain rule and the definition of β, we have $D_i(\beta \circ x_n)(s) = d(x_n(s))D_i x_n(s)$, hence $\|D_i(\beta \circ x_n)\|_q \le \|d \circ x_n\|_{pq/(p-q)}\|D_i x_n\|_p$. Using the estimate (9.16) for $x = x_n$ and $y = y_n$, Lemma 9.3 for an open subdomain Ω' with $\bar{\Omega}' \subset \Omega$, and the growth conditions (9.41) and (9.42), we get

$$\|D_i y_n\|_q' \le \|a\|_q + \|b \circ x_n\|_q + \|c\|_{pq/(p-q)}'\|D_i x_n\|_p$$
$$+ \|D_i(\beta \circ x_n)\|_q \le \|a\|_q + C[1 + \|x_n\|_{Np/(N-p)}^{Np/q(N-p)}]$$
$$+ \|c\|_{pq/(p-q)}'\|x_n\|_{1,p} + C[1 + \|x_n\|_{Np/(N-p)}^{N(p-q)/q(N-p)}]\|x_n\|_{1,p}.$$

This shows that the sequence $D_i y_n$ is bounded in L_q for $i = 1, \cdots, N$, and so we are done. $\Leftarrow$

9.5 Boundedness and continuity conditions in W_p^k

In view of applications to higher–order (especially second order) partial differential equations, it would be very useful to generalize the necessary and sufficient conditions under which the superposition operator F acts in the Sobolev space W_p^k (especially W_p^2) and has the necessary properties. Unfortunately, only sufficient conditions are known up to the present. We begin by stating a sufficient acting condition for a nonlinear *higher-order differential operator*

$$\Phi x(s) = \phi(s, x(s), Dx(s), \cdots, D^k x(s)) \qquad (9.43)$$

$(D^k x = \{D^\alpha x : |\alpha| = k\})$ between W_p^k and L_q. To this end, we use the following notation. If $N(j) = (N + j - 1)![j!(N - 1)!]^{-1}$ denotes the number of all multi-indices α with $|\alpha| = j$ $(j = 0, 1, \cdots, k)$, the function ϕ in (9.43) is defined on $\Omega \times \prod_{j=0}^{k} \mathbb{R}^{N(j)}$, and we write $\phi = \phi(s, \xi)$, where $\xi = (\xi_0, \xi_1, \cdots, \xi_k)$ with $\xi_j = \{\xi_\alpha : |\alpha| = j\}$ $(j = 0, 1, \cdots, k)$. For a fixed multi-index α with $0 \le |\alpha| \le k$, let

$$r(\alpha) \begin{cases} = \frac{Np}{N-(k-|\alpha|)p} & \text{if } |\alpha| > k - N/p, \\ \ge 1 \text{ arbitrary} & \text{if } |\alpha| = k - N/p. \end{cases} \qquad (9.44)$$

The following is a direct generalization of Lemma 9.5.

Lemma 9.6   Let $\Omega \subset \mathbb{R}^N$ be bounded, let ϕ be a Carathéodory function on $\Omega \times \mathbb{R}^{N(0)} \times \cdots \times \mathbb{R}^{N(k)}$, and let $1 \le p, q < \infty$. Suppose that the function ϕ satisfies the growth condition

$$|\phi(s, \xi)| \le c(\sum_{0 \le |\alpha| < k-N/p} |\xi_\alpha|)\{a(s) + b \sum_{k-N/p \le |\alpha| \le k} |\xi_\alpha|^{r(\alpha)/q}\}, \qquad (9.45)$$

where $a \in L_q(\Omega)$, $b \ge 0$, $c \in C([0, \infty))$, and $r(\alpha)$ is given by (9.44). Then the differential operator (9.43) generated by ϕ maps the space W_p^k into the space L_q and is bounded and continuous.

The proof of Lemma 9.6 builds on the imbedding lemma 9.1 and is somewhat cumbersome, but straightforward. In the case $k = 1$, $q = 1$, and $1 \le p < N$, in particular, we have $k - N/p < 0$, $r(\alpha) = Np/(N - p)$ for $|\alpha| = 0$, and $r(\alpha) = p$ for $|\alpha| = 1$, and thus (9.45) coincides precisely with (9.18). More generally, (9.45) contains all three conditions (9.36), (9.37) and (9.38) in case $k = 1$. In the case $k = 2$, $q = 1$, and $1 \leq p < N/2$, on the other hand, we have $k - N/p < 0$, $r(\alpha) = Np/(N - 2p)$ for $|\alpha| = 0$, $r(\alpha) = Np/(N - p)$ for $|\alpha| = 1$, and $r(\alpha) = p$ for $|\alpha| = 2$, and thus (9.45) becomes

$$|\phi(s, u, \eta, \zeta)| \leq a(s) + b_0|u|^{Np/(N-2p)} + b_1|\eta|^{Np/(N-p)} + b_2|\zeta|^p \qquad (9.46)$$

($s \in \Omega$, $u \in \mathbb{R}$, $\eta \in \mathbb{R}^N$, $\zeta \in \mathbb{R}^{N(N+1)/2}$). Unfortunately, it is not clear whether or not (9.46) is also necessary for the differential operator

$$\Phi x(s) = \phi(s, x(s), Dx(s), D^2(x(s))) \qquad (9.47)$$

to map W_p^2 into L_1, where $D^2 x = \{D_{ij} x : i, j = 1, \cdots, N\}$.

In order to get (sufficient) acting conditions for the superposition operator F between the Sobolev spaces W_p^k and W_q^k, we begin with the simplest case when F is a multiplication operator generated by the function

$$f(s, u) = a(s)u \qquad (9.48)$$

with a fixed multiplicator $a \in W_r^k$.

Lemma 9.7   *Let $\Omega \subset \mathbb{R}^N$ be bounded, $1 \leq q \leq p < \infty$, $r \geq q$, and*

$$\frac{k}{N} > \frac{1}{p} - \frac{1}{q} + \frac{1}{r}. \qquad (9.49)$$

Then the superposition operator F generated by the function (9.48) with $a \in W_r^k(\Omega)$ maps $W_p^k(\Omega)$ into $W_q^k(\Omega)$ and is bounded and continuous, hence

$$\|Fx\|_{k,q} \leq C\|a\|_{k,r}\|x\|_{k,p}. \qquad (9.50)$$

$\Rightarrow$ We prove the lemma by induction on k. Let first $k = 1$, hence $\frac{1}{N} + \frac{1}{q} > \frac{1}{p} + \frac{1}{r}$. We distinguish the following four cases.

    <u>Case 1</u>   If $p \leq N$ and $r \leq N$, then $p > q$ and $r > q$, and, by Lemma 9.1, $W_p^1 \subseteq L_{qr/(r-q)}$ and $W_r^1 \subseteq L_{pq/(p-q)}$. Since $D_i x \in L_p$ and $D_i a \in L_r$, by Hölder's inequality we get for $y = Fx$

$$D_i y = aD_i x + xD_i a \in L_q \qquad (9.51)$$

for $i = 1, \cdots, N$, hence $y \in W_q^1$.

<u>Case 2</u> If $p \leq N$ and $r > N$, then $p = q$, and, by Lemma 9.1 again, $W_p^1 \subseteq L_{qr/(r-q)}$ and $W_r^1 \subseteq L_\infty$. Thus (9.51) follows again from $D_i x \in L_p$ and $D_i a \in L_r$.

<u>Case 3</u> If $p > N$ and $r \leq N$, the result follows as in Case 2, by symmetry.

<u>Case 4</u> If $p > N$ and $r > N$, then necessarily $p = q = r$, and (9.51) follows from the fact that $W_p^1 \subseteq L_\infty$.

Suppose now that the statement is proved for fixed k; we show that $F(W_p^{k+1}) \subseteq W_q^{k+1}$, provided that $a \in W_r^{k+1}$, with $\frac{k+1}{N} + \frac{1}{q} > \frac{1}{p} + \frac{1}{r}$. Consider first the case when $kp \leq N$ and $kr \leq N$, and set $p_* = Np/(N - p)$ and $r_* = Nr/(N - r)$.
Again by Lemma 9.1, we then have $x \in W_{p_*}^k$ and $a \in W_{r_*}^k$; moreover, $D_i x \in W_p^k$ and $D_i a \in W_r^k$, by hypothesis $(i = 1, \cdots, N)$. Now, our choice of p_* and r_* implies that $\frac{k}{N} + \frac{1}{q} > \max\left\{\frac{1}{p} + \frac{1}{r_*}, \frac{1}{p_*} + \frac{1}{r}\right\}$. By the induction hypothesis, both products $a D_i x$ and $x D_i a$ belong to W_q^k, and thus $y = ax \in W_q^{k+1}$ as claimed. The case when $kp > N$ or $kr > N$ is treated similarly.
The formula (9.50) is proved by approximating the functions a and x by appropriate sequences $a_n \in C^\infty(\Omega) \cap W_r^k(\Omega)$ and $x_n \in C^\infty(\Omega) \cap W_p^k(\Omega)$, respectively, and calculating the derivatives of $y_n = a_n x_n$ by means of the classical chain rule. $\Leftarrow$

Lemma 9.7 states that, in case of a bounded domain $\Omega \subset \mathbb{R}^N$, the space $W_r^k(\Omega)$ is continuously imbedded in the multiplicator space $W_q^k(\Omega)/W_p^k(\Omega)$ (see (2.40)), provided (9.49) holds. We point out that this is true also for unbounded domains Ω under the additional requirement that $kp \leq N$ if $r \neq q$, that $kr \leq N$ if $p \neq q$, and that (9.49) is replaced by

$$\frac{k-1}{N} \leq \frac{1}{p} - \frac{1}{q} + \frac{1}{r} < \frac{k}{N}$$

if $r \neq q$ and $p \neq q$.

Theorem 9.7   Let $\Omega \subset \mathbb{R}^N$ be bounded, and let $kp \geq N$. Suppose that the function f is k-times continuously differentiable on $\bar{\Omega} \times \mathbb{R}$. Then the superposition operator F generated by f maps $W_p^k(\Omega)$ into itself and is bounded and continuous.

$\Rightarrow$ The proof proceeds again by induction on k. In case $k = 1$ the space $W_p^1(\Omega)$ is, by Lemma 9.1 (c), imbedded in the space $C(\bar{\Omega})$. Since the function f is C^1 on $\bar{\Omega} \times \mathbb{R}$, all partial derivatives $D_i f$ $(i = 1, \cdots, N)$ and $D_u f$ are continuous on $\bar{\Omega} \times \mathbb{R}$, and thus it suffices to apply the chain rule (9.3) to $y = Fx$ to ensure that $y \in W_p^1(\Omega)$.

Suppose now that the assertion holds for fixed k, and let f be of class C^{k+1} on $\bar{\Omega} \times \mathbb{R}$, where $(k+1)p \geq N$. We distinguish three cases.

Case 1 If $p > N$, then the functions $D_i f(s, x(s))$ and $D_u f(s, x(s))$ belong to W_p^k for each $x \in W_p^{k+1}$ (even for each $x \in W_p^k$), by the induction hypothesis. Moreover, from Lemma 9.7 (for $p = q = r$) it follows that the product $D_u f(s, x(s)) D_i x(s)$ also belongs to W_p^k for $i = 1, \cdots, N$. Thus, $D_i y = D_i F x \in W_p^k$, hence $y = F x \in W_p^{k+1}$.

Case 2 If $p = N$, fix $r > N$ and observe that $W_N^{k+1} \subseteq W_r^k$, by Lemma 9.1 (a). Again by the induction hypothesis, the functions $D_i f(s, x(s))$ and $D_u f(s, x(s))$ belong to W_r^k for each $x \in W_N^{k+1}$ (even for each $x \in W_r^k$). By Lemma 9.7, applied to $p = q = N$, we get that also the product $D_u f(s, x(s)) D_i x(s)$ belongs to W_N^k for $i = 1, \cdots, N$, and the statement follows as above.

Case 3 If $p < N$, the hypothesis $(k+1)p \geq N$ implies that $\frac{1}{p} - \frac{1}{N} < \frac{k+1}{N} - \frac{1}{N} = \frac{k}{N}$, and thus $W_p^{k+1} \subseteq W_{Np/(N-p)}^k$, by Lemma 9.1 (a). Applying again Lemma 9.7 to $p = q$ and $r = Np/(N-p)$, we conclude that $D_i y = D_i F x \in W_p^k$, hence $y = F x \in W_p^{k+1}$.

The continuity of F follows from Lemma 9.7 as well. The boundedness may be proved by a straightforward calculation. Moreover, one may obtain an estimate for the growth function (2.24) of F, in terms of the partial derivatives of F, which is analogous to that obtained in the space C^k (see e.g. Theorem 8.2). $\Leftarrow$

In most of the preceding statements we restricted ourselves to the case of a *bounded* domain $\Omega \subset \mathbb{R}^N$. Now we shall give two results in the case when Ω is the whole space $\mathbb{R}^N$. It seems reasonable that in this case the restrictions on the growth of the function f have to be strengthened. We will do so by requiring that $f = f(u)$ is a C^∞ function which has either a compact support, or whose derivatives are rapidly decreasing (see (9.52) below). The first result is a direct consequence of Lemma 9.1 (see also Theorem 9.7) and will be stated without proof.

Theorem 9.8 *Let $f = f(u)$ be a C^∞-function with compact support in $\mathbb{R}$. Let $p \geq N/k$. Then the superposition operator F generated by f maps $W_p^k(\mathbb{R}^N)$ into itself and is bounded and continuous.*

Of course, for applying Lemma 9.1 to the various terms that arise in estimating the L_p norms of the derivatives $D^\alpha y$ of $y = Fx$, the assumption $p \geq N/k$ is crucial. The question arises what happens if $p < N/k$. The following theorem gives a partial answer in case $k = 2$. As we shall see later (Theorems 9.10 and 9.11), this is not accidental. Here the function f is supposed to be defined only on the half-axis $[0, \infty)$, and thus the corresponding superposition operator F applies only to functions in the "cone"

$$W_{p,+}^k(\mathbf{R}^N) = \{x : x \in W_p^k(\mathbf{R}^N),\ x \geq 0 \text{ almost everywhere in } \mathbf{R}^N\}.$$

Theorem 9.9   Let $f = f(u)$ be a C^∞ function on $[0, \infty)$ which satisfies the growth condition

$$\sup_{u>0} |u^{j-1} f^{(j)}(u)| \leq M < \infty \quad (j = 0, 1, \cdots). \tag{9.52}$$

Let $1 < p < N/2$. Then the superposition operator F generated by f maps $W_{p,+}^2(\mathbf{R}^N)$ into $W_p^2(\mathbf{R}^N)$ and is bounded and continuous.

$\Rightarrow$ Given $x \in W_{p,+}^2$, we have to show that the second-order derivatives of $y = Fx$, i.e.

$$D_{ij}y(s) = f'(x(s))D_{ij}x(s) + f''(x(s))D_ix(s)D_jx(s) \tag{9.53}$$

belong to $L_p = L_p(\mathbf{R}^N)$. For the first term in (9.53), this is a direct consequence of (9.52) and the fact that $D_{ij}x \in L_p$; it is the second term that requires attention. By Hölder's inequality, it suffices to show that

$$\int_{\mathbf{R}^N} |f''(x(s))|^p |D_ix(s)|^{2p} ds < \infty. \tag{9.54}$$

To prove (9.54), let first $x \in C_0^\infty(\mathbf{R}^N)$, $x \geq 0$ almost everywhere in $\mathbf{R}^N$, and consider the integral

$$I = \int_{\mathbf{R}^N} \frac{|D_ix(s))|^{2p}}{[1 + x(s)]^p} ds.$$

Integrating by parts we get

$$I = \int_{\mathbf{R}^N} \frac{\text{sgn } D_ix(s)|D_ix(s)|^{2p-1} D_i[1 + x(s)]}{[1 + x(s)]^p} ds$$

$$= -(2p - 1) \int_{\mathbf{R}^N} \frac{|D_ix(s)|^{2p-2} D_{ii}x(s)}{[1 + x(s)]^{p-1}} ds$$

$$+ p \int_{\mathbf{R}^N} \frac{\text{sgn } D_ix(s)|D_ix(s)|^{2p-1} D_ix(s)}{[1 + x(s)]^p} ds.$$

Since the last integral is just I again we conclude that

$$I = \frac{2p - 1}{p - 1} \int_{\mathbf{R}^N} \frac{|D_ix(s)|^{2p-2} D_{ii}x(s)}{[1 + x(s)]^{p-1}} ds;$$

by Hölder's inequality we get

$$I \leq \frac{2p-1}{p-1} I^{1-1/p} \|D_{ii}x\|_p^p,$$

hence

$$I = \int_{\mathbb{R}^N} \frac{|D_i x(s)|^{2p}}{[1+x(s)]^p} ds \leq C \|D_{ii}x\|_p^p,$$

where the constant C depends on p, but not on x. Now approximating $x \in W_{p,+}^2$ by a sequence $x_n \in C_0^\infty$ and replacing x by x/ε $(\varepsilon > 0)$ gives

$$\int_{\mathbb{R}^N} \frac{|D_i x(s)|^{2p}}{[\varepsilon + x(s)]^p} ds \leq C \|D_{ii}x\|_p^p;$$

finally, letting ε tend to 0 we conclude that

$$\int_{\mathbb{R}^N} |f''(x(s))|^p |D_i x(s)|^{2p} ds \leq M^p \int_{\mathbb{R}^N} \frac{|D_i x(s)|^{2p}}{x(s)^p} ds \leq M^p C \|D_{ii}x\|_p^p < \infty,$$

and this is (9.54). The boundedness and continuity of F follows from the fact that $\|Fx\|_{2,p}$ is linearly bounded by $\|x\|_{2,p}$. $\Leftarrow$

The Theorems 9.8 and 9.9 give sufficient acting conditions for F in the space $W_p^k(\mathbb{R}^N)$ only in case $p \geq N/k$ and $k \geq 1$, on the one hand, and $1 < p < N/k$ and $k = 2$, on the other. We shall see later (Theorems 9.10 and 9.11) that these are essentially the only cases for which $F(W_p^k(\mathbb{R}^N)) \subseteq W_p^k(\mathbb{R}^N)$.

9.6 Degeneracy results

In previous chapters we have seen that, if the superposition operator F acts in some function space X, rather mild regularity conditions on F may lead to a strong degeneracy of the corresponding function f. To recall some examples, the differentiability of F in $X = L_p$ (Theorem 3.12) or $X = L_M$ (Theorem 4.11), or the (global) Lipschitz continuity of F in $X = NBV$ (Theorem 6.15), $X = H_\phi$ (Theorem 7.8), $X = C^1$ (Theorem 8.4), $X = H_\phi^1$ (Theorem 8.7), or $X = R_\mu(L)$ (Theorem 8.10), lead necessarily to affine functions $f(s,u) = a(s) + b(s)u$. We shall see now that merely the fact that F *acts* in the space $W_p^k(\mathbb{R}^N)$ in the cases not covered by Theorems 9.8 and 9.9, leads to the degeneracy

$$f(u) = a + bu. \tag{9.55}$$

The following result complements Theorem 9.8.

Theorem 9.10 *Let $f = f(u)$ be a C^∞ function with compact support in $\mathbb{R}$. Let $1 \leq p < N/k$ and $k \geq 3$. Suppose that the superposition operator F generated by f maps $W_p^k(\mathbb{R}^N)$ into itself. Then f has the form (9.55).*

$\Rightarrow$ Suppose first that $f(0) = 0$, and let φ be a C^∞ function with compact support in $\mathbb{R}^N$ such that $\varphi(s) = \varphi(s_1, \cdots, s_N) = s_1$ for $|s| \leq 1$, and $\varphi(s) \equiv 0$ for $|s| \geq 2$. Fix a sequence $\sigma_n \in \mathbb{R}^N$ with $|\sigma_i - \sigma_j| \geq 10$ for $i \neq j$, and consider the function $x = \sum_{n=1}^\infty \varphi_n$, where $\varphi_n(s) = n^\beta \varphi[n^\alpha(s - \sigma_n)]$ $(n = 1, 2, \cdots)$, with $\alpha = \frac{2}{N-kp}$ and $\beta = \frac{1}{kp-1}$. We claim that $x \in W_p^k$. In fact, $x \in C^\infty(\mathbb{R}^N)$ and

$$\sum_{|\alpha| \leq k} \int_{\mathbb{R}^N} |D^\alpha x(s)|^p ds \leq C \sum_{n=1}^\infty n^{\beta p} n^{-\alpha(N-kp)} = C \sum_{n=1}^\infty n^{(p-2kp+2)/(kp-1)} < \infty,$$

since $(p - 2kp + 2)/(kp - 1) < -1$. We shall now prove that f must be a polynomial of degree less than k. If this is not the case, we find some interval $[a, b] \subset \mathbb{R}$ such that $|f^{(k)}(u)| \geq \gamma > 0$ for $a \leq u \leq b$. Let

$$S_n = \{s : s \in \mathbb{R}^N, |s - \sigma_n| < n^{-\alpha}, an^{-(\alpha+\beta)} < s_1 - \sigma_{n,1} < bn^{-(\alpha+\beta)}\}.$$

By the definition of α and β, the area of the circular section S_n is bigger than the constant times $n^{-(N\alpha+\beta)}$ for n large enough. Moreover, since the functions φ_n are disjoint it follows that $x = \varphi_n$ on S_n; consequently, $y = Fx$ satisfies

$$\int_{\mathbb{R}^N} |\frac{\partial^k}{\partial s_1^k} y(s)|^p ds \geq C \sum_{n=1}^\infty |S_n| n^{\beta pk} n^{\alpha pk}$$

$$= C \sum_{n=1}^\infty n^{-(N\alpha+\beta)+(\alpha+\beta)pk} = C \sum_{n=1}^\infty n^{-1} = \infty,$$

and thus y cannot belong to W_p^k. This shows that f is a polynomial of degree at most k. By considering functions of the form $x(s) = |s|^{-\alpha}\psi(s)$ with $\psi \in C_0^\infty(\mathbb{R}^N)$ and $\psi(0) \neq 0$, one sees that f is even linear, i.e. $f(u) = bu$ for some $b \in \mathbb{R}$. The representation (9.55) follows then with $a = f(0)$. $\Leftarrow$

The following result complements Theorem 9.9.

Theorem 9.11 *Let $f = f(u)$ be a C^∞ function on $[0, \infty)$ which satisfies (9.52). Let $1 < p < N/k$ and $k \geq 3$. Suppose that the superposition operator F generated by f maps $W_{p,+}^k(\mathbb{R}^N)$ into $W_p^k(\mathbb{R}^N)$. Then f has the form (9.55).*

$\Rightarrow$ As above the result follows if we can show that f is a polynomial on $(0, \infty)$. We claim that $f^{(\ell)}(u) \equiv 0$ for $u > 0$, where $\ell = \lceil \frac{k}{2} \rceil$. Suppose that this is not the case, fix $u_0 > 0$ with $f^{(\ell)}(u_0) \neq 0$, and let $g(u) = f(u + u_0)$. Since $g^{(\ell)}(0) \neq 0$, there are numbers $c > 0$ and $a > 0$ such that $|g^{(k)}(u^2)| \geq cu^{<k>}$ for $0 < u < a$, where $< k >= 0$ if k is even, and $< k >= 1$ if k is odd. Choose sequences $\omega_n > 1$ and $\varepsilon_n > 0$ such that

$$\sum_{n=1}^{\infty} \varepsilon_n < \infty, \quad \sum_{n=1}^{\infty} \omega_n^p \varepsilon_n^{N-kp} < \infty, \quad \sum_{n=1}^{\infty} \omega_n^{(kp-1)/2} \varepsilon_n^{N-kp} = \infty. \quad (9.56)$$

Putting $\sigma_n = (0, 5\sum_{j=1}^{n} \varepsilon_j, 0, \cdots, 0) \in \mathbb{R}^N$, the balls $B_n = \{s : s \in \mathbb{R}^N, |s - \sigma_n| < 2\varepsilon_n\}$ are disjoint and contained in some large common ball $B_R = \{s : s \in \mathbb{R}^N, |s| < R\}$. Let φ be a C_0^{∞} function in $\mathbb{R}^N$ such that $\varphi(s) = \varphi(s_1, \cdots, s_N) = s_1^2$ for $|s| \leq 1$, and $\varphi(s) \equiv 0$ for $|s| \geq 2$. Consider the function $x = (u_0 + \sum_{n=1}^{\infty} \varphi_n)\psi$, where $\varphi_n(s) = \omega_n \varphi[\varepsilon_n^{-1}(s - \sigma_n)]$, and ψ is some nonnegative C_0^{∞} function with $\psi(s) \equiv 1$ for $|s| \leq R$. As in the proof of Theorem 9.10, it follows from (9.56) that $x \in W_{p,+}^k$, but $y = Fx \notin W_p^k$, contradicting the hypothesis. $\Leftarrow$

9.7 The superposition operator in Sobolev–Orlicz spaces

Sobolev–Orlicz spaces generalize the classical Sobolev spaces in rather the same way as Orlicz spaces generalize the classical Lebesgue spaces. The aim of this section is to recall the definition and the basic facts on Sobolev–Orlicz spaces, and to give some results on the superposition operator between them.

Throughout this section we consider a domain $\Omega \subset \mathbb{R}^N$ and an autonomous Young function $M = M(u)$ (see Section 4.1) on $\mathbb{R}$ with the additional property that

$$\int_0^1 \frac{M^{-1}(t)}{t^{1+1/N}} dt < \infty. \quad (9.57)$$

The last requirement may always be fulfilled by passing, if necessary, to an *equivalent* Young function $\bar{M} \sim M$, i.e. with the same behaviour as M near infinity. For $k \in \mathbb{N}$ and M a Young function with (9.57), the *Sobolev–Orlicz space* $W^k L_M = W^k L_M(\Omega)$ consists, by definition, of all functions $x \in L_M(\Omega)$ for which all distributional derivatives $D^{\alpha} x$ ($|\alpha| \leq k$) belong to $L_M(\Omega)$, equipped with the norm

$$\|x\|_{k,M} = \sum_{|\alpha| \leq k} \|D^{\alpha} x\|_M, \quad (9.58)$$

where $\|z\|_M$ denotes the Luxemburg norm (4.8) of z. The closed subspace $W^k E_M = W^k E_M(\Omega)$ is defined analogously. In particular, $W^0 L_M$ and $W^0 E_M$ are just the spaces L_M and E_M defined by (4.7) and (4.10), respectively.

It is clear that, for bounded Ω, the space $C^k(\bar{\Omega})$ is imbedded in $W^k L_M(\Omega)$ for each $k \geq 0$. Moreover, one may prove some density theorems which are similar to those for the usual Sobolev spaces W_p^k. For instance, $C^\infty(\bar{\Omega}) \cap W^k E_M(\Omega)$ is dense in $W^k E_M(\Omega)$ for Ω bounded, and the space $W^{k,0} L_M(\Omega)$ (i.e. the closure of $C_0^\infty(\Omega)$ with respect to the norm (9.58)) coincides with $W^k E_M(\Omega)$ if $\Omega = \mathbb{R}^N$.

We do not repeat the general discussion on the various properties of the Sobolev spaces W_p^k in Section 9.1, but just make some remarks on *imbedding theorems* in the spirit of Lemma 9.1. As we have seen several times in the previous sections, the number $p_* = Np/(N-p)$ plays a particular role as the "limiting" exponent of the space W_p^k; p_* is sometimes called the *Sobolev conjugate* of p. Now, given a Young function M, we define the *Sobolev-Young conjugate* M_* of M by

$$M_*^{-1}(u) = \int_0^{|u|} \frac{M^{-1}(t)}{t^{1+1/N}} dt.$$

Obviously, if $M(u) \sim |u|^p$ $(1 \leq p \leq N)$, then $M_*(u) \sim |u|^{p_*}$ with p_* as above.

Let us take a closer look at two special cases of Lemma 9.1. First, choosing $p < N$, $q = p_* = Np/(N-p)$, $s = 1$, and $r = 0$ in Lemma 9.1 (a) yields the imbedding $W_p^1(\Omega) \subseteq L_{p_*}(\Omega)$. Second, choosing $p > N$, $k = 0$, $\sigma = 1 - N/p$, and $s = 1$ in Lemma 9.1 (c) yields the imbedding $W_p^1(\Omega) \subseteq H_{1-N/p}(\bar{\Omega})$. Both imbeddings may be generalized to Sobolev-Orlicz spaces as follows. If, on the one hand,

$$\int_1^\infty \frac{M^{-1}(t)}{t^{1+1/N}} dt = \infty$$

(which corresponds to the case $p < N$ if $M(u) \sim |u|^p$), then the imbedding $W^1 L_M(\Omega) \subseteq L_{M_*}(\Omega)$ holds. If, on the other hand

$$\int_1^\infty \frac{M^{-1}(t)}{t^{1+1/N}} dt < \infty$$

(which corresponds to the case $p > N$ if $M(u) \sim |u|^p$), then the imbedding $W^1 L_M(\Omega) \subseteq H_\phi(\bar{\Omega})$ holds, where H_ϕ is the Hölder space (see Section 7.1) generalized by the Hölder function

$$\phi(t) = \int_{t^{-N}}^{\infty} \frac{M^{-1}(\tau)}{\tau^{1+1/N}} d\tau \,.$$

Of course, in case $M(u) \sim |u|^p$ with $p > N$ we get again the classical Hölder space $H_\sigma(\Omega)$ with $\sigma = 1 - N/p$.

We turn now to the superposition operator in Sobolev–Orlicz spaces. Apparently, the only systematic results presently known are generalizations, due to G. Hardy, of three theorems of M. Marcus and V.J. Mizel given in Sections 9.2 and 9.4. Since the proofs are essentially parallel to those of the corresponding statements in Sobolev spaces, we shall drop them.

The first result is a natural generalization of Theorem 9.1.

Theorem 9.12 *Suppose that f is locally Lipschitz continuous on $\Omega \times \mathbb{R}$, where $\Omega \subset \mathbb{R}^N$ is bounded. Let M, N, Q, and R be Young functions such that $Q \in \Delta_2$, the derivative Q' of Q is strictly increasing on $(0, \infty)$, the Sobolev–Young conjugate N_* of N satisfies $N_*^{-1}(u) \to \infty$ as $|u| \to \infty$, and*

$$N_* \sim R_* \circ Q^{-1}, \quad R \sim M_* \circ (Q')^{-1} \,. \tag{9.59}$$

Assume that the estimates

$$|D_i f(s, u)| \le a(s) + bQ(|u|)$$

$(i = 1, \cdots, N)$ and

$$|D_u f(s, u)| \le c(s) + dQ'(|u|)$$

hold for some $a \in L_N$, $c \in L_R$, $b \ge 0$, and $d \ge 0$, where the partial derivatives exist. Then the superposition operator F generated by f maps $W^1 L_M(\Omega)$ into $W^1 L_N(\Omega)$.

Choosing, in particular, $M(u) \sim |u|^p$, $N(u) \sim |u|^q$, $Q(u) \sim |u|^{p(N-q)/q(N-p)}$, and $R(u) \sim |u|^{pq/(p-q)}$, where $1 < q \le p < N$, we get Theorem 9.1 as a special case of Theorem 9.12.

The next result generalizes Theorem 9.2.

Theorem 9.13 *Suppose that f is a locally absolutely continuous Carathéodory function on $\Omega \times \mathbb{R}$, where $\Omega \subset \mathbb{R}^N$ is bounded. Let M, N, P, and R be Young functions such that $M, N \in \Delta_2$, $N \sqsubseteq M$, and*

$$P \sqsubseteq N^{-1} \circ M, \quad \tilde{P} \sqsubseteq N^{-1} \circ R, \tag{9.60}$$

where the relation $\sqsubseteq$ is defined in (4.19), and $\tilde{P}$ denotes the associated Young function (4.16) of P. Assume that the estimates (9.9) and (9.10) hold for almost all $s \in \Omega$ and all $u \in \mathbb{R}$, where $a \in L_N$, $b \in C(\mathbb{R})$, $c \in L_R$, and $d \in L_1(\mathbb{R})$ are nonnegative functions. Given $x \in W^1 L_M$, suppose that $b \circ x \in L_N$, $d \circ x \in L_R$, and $(d \circ x) D_i x \in L_N$ $(i = 1, \cdots, N)$. Then $y = Fx \in W^1 L_N$.

If we choose, in particular, $M(u) \sim |u|^p$, $N(u) \sim |u|^q$, $P(u) \sim |u|^{p/q}$, and $R(u) \sim |u|^{pq/(p-q)}$, where $1 < q \leq p < N$, we get Theorem 9.2 as a special case of Theorem 9.13.

Finally, the demi-continuity result given in Theorem 9.6 is generalized by the following:

Theorem 9.14 *Suppose that f is a locally absolutely continuous Carathéodory function on $\Omega \times \mathbb{R}$, where $\Omega \subset \mathbb{R}^N$ is bounded. Let M, N, P, and R be Young functions such that $M, N, \tilde{N} \in \Delta_2$, $R^{-1} \circ M_* \in \Delta_2$, $N^{-1} \circ M_* \in \Delta_2$, $M_*^{-1}(u) \to \infty$ as $|u| \to \infty$, $N \sqsubseteq M$, and (9.60) holds. Assume that the estimates (9.9) and (9.10) hold for almost all $s \in \Omega$ and all $u \in \mathbb{R}$, where $a \in L_N$, $c \in L_R$, and b and d are nonnegative continuous functions on $\mathbb{R}$ which generate continuous superposition operators from L_{M_*} into L_N, and from L_{M_*} into L_R, respectively. Then the superposition operator F generated by f is demi-continuous between $W^1 L_M$ and $W^1 L_N$.*

Theorem 9.6 is obtained as a special case of Theorem 9.14 by the same choice of M, N, P, and R as after Theorem 9.13. Observe, in particular, that the requirement $\tilde{N} \in \Delta_2$ corresponds to the requirement $q > 1$.

9.8 Notes, remarks and references

1. Sobolev spaces have become a well established and largely used tool in functional analysis, numerical analysis, and the theory of distributions and partial differential equations. The classical monograph of S.L. Sobolev [326] is of historical interest. Nowadays there are many excellent textbooks on both the theory and the applications of Sobolev spaces; we just mention the books [5] and [185], see also Sections 2.6 and 2.7 of [64] for a concise introduction with some illuminating examples.

Lemma 9.1 is generally wellknown as the *Sobolev imbedding theorem* and may be found in many textbooks. The characterization of W_p^1 as space of absolutely continuous functions (see Lemma 9.2) is due to E. Gagliardo [117]. This characterization is crucial in analyzing the chain rule for functions which are not C^1 everywhere. The classical chain rule for absolutely continuous functions goes back to De la Vallée–Poussin; a stronger result was later given by J. Serrin; see [208] and [209] for details.

2. The sufficient acting conditions discussed in Section 9.2 build essentially on a very delicate analysis of the chain rule. All results presented in this and the following section are due to M. Marcus and V.J. Mizel [208–210], [214–217]. We point out again that the scope of these papers is far beyond the rather particular results that we presented in Sections 9.2 and 9.3, inasmuch as *vector valued functions* are also considered. In particular, [208] and [209] discuss the chain rule for vector functions which are locally absolutely continous "on tracks". Further, apart from acting conditions in the Sobolev space W_p^k functions in the spaces $W_{p,\mathrm{loc}}^k$ are also considered. All these refinements are beyond the scope of the present chapter.

Theorem 9.1 is Theorem 2.2 in [208], while our Theorem 9.2 is Theorem 2.1 in [209], where also the notion of a locally absolutely continuous Carathéodory function is introduced. Observe that we used the hypothesis $q > 1$ in Theorem 9.2 only to apply Lemma 9.3; the proof in case $q = 1$ requires a slightly different technique and may be found as Theorem 3.1 in [209]. The proof of Lemma 9.3 may be found, for example, in [119].

3. The crucial part in obtaining necessary acting conditions for the superposition operator F between W_p^1 and W_q^1 is Lemma 9.4, which is of independent interest. This lemma is formulated in [214] (for both cases $p < N$ and $p > N$) and proved in [216], as well as Theorems 9.3. and 9.4. The counterexample (9.34) is due to V.J. Mizel [228].

Unfortunately, a parallel result for *second-order Sobolev spaces* seems to be unknown. As already pointed out after Lemma 9.6, one does not know whether the acting condition $\Phi(W_p^2) \subseteq L_1$ for the differential operator (9.47) (in the autonomous case) implies the growth condition

$$|\phi(u,\eta,\zeta)| \leq C(1 + |u|^{Np/(N-2p)} + |\eta|^{Np/(N-p)} + |\zeta|^p)$$

for $u \in \mathbb{R}$, $\eta \in \mathbb{R}^N$, $\zeta \in \mathbb{R}^{N(N+1)/2}$; this would be the second-order analogue to (9.18). In fact, in this case the superposition operator F generated by $f = f(u)$ would map W_p^2 into W_q^2 ($1 \leq q \leq p < N/2$) if and only if the growth conditions

$$|f'(u)| \leq C(1 + |u|^{N(p-q)/q(N-2p)})$$

and

$$|f''(u)| \leq C(1 + |u|^{(Np-2Nq+2pq)/q(N-2p)})$$

hold.

4. Boundedness and continuity properties of the superposition operator between Sobolev spaces are discussed, for instance, in [210] and [215]. Theorem 9.5 is given as Theorem 4.2 in [210]; this shows that, roughly speaking, the acting condition $F(W_p^1) \subseteq W_q^1$ implies the continuity of F if the partial derivatives of the function f are also Carathéodory functions generating superposition operators between appropriate Lebesgue spaces. In [215] it is shown that F is always continuous in the autonomous case $f = f(u)$.

We point out again that the situation becomes more complicated in the vector valued setting, i.e. when f maps $\mathbb{R}^m$ into $\mathbb{R}^m$ for some $m > 1$. For instance, the continuity of F does not follow merely from the acting condition $F(W_p^1) \subseteq (W_q^1)^m$ (as it does in case $m = 1$), but more regularity of f is required in case $m > 1$ [214].

The sufficient condition for the demi-continuity of F given in Theorem 9.6 is taken from [209]. Moreover, one may show that, under the hypotheses of Theorem 9.6, F is continuous from W_p^1 into L_q.

Continuity properties of the differential operator (9.17) are particularly important in *variational problems* of nonlinear analysis, and are dealt with in a vast literature. As a sample reference, we mention [74] in the autonomous case $\phi = \phi(u, \eta)$, and [10] in the non-autonomous case $\phi = \phi(s, u, \eta)$.

5. In higher-order Sobolev spaces, only *sufficient* conditions for the acting, boundedness, continuity, etc., of the superposition operator are known. Lemma 9.6 follows directly from the Sobolev imbedding theorem and may be found in several books on nonlinear differential equations, e.g. [116]. The acting condition for the multiplication operator (9.48) given in Lemma 9.7, is due to T. Valent [355], as well as Theorem 9.7, which apparently was used by many authors.

In some applications, it is often tacitly assumed that the nonlinearity involved is sufficiently smooth (e.g. $f = f(u)$ is C_0^∞ in $\mathbb{R}$), in order to ensure that the corresponding superposition operator F acts in the space $W_p^k(\mathbb{R}^N)$ and has "nice" properties. As Theorems 9.8 – 9.11 show, however, one has to be careful with the right choice of k and p. Theorems 9.8 and 9.9 are proved in [4] in connection with norm estimates for *Riesz potentials*. The question of whether Theorem 9.9 is valid also in $W_{p,+}^k$ for $k > 2$ is stated as an open problem in [4].

We remark that there are other (sufficient) acting conditions for the superposition operator in higher-order Sobolev spaces, which we did not discuss here. It is clear from the classical imbedding relations and density results that the case $kp > N$ is particularly pleasant, since $W_p^k(\Omega) \subseteq C(\bar{\Omega})$ in this case. A growth condition on the derivatives of the function $f = f(s, u)$ (in the distributional sense for the derivatives $D_i f$, and in the classical sense for the derivative $D_u f$) which ensures the boundedness of F between $W_p^k(\Omega) \cap C(\bar{\Omega})$ and $W_p^k(\Omega)$ may be found in [229]; this condition in turn generalizes some results of [230], [231].

Most sufficient conditions for the superposition operator to act between

Sobolev spaces are expressed in terms of smoothness properties (see e.g. Theorem 9.7) of the function f. There are also some papers, e.g. [330–335], where the function $f = f(u)$ is supposed to belong to some Sobolev space, and the relations between this space, on the one hand, and the "acting spaces" for F, on the other hand, are studied. For example, it is shown in [330] (in the scalar case $N = 1$) that every function $f \in W_2^k$ generates a superposition operator F from W_2^ℓ into W_2^m, provided

$$m - \frac{1}{2} < (k - \frac{1}{2})(\ell - \frac{1}{2}). \tag{9.61}$$

Moreover, it is shown by a counterexample that the condition (9.61) is sharp. A certain analogue of this in the case $N > 1$ is proved in [332], together with applications to Fourier series of W_2^k functions. These results employ the definition of Sobolev spaces W_p^s with non-integer index s which we briefly discussed in Section 9.1. The following result is parallel to Theorem 9.7 and may be found in Chapter 2 of [356]: if the function f is of class C^{k+1} on $\bar{\Omega} \times \mathbb{R}$, then the superposition operator F generated by f maps $W_p^k(\Omega)$ into itself and is continuously Fréchet differentiable. A sufficient condition for analyticity of F on W_p^k ($k \geq 1$) is also given in [356]; of course, the case $k = 0$ must be excluded, by Theorem 3.16.

6.   The degeneracy results given in Theorems 9.10 and 9.11 show that it is reasonable to consider the superposition operator F in the Sobolev space W_p^k only in case $kp \geq N$ for $k \geq 3$, if F is generated by a C_0^∞ function $f = f(u)$ in $\mathbb{R}$. The proofs of these theorems were taken from [87]. There are also other degeneracy results which are related to additional analytical properties of F. For instance, it is shown in [352] (see also [353], [354], [357]) that, whenever the *differential operator* $\Phi x(s) = \phi(s, Dx(s))$ maps the Sobolev space $W_p^{1,0}(\Omega)$ into the Lebesgue space $L_p(\Omega)$ ($\Omega \subset \mathbb{R}^N$ bounded) and is differentiable at some point $x_0 \in W_p^{1,0}(\Omega)$, then the function ϕ is necessarily of the form $\phi(s, \eta) = a(s) + b(s) \sum_{i=1}^N \beta_i \eta_i$ ($s \in \Omega$, $\eta \in \mathbb{R}^N$), i.e. is *affine* in η. In view of the degeneracy results for the superposition operator in Lebesgue spaces (see Theorem 3.12), this is not at all surprising. There is a degeneracy result for *compact* differential operators which is similar to Theorem 1.8 and its analogues in ideal spaces: if the operator (9.37) is compact as an operator between W_p^2 and W_p^1 on $\Omega = (a, b)$, then the corresponding function $\phi = \phi(s, u, \eta)$ is independent of η [96].

We still mention the following representation theorems. In [211] it is shown that every continuous disjointly additive operator F from $W_p^{k,0}(a, b)$ into $L_1(a, b)$ is of the form $Fx(s) = f(s, D^k x(s))$ with some Carathéodory function f. Similar characterizations of the *integral functional*

$$\Phi x = \int_\Omega f(s, D^k x(s)) ds$$

may be found in [213]. Finally, the integral functional

$$\Phi(x, D) = \int_D f(s, x(s), Dx(s)) ds \quad (D \subseteq \Omega)$$

on $W_p^1(\Omega)$ (which is of course analogous to the functional (1.22) on $L_p(\Omega)$) is studied in detail in [73].

7. Sobolev–Orlicz spaces arise quite naturally in applications to nonlinear problems (e.g. boundary value problems for *nonlinear elliptic equations*, see [122]) involving *strong* (i.e. non-polynomial) *nonlinearities*. The first systematic account on imbedding theorems between Sobolev–Orlicz spaces is [97]. A detailed description of the theory of Sobolev–Orlicz spaces may be found, for example, in Chapter 8 of [5] or Chapter 7 of [185]. We remark that there exist also higher–order analogues of the imbeddings $W^1 L_M \subseteq L_{M_*}$ and $W^1 L_M \subseteq H_\phi$, generalizing Lemma 9.1.

All results on the superposition operator in Sobolev–Orlicz spaces given in Section 9.7 are due to G. Hardy [140-143]. More precisely, Theorem 9.12 is taken from [141], Theorem 9.13 from [140], and Theorem 9.14 from [143]. The statements of Lemma 9.7 and Theorem 9.7 carry over to the higher–order Sobolev–Orlicz spaces $W^k E_M$ as well [26].

Bibliography

[1] Ju. A. Abramovich: *Multiplicative representations of disjointness preserving operators*, Indag. Math. (45) 86 (1983), 265-279
Zbl. 527.47025　　　R.Zh. 2B1015(84)　　　M.R. 85f47040

[2] Ju. A. Abramovich, A. I. Veksler, A. V. Koldunov: *On operators preserving disjointness* (Russian), Doklady Akad. Nauk SSSR 248, 5 (1979), 1033-1036 [= Soviet Math., Doklady 20, 5 (1979), 1089-1093]
Zbl. 445.46017　　　R.Zh. 3B613(80)　　　M.R. 81e47034

[3] Ju. A. Abramovich, A. I. Veksler, A. V. Koldunov: *Operators preserving disjointness, their continuity and multiplicative representation* (Russian), Lin. Oper. Prilozh., Leningrad (1981), 13-34

[4] D. R. Adams: *On the existence of capacitary strong type estimates in* $\mathbb{R}^n$, Ark. Mat. 14 (1976), 125-140
Zbl. 325.31008　　　R.Zh. 12B647(76)　　　M.R. 54♯5822

[5] R. Adams: *Sobolev spaces*, Academic Press, New York 1976
Zbl. 314.46030　　　R.Zh. 9B541(76)　　　M.R. 56♯9247

[6] M. L. Agranovskij, R. D. Baglaj, K. K. Smirnov: *Identification of a class of nonlinear operators* (Russian), Zhurn. Vychisl. Mat. Mat. Fiz. 18, 2 (1978), 284-293 [= USSR Comput. Math. Math. Phys. 18 (1978), 7-15]
Zbl. 436.47059　　　R.Zh. 8B907(78)　　　M.R. 58♯9508

[7] R. R. Ahmerov, M. I. Kamenskij, A. S. Potapov, B. N. Sadovskij: *Condensing operators* (Russian), Itogi Nauki Tehniki 18 (1980), 185-250 [=J. Soviet Math. 18 (1982), 551-592]
Zbl. 443.47056　　　R.Zh.1B1123(81)　　　M.R. 83e47039

[8] R. R. Ahmerov, M. I. Kamenskij, A. S. Potapov, A. Je. Rodkina, B. N. Sadovskij: *Measures of noncompactness and condensing operators* (Russian), Nauka, Novosibirsk 1986
Zbl. 623.47070　　　R.Zh. 3B885(87)　　　M.R. 88f47048

[9] R.A. Alò: *Non-linear operators on Lebesgue spaces*, Karachi J. Math. 2 (1984), 1-26
Zbl. 668.47048

[10] L. Ambrosio, G. Buttazzo, A. Leaci: *Continuous operators of the form* $T_f(u) = f(x, u, Du)$, Boll. Unione Mat. Ital. (7) 2-B (1988), 99-108
Zbl. 639.47051　　　R.Zh. 10B1140(88)

[11] T. Andô: *On products of Orlicz spaces*, Math. Ann. 140 (1960), 174-186
Zbl. 91.278   R.Zh. 4B332(62)   M.R. 22♯3965

[12] J. Appell: *Implicit functions, nonlinear integral equations, and the measure of noncompactness of the superposition operator*, J. Math. Anal. Appl. 83, 1 (1981), 251-263
Zbl. 495.45007   R.Zh. 4B1035(82)   M.R. 83c47084

[13] J. Appell: *On the solvability of nonlinear noncompact problems in function spaces with applications to integral and differential equations*, Boll. Unione Mat. Ital. (6) 1-B (1982), 1161-1177
Zbl. 511.47045   R.Zh. 5B909(83)   M.R. 84j47099

[14] J. Appell: *Upper estimates for superposition operators and some applications*, Ann. Acad. Sci. Fenn., Ser. A I 8 (1983), 149-159
Zbl. 489.47017   R.Zh. 3B1109(84)   M.R. 84i47079

[15] J. Appell: *"Genaue" Fixpunktsätze und nichtlineare Sturm-Liouville-Probleme*, Proc. Int. Conf. "Equadiff 82" Würzburg, Lect. Notes Math. 1071 (1983), 38-42
Zbl. 533.47058   R.Zh. 5B920(84)   M.R. 85d47061

[16] J. Appell: *On the differentiability of the superposition operator in Hölder and Sobolev spaces*, Nonlinear Anal., Theory, Methods Appl. 8, 10 (1984), 1253-1254
Zbl. 555.47037   R.Zh. 5B1055(85)

[17] J. Appell: *Über eine Klasse symmetrischer idealer Funktionenräume nebst Anwendungen*, Comm. Math. Univ. Carolinae 25, 2 (1984), 337-354
Zbl. 554.46009   R.Zh. 4B951(85)   M.R. 86c46024

[18] J. Appell: *Deux méthodes topologiques pour la résolution des équations elliptiques non linéaires sans compacité*, Presses Univ. Montréal 95 (1985), 9-19
Zbl. 578.47044   M.R. 87h35092

[19] J. Appell: *Über die Differenzierbarkeit des Superpositionsoperators in Orliczräumen*, Math. Nachr. 123 (1985), 335-344
Zbl. 537.47031   R.Zh. 1B1206(86)   M.R. 87a47096

[20] J. Appell: *Untersuchungen zur Theorie nichtlinearer Operatoren und Operatorgleichungen*, Habilitationsschrift, Univ. Augsburg 1985
Zbl. 573.47049

[21]   J. Appell: *Misure di non compattezza in spazi ideali*, Rend. Ist. Lomb. Accad. Sci. Milano A - 119 (1985), 157-174
Zbl. 619.47043   M.R. 89c46042

[22]   J. Appell, E. De Pascale: *Su alcuni parametri connessi con la misura di non compattezza di Hausdorff in spazi di funzioni misurabili*, Boll. Unione Mat. Ital. (6) 3-B (1984), 497-515
Zbl. 507.46025   R.Zh. 4B68(85)   M.R. 86f46024

[23]   J. Appell, E. De Pascale: *Théorèmes de bornage pour l'opérateur de Nemyckii dans les espaces idéaux*, Canadian J. Math. 38, 6 (1986), 1338-1355
Zbl. 619.47051   R.Zh. 10B1203(87)   ' M.R. 88c47134

[24]   J. Appell, E. De Pascale: *Lipschitz and Darbo conditions for the superposition operator in some non-ideal spaces of smooth functions*, Annali Mat. Pura Appl. (to appear)

[25]   J. Appell, E. De Pascale, P. P. Zabrejko: *An application of B. N. Sadovskij's fixed point principle to nonlinear singular equations*, Zeitschr. Anal. Anw. 6 (3) (1987), 193-208
Zbl. 628.45003   R.Zh. 2B600(88)   M.R. 88m47086

[26]   J. Appell, G. Hardy: *On products of Sobolev–Orlicz spaces*, Preprint Univ. Würzburg 1989

[27]   J. Appell, I. Massabò, A. Vignoli, P. P. Zabrejko: *Lipschitz and Darbo conditions for the superposition operator in ideal spaces*, Annali Mat. Pura Appl. 152 (1988), 123-137
Zbl. 638.47063   R.Zh. 10B1166(89)

[28]   J. Appell, V. I. Nazarov, P. P. Zabrejko: *Composing infinitely differentiable functions*, Preprint Univ. Calabria 1988 R.Zh. 10B1166(89)

[29]   J. Appell, Je. M. Semjonov: *Misure di noncompattezza debole in spazi ideali simmetrici*, Rend. Ist. Lomb. Accad. Sci. Milano A-122 (1988), 1-18

[30]   J. Appell, P. P. Zabrejko: *On a theorem of M. A. Krasnosel'skij*, Nonlinear Anal., Theory, Methods Appl. 7, 7(1983), 695-706
Zbl. 522.47056   R.Zh. 12B1175(83)   M.R. 84m47088

[31]   J. Appell, P. P. Zabrejko: *Sharp upper estimates for the superposition operator* (Russian), Doklady Akad. Nauk BSSR 27, 8 (1983), 686-689
Zbl. 525.47020   R.Zh. 1B52(84)   M.R. 84j47105

[32]   J. Appell, P. P. Zabrejko: *On analyticity conditions for the superposition operator in ideal function spaces*, Boll. Unione Mat. Ital. (6)
4-C (1985), 279-295
Zbl. 583.47057   R.Zh. 6B1221(86)   M.R. 87g47107

[33]   J. Appell, P. P. Zabrejko: *Analytic superposition operators* (Russian),
Doklady Akad. Nauk BSSR 29, 10 (1985), 878-881
Zbl. 591.47048   R.Zh. 4B1169(86)   M.R. 87e47086

[34]   J. Appell, P. P. Zabrejko: *On the degeneration of the class of differentiable superposition operators in function spaces*, Analysis 7 (1987),
305-312
Zbl. 627.47033   R.Zh. 7B979(88)   M.R. 89c47082

[35]   J. Appell, P. P. Zabrejko: *Boundedness properties of the superposition operator*, Bull. Polish Acad. Sci. Math. (to appear)

[36]   J. Appell, P. P. Zabrejko: *Continuity properties of the superposition*
operator, J. Austral. Math. Soc. 46 (1989), 1-25

[37]   J. Appell, P. P. Zabrejko: *Über die L-Charakteristik nichtlinearer*
Operatoren in Räumen integrierbarer Funktionen, Manuscripta Math.
62, 3 (1988), 355-367
R.Zh. 4B1293(89)

[38]   L. Ardizzone, D. Averna: *Misurabilità delle funzioni astratte di più*
variabili, Atti Accad. Sci. Lett. Arti Palermo (1986), 1-11

[39]   D. Averna, A. Fiacca: *Sulla proprietà di Scorza-Dragoni*, Atti Sem.
Mat. Fis. Univ. Modena 33 (1984), 313-318
Zbl. 599.28017   M.R. 88d28015

[40]   A. A. Babajev: *The structure of a certain nonlinear operator and its*
application (Russian), Azerbajdzh. Gos. Univ. Uchen. Zapiski 4
(1961), 13-16
R.Zh. 3B300(63)   M.R. 36♯4395

[41]   R. J. Bagby, J. D. Parsons: *Orlicz spaces and rearranged maximal*
functions, Math. Nachr. 132 (1987), 15-27
Zbl. 662.46031   R.Zh. 3B714(88)   M.R. 88i46039

[42]   J. Banaś: *On the superposition operator and integrable solutions of*
some functional equations, Nonlinear Anal., Theory, Methods Appl.
12, 8 (1988), 777-784
Zbl. 656.47057   M.R. 89h47096

[43] R. Bellman: *Dynamic programming*, Princeton Univ. Press, Princeton 1957

[44] V. Benci, D. Fortunato: *Weighted Sobolev spaces and the nonlinear Dirichlet problem in unbounded domains*, Annali Mat. Pura Appl. 121 (1979), 319-336
Zbl. 422.35038   R.Zh. 8B267(80)   M.R. 81f35040

[45] Je. I. Berezhnoj: *On differentiable maps in finite-dimensional spaces* (Russian), Jaroslav. Gos. Univ. Vestnik 12 (1975), 3-10
R.Zh. 5B55(76)

[46] M. Z. Berkolajko: *On a nonlinear operator acting in generalized Hölder spaces* (Russian), Voronezh. Gos. Univ. Trudy Sem. Funk. Anal. 12 (1969), 96-104
Zbl. 266.47053   R.Zh. 1B699(70)

[47] M. Z. Berkolajko: *On the continuity of the superposition operator acting in generalized Hölder spaces* (Russian), Voronezh. Gos. Univ. Sbornik Trudov Aspir. Mat. Fak. 1 (1971), 16-24
R.Zh. 1B629(73)

[48] M. Z. Berkolajko: *On a property of the superposition operator in the Hölder space H_φ* (Russian), Voronezh. Gos. Univ. Sbornik Nauchn. Statjej (1975), 3-8
R.Zh. 12B923(76)   M.R. 57♯116

[49] M. Z. Berkolajko: *Estimates for the moduli of continuity of functions belonging to the space $B_{p,\theta}^{a,\phi}$ and $H_{\phi,\theta}$ and their applications* (Russian), Doklady Akad. Nauk SSSR 233, 5 (1976), 761-764 [= Soviet Math., Doklady 18, 2 (1977), 469-473]
Zbl. 379.46020   R.Zh. 8B64(76)   M.R. 58♯12330

[50] M. Z. Berkolajko: *Estimates for the modulus of continuity of functions from Besov and Hölder spaces and their applications* (Russian), Oper. Metody Diff. Uravn., Izdat. Voronezh. Gos. Univ.(1979), 3-18
R.Zh. 3B655(80)

[51] M. Z. Berkolajko, Ja. B. Rutitskij: *On operators in Hölder spaces* (Russian), Doklady Akad. Nauk SSSR 192, 6 (1970), 1199-1201 [= Soviet Math., Doklady 11, 3 (1970), 787-789]
Zbl. 231.46051   R.Zh. 12B733(70)   M.R. 57♯17365

[52] M. Z. Berkolajko, Ja. B. Rutitskij: *Operators in generalized Hölder spaces* (Russian), Sibir. Mat. Zhurn. 12, 5 (1971), 1015-1025 [= Siber. Math. J. 12, 5 (1971), 731-738]
Zbl. 247.47045   R.Zh. 1B977(72)   M.R. 45♯4176

[53] V. A. Bondarenko: *The superposition operator in the space of Lipschitzian functions* (Russian), Jaroslav. Gos. Univ. Vestnik 2 (1973), 5-10
R.Zh. 12B743(73)   M.R. 57♯1217

[54] V. A. Bondarenko: *Cases of degeneracy of the superposition operator in function spaces* (Russian), Jaroslav. Gos. Univ. Kach. Priblizh. Metody Issled. Oper. Uravn. 1 (1976), 11-16
R.Zh. 9B900(77)   M.R. 58♯30579

[55] V. A. Bondarenko, P. P. Zabrejko: *The superposition operator in Hölder functions spaces* (Russian), Doklady Akad. Nauk SSSR 222, 6 (1975), 1265-1268 [= Soviet Math., Doklady 16, 3 (1975), 739-743]
Zbl. 328.47040   R.Zh. 12B1005(75)   M.R. 53♯6384

[56] V. A. Bondarenko, P. P. Zabrejko: *The superposition operator in Hölder spaces* (Russian), Jaroslav. Gos. Univ. Kach. Priblizh. Metody Issled. Oper. Uravn. 3 (1978), 17-29
Zbl. 441.47064   R.Zh. 10B1024(79)   M.R. 81d47041

[57] A. Bottaro Aruffo: *Su alcune estensioni del teorema di Scorza Dragoni*, Rend. Accad. Naz. Sci. XL 9 (1985), 87-202
Zbl. 599.28018   R.Zh. 7B1374(86)   M.R. 88f49012

[58] A. Bottaro Aruffo: *($\mathcal{L} \times \mathcal{B}(\rho)$)-misurabilità e convergenza in misura*, Rend. Sci. Mat. Appl. A - 118 (1987), 203-246
Zbl. 648.47049   M.R. 89c28013

[59] A. Bottaro Aruffo: *Maggiorazioni integrali*, Preprint Univ. Genova 1984

[60] A. Bottaro Aruffo: *Proprietà di inclusione e di uniforme continuità dell'operatore di Nemytskii*, Ricerche Mat. 34, 1 (1985), 59-104
Zbl. 606.47068   R.Zh. 10B1176(86)   M.R. 88a47059a

[61] A. Bottaro Aruffo: *Condizioni necessarie e sufficienti per la continuità dell'operatore di sovrapposizione*, Ricerche Mat. 34, 1 (1985), 105-145
Zbl. 606.47069   R.Zh. 10B1177(86)   M.R. 88a47059b

[62] G. Bouchitte: *Maximale monotonie et pseudomonotonie renforcées de l'opérateur de Nemickij associé au sous-différentiel d'un intégrande convexe normal*, Travaux Sém. Anal. Conv. 8, 2 (1978), 27 p.
M.R. 80b47067

[63]   H. Brézis: *Opérateurs maximaux monotones et semi-groupes de con-*
tractions dans les espaces de Hilbert, North – Holland, New York
1973
Zbl. 252.47055   R.Zh. 6B928(74)   M.R. 50♯1060

[64]   F. Brezzi, G. Gilardi: *Functional spaces*, in: H. Kardestuncer (Ed.),
Finite element handbook, McGraw-Hill, New York 1987
Zbl. 638.65076   M.R. 89g65001

[65]   G. Bruckner: *On local Lipschitz continuity of superposition operators*,
Comm. Math. Univ. Carolinae 17, 3 (1976), 593-608
Zbl. 346.47053   R.Zh. 6B822(77)   M.R. 57♯13608

[66]   G. Bruckner: *Über lokale Hölder-Stetigkeit und lokale gleichmässige*
Monotonie von Superpositionsabbildungen zwischen Lebesgueschen
Räumen, Math. Nachr. 80 (1977), 337-352
Zbl. 386.47043   R.Zh. 11B1110(78)   M.R. 58♯17998

[67]   G. Bruckner: *A characterization of superposition operators with local*
properties, Math. Nachr. 102 (1981), 137-140
Zbl. 485.47035   R.Zh. 6B1007(82)   M.R. 83h47039

[68]   G. Bruckner: *On a general principle and its application to superpo-*
sition operators in Orlicz spaces, Math. Nachr. 104 (1981), 183-199
Zbl. 495.47042   R.Zh. 11B1020(82)   M.R. 84f47073

[69]   Ju. A. Brudnyj, S. G. Krejn, Je. M. Semjonov: *Interpolation of linear*
operators (Russian), Itogi Nauki Tehniki 24 (1986), 3-163 [= J. Soviet
Math. 42 (1988), 2009-2113]
Zbl. 631.46070   R.Zh. 4B953(87)   M.R. 88e46056

[70]   J. Brüning: *Personal communication* 1984

[71]   G. Buttazzo: *Semicontinuity, relaxation and integral representation*
problems in the calculus of variations, Textas e Notas 34, Centro Mat.
Aplic. Fund. Univ. Lisboa 1986

[72]   G. Buttazzo, G. Dal Maso: *On Nemyckii operators and integral rep-*
resentations of local functionals, Rend. Mat. 3 (1983), 491-509
Zbl. 536.47027   R.Zh. 10B674   M.R. 85e47093

[73]   G. Buttazzo, G. Dal Maso: *A characterization of nonlinear function-*
als on Sobolev spaces which admit an integral representation with a
Carathéodory integrand, J. Math. Pures Appl. 64 (1985), 337-361
Zbl. 536.46022   R.Zh. 10B643(86)   M.R. 88f46069

[74] G. Buttazzo, A. Leaci: *A continuity theorem for operators from*
      $W^{1,q}(\Omega)$ *into* $L^r(\Omega)$, J. Funct. Anal. 58,2 (1984), 216-224
      Zbl. 548.46027   R.Zh. 1B51(85)   M.R. 85k49036

[75] P. L. Butzer, H. Berens: *Semigroups of operators and approximation*,
      Springer, New York 1967
      Zbl. 164.437   R.Zh. 12B612(68)   M.R. 37♯5588

[76] A. P. Calderón: *Intermediate spaces and interpolation, the complex
      method*, Studia Math. 24 (1964), 113-190
      Zbl. 204.137   R.Zh. 2B504(65)   M.R. 29♯5097

[77] B. Calvert: *Perturbation by Nemytskii operators of m-T-accretive
      operators in* $L^q, q \in (1, \infty)$, Revue Roumaine Math. Pure Appl. 22
      (1977), 883-906
      Zbl. 386.47034   R.Zh. 5B715(78)   M.R. 58♯17978

[78] K. Carathéodory: *Vorlesungen über reelle Funktionen*, De Gruyter,
      Leipzig-Berlin 1918

[79] L. Cesari: *Un criterio per la misurabilità degli insiemi*, Rend. Accad.
      Lincei (8) 1 (1946), 1256-1263

[80] L. Cesari, M. B. Suryanarayana: *Nemitsky's operators and lower clo-
      sure theorems*, J. Optim. Theory Appl. 19 (1976), 165-183
      Zbl. 305.49016   R.Zh. 1B603(77)   M.R. 56♯1169

[81] R. V. Chacon, N. Friedman: *Additive functionals*, Arch. Rat. Mech.
      Analysis 18 (1965), 230-240
      Zbl. 138.93   R.Zh. 10B343(65)   M.R. 30♯2329

[82] M. Chaika, D. Waterman: *On the invariance of certain classes of
      functions under composition*, Proc. Amer. Math. Soc. 43, 2 (1974),
      345-348
      Zbl. 287.42011   R.Zh. 2B42(75)   M.R. 48♯8704

[83] J. Chatelain: *Quelques propriétés de type Orlicz de certains intégran-
      des convexes normaux*, Travaux Sém. Anal. Conv. 5, 10 (1975), 10 p.
      Zbl. 353.46016   M.R. 58♯30111

[84] J. Chatelain: *Espaces d'Orlicz à paramètre et plusieurs variables:
      Application à la surjectivité d'un opérateur non linéaire*, C. R. Acad.
      Sci. Paris 283 (1976), 763-766
      Zbl. 342.46017   R.Zh. 5B351(77)   M.R. 55♯1147

[85]   J. Chatelain: *Surjectivité de l'opérateur de Nemickii dans les espaces d'Orlicz à paramètre et plusieurs variables*, Travaux Sém. Anal. Conv. 6, 9 (1976), 11 p.
Zbl. 409.46026   M.R. 58♯30113

[86]   J. Ciemnoczołowski, W. Orlicz: *Composing functions of bounded φ-variation*, Proc. Amer. Math. Soc. 96, 3 (1986), 431-436
Zbl. 603.26004   M.R. 87k26012

[87]   B. E. J. Dahlberg: *A note on Sobolev spaces*, Proc. Symp. Pure Math. 35, 1 (1979), 183-185
Zbl. 421.46027   R.Zh. 6B815(80)   M.R. 81h46030

[88]   J. Daneš: *Fixed points theorems, Nemytskij and Uryson operators, and continuity of nonlinear mappings*, Comm. Math. Univ. Carolinae 11, 3 (1970), 481-500
Zbl. 202.148   R.Zh. 6B801(71)   M.R. 44♯877

[89]   G. Darbo: *Punti uniti in trasformazioni a codominio non compatto*, Rend. Sem. Mat. Univ. Padova 24 (1955), 84-92

[90]   R. B. Darst: *A characterization of universally measurable sets*, Proc. Cambridge Philos. Soc. 65 (1969), 617-618
Zbl. 176.338   R.Zh. 11B69(69)   M.R. 39♯1615

[91]   R. O. Davies: *Separate approximate continuity implies measurability*, Proc. Cambridge Philos. Soc. 73 (1973), 461-465
Zbl. 254.26011   R.Zh. 11B48(73)   M.R. 48♯4216

[92]   F. De Blasi: *On a property of the unit sphere in a Banach space*, Bull. Math. Soc. Sci. Math. R. S. Roumanie 21 (69) (1977), 259-262
                            R.Zh. 8B887(78)   M.R. 58♯2475

[93]   F. Dedagić: *O operatorima superpozicije u Banahovim prostorima nizova*, Doct. Diss., Univ. Priština 1986

[94]   F. Dedagić, P. P. Zabrejko: *On the superposition operator in ℓ_p spaces* (Russian), Sibir. Mat. Zhurn. 28,1 (1987), 86-98 [= Siber. Math. J. 28, 1 (1987), 63-73]
Zbl. 632.47046   R.Zh. 6B1044(87)   M.R. 88d47077

[95]   K. Deimling: *Nonlinear functional analysis*, Springer, New York 1985
                                                                 M.R. 86j47001

[96] P. Dierolf, J. Voigt: *Weak compactness of Fréchet derivatives: Application to composition operators,* J. Austral. Math. Soc. A 29, 4 (1980), 399-406
Zbl. 408.58006   R.Zh. 12B819(80)   M.R. 81h47062

[97] T. K. Donaldson, N. S. Trudinger: *Orlicz-Sobolev spaces and imbedding theorems,* J. Funct. Anal. 8 (1971), 52-75
Zbl. 216.157   R.Zh. 4B805(72)   M.R. 46♯658

[98] P. Drábek: *Continuity of Nemyckii's operator in Hölder spaces,* Comm. Math. Univ. Carolinae 16, 1 (1975), 37-57
Zbl. 302.26008   R.Zh. 10B814(75)   M.R. 52♯1447

[99] J. Dravecký: *On measurability of superpositions,* Acta Math. Univ. Comenianae 44/45 (1984), 181-183
Zbl. 586.28004   R.Zh. 4B1168(87)   M.R. 86d28005

[100] L. Drewnowski, W. Orlicz: *A note on modular spaces XI,* Bull. Acad. Polon. Sci., Sér. Sci. Math. Astron. Phys. 16 (1968), 877-882
Zbl. 174.439   R.Zh. 4B620(70)   M.R. 39♯6068

[101] L. Drewnowski, W. Orlicz: *On orthogonally additive functionals,* Bull. Acad. Polon. Sci., Sér. Sci. Math. Astron. Phys. 16 (1968), 883-888
Zbl. 172.420   M.R. 39♯6069

[102] L. Drewnowski, W. Orlicz: *On representation of orthogonally additive functionals,* Bull. Acad. Polon. Sci., Sér. Sci. Math. Astron. Phys.17 (1969), 167-173
Zbl. 174.462   M.R. 40♯3296

[103] L. Drewnowski, W. Orlicz: *Continuity and representation of orthogonally additive functionals,* Bull. Acad. Polon. Sci., Sér. Sci. Math. Astron. Phys. 17 (1969), 647-653
Zbl. 198.193   M.R. 41♯806

[104] N. Dunford, J. Schwartz: *Linear operators I,* Int. Publ., Leyden 1963
Zbl. 84.104   R.Zh. 9B485(61)   M.R. 35♯7138

[105] J. Durdil: *On the differentiability of Urysohn and Nemyckij operators,* Comm. Math. Univ. Carolinae 8, 3 (1967), 515-554
Zbl. 165.489   R.Zh. 9B657(69)   M.R. 37♯3406

[106] A. S. Dzhafarov: *Imbedding theorems for classes of functions with differential properties in norms of special spaces* (Russian), Doklady Akad. Nauk Azerb. SSR 21, 2 (1965), 10-14
Zbl. 152.127   R.Zh. 1B74(66)   M.R. 32♯4535

[107] G.A. Dzhanashija: *On the Carleman problem for Gevrey function classes* (Russian), Doklady Akad. Nauk SSSR 145, 2 (1962), 259-262 [= Soviet Math., Doklady 3, 2 (1962), 969-972]
Zbl. 136.40   R.Zh. 7B87(63)   M.R. 26♯1410

[108] G. A. Dzhanashija: *On the superposition of two functions from Gevrey classes* (Russian), Soobsh. Akad. Nauk Gruzin. SSR 33 (1964), 257-261
Zbl. 148.287   M.R. 29♯2349

[109] M. A. Efendijev: *On the F-QL property of some class of nonlinear operators* (Russian), Izvestija Akad. Nauk Azerbajdzh. SSR, Ser. Fiz.-Teh. Mat. Nauk 7 (1986), 11-14
R.Zh. 5B1002(88)

[110] S. R. Firshtejn: *On a compactness criterion in the Banach space $C^{l+\alpha}[0,1]$*, Izvestija Vyssh. Uchebn. Zaved. Mat. 8 (1969), 117-118
Zbl. 186.448   M.R. 40♯1761

[111] H. Flaschka: *A nonlinear spectral theorem for abstract Nemytskii operators*, Indiana Univ. Math. J. 21 (1971), 265-276
Zbl. 215.213   R.Zh. 7B790(73)   M.R. 44♯4595

[112] A. Fougères: *Intégrandes de Young et cônes d'Orlicz associés*, C. R. Acad. Sci. Paris 283 (1976), 759 - 762
Zbl. 368.46034   R.Zh. 5B349(77)   M.R. 55♯1146

[113] A. Fougères: *Intégrandes de Young et cônes d'Orlicz associés*, Travaux Sém. Anal. Conv. 6, 10 (1976), 10 p.
Zbl. 356.46033   M.R.58♯30114

[114] L. E. Fraenkel: *Formulae for high derivatives of composite functions*, Math. Proc. Cambridge Philos. Soc. 83 (1978), 159-165
Zbl. 388.46032   M.R. 58♯6124

[115] A. Friedman: *Generalized functions and partial differential equations*, Prentice-Hall, New York 1963
Zbl. 116.070   R.Zh. 7B269(64)   M.R. 29♯2672

[116] S. Fučik, A. Kufner: *Nonlinear differential equations*, Elsevier, Amsterdam-Oxford-New York 1980
Zbl. 474.35001   R.Zh. 11B495(81)   M.R. 81e35001

[117] E. Gagliardo: *Proprietà di alcune classi di funzioni di più variabili*, Ricerche Mat. 7, 1 (1958), 102-137

[118] M. Gevrey: *Sur la nature analytique des solutions des équations aux dérivées partielles*, Ann. Sci. École Norm. Sup. Paris 35 (1918), 129-190

[119] D. Gilbarg, N.S. Trudinger: *Elliptic partial differential equations of second order*, Springer, Berlin 1977
Zbl. 361.35003   R.Zh. 10B328(78)   M.R. 57♯13109

[120] M. Goebel: *On Fréchet-differentiability of Nemytskij operators acting in Hölder spaces*, Preprint Bergakad. Freiberg 1988

[121] M. L. Gol'dman: *Imbedding of generalized Hölder classes* (Russian), Mat. Zametki 12 (1972), 325-336 [= Math. Notes 12 (1972), 626-631]
Zbl. 253.46073   R.Zh. 12B80(72)   M.R. 48♯2746

[122] J.-P. Gossez: *Nonlinear elliptic boundary value problems for equations with rapidly (or slowly) increasing coefficients*, Trans. Amer. Math. Soc. 190 (1974), 163 – 205
Zbl. 277.35052   M.R. 49♯7598

[123] Z. Grande: *Sur la mesurabilité des fonctions de deux variables*, Bull. Acad. Polon. Sci., Sér. Sci. Math. Astron. Phys. 21 (1973), 813-816
Zbl. 267.26005   R.Zh. 4B102(74)   M.R. 48♯8736

[124] Z. Grande: *La mesurabilité des fonctions de deux variables*, Bull. Acad. Polon. Sci., Sér. Sci. Math. Astron. Phys. 22 (1974), 657-661
Zbl. 287.28003   R.Zh. 3B44(75)   M.R. 50♯2433

[125] Z. Grande: *On the measurability of functions of two variables*, Proc. Cambridge Philos. Soc. 77 (1975), 335-336
Zbl. 296.28007   R.Zh. 9B58(75)   M.R. 51♯841

[126] Z. Grande: *Les fonctions qui ont la propriété (K) et la mesurabilité des fonctions de deux variables*, Fundamenta Math. 93 (1976), 155-160
Zbl. 347.28007   R.Zh. 8B50(77)   M.R. 55♯5827

[127] Z. Grande: *Quelques remarques sur un théorème de Kamke et les functions sup-mesurables*, Real Anal. Exch. 4, 2 (1979), 167-177
Zbl. 407.28003   M.R. 80j28009

[128] Z. Grande: *Un exemple de fonction non mesurable*, Revue Roumaine Math. Pure Appl. 24 (1979), 101-102
Zbl. 401.28006   R.Zh. 10B45(79)   M.R. 80i28011

[129] Z. Grande: *La mesurabilité des fonctions de deux variables et de la superposition* $F(x, f(x))$, Dissertationes Math., Warszawa 159 (1979), 45 p.
Zbl. 399.28005   R.Zh. 10B46(79)   M.R. 80j28013

[130] Z. Grande: *Semiéquicontinuité approximative et mesurabilité*, Colloquium Math. 45, 1 (1981), 133-135
Zbl. 492.26004   R.Zh. 11B66(82)   M.R. 83f28003

[131] Z. Grande: *Les problèmes concernant les fonctions réelles*, Problemy Mat. 3 (1982), 11-27
M.R. 85a26020

[132] E. Grande, Z. Grande: *Quelques remarques sur la superposition* $F(x, f(x))$, Fundamenta Math. 121 (1984), 199-211
Zbl. 573.28007   R.Zh.7B52(85)   M.R. 86c28011

[133] Z. Grande, J. S. Lipiński: *Un exemple d'une fonction sup-mesurable qui n'est pas mesurable*, Colloquium Math. 39, 1 (1978), 77-79
Zbl. 386.28006   R.Zh. 4B56(79)   M.R. 80b28006

[134] Ju. I. Gribanov: *Nonlinear operators in Orlicz spaces* (Russian), Uchen. Zapiski Kazansk. Univ. 115, 7 (1955), 5-13

[135] J. J. Grobler: *Orlicz spaces – a survey of certain aspects*, Math. Centrum Amsterdam 149 (1982), 1-12
Zbl. 492.46024   R.Zh. 11B780(82)   M.R. 83i46033

[136] D.J. Guo: *Properties and applications of V.V. Nemytskij operators* (Chinese), Shuxue Jinzhan 6 (1963), 7-91
M.R. 32♮4494

[137] A. I. Gusejnov, H. Sh. Muhtarov: *On the structure of a certain nonlinear operator and existence theorems for bounded solutions to nonlinear singular integral equations with Cauchy kernel* (Russian), Izvestija Akad. Nauk Azerbajdzh. SSR, Ser. Fiz.-Teh. Mat. Nauk 6 (1963), 107-113
Zbl. 178.471   R.Zh. 11B295(64)

[138] A. I. Gusejnov, H. Sh. Muhtarov: *Introduction to the theory of nonlinear singular integral equations* (Russian), Nauka, Moskva 1980
Zbl. 474.45003   R.Zh. 10B480(80)   M.R. 82g45013

[139] P. R. Halmos: *Measure Theory*, Springer, New York 1974
Zbl. 283.28001   R.Zh. 6B1061(75)

[140] G. Hardy: *Extensions of theorems of Gagliardo and Marcus and Mizel to Orlicz spaces,* Bull. Austral. Math. Soc. 23, 1 (1981), 121-138
Zbl. 451.46022   R.Zh. 12B942(81)   M.R. 82f46035

[141] G. Hardy: *Nemitsky operators between Orlicz-Sobolev spaces,* Bull. Austral. Math. Soc. 30, 2 (1984), 251-269
Zbl. 548.46028   R.Zh. 11B1221(85)   M.R. 86d46028

[142] G. Hardy: *Some theorems on Orlicz-Sobolev spaces, and an application to Nemitsky operators,* Proc. Centre Math. Anal. Austral. Nat. Univ. 9 (1985), 200-205

[143] G. Hardy: *Demicontinuity of Nemitsky operators on Orlicz-Sobolev spaces,* Bull. Austral. Math. Soc. 37, 1 (1988), 29-42
Zbl. 638.47064   R.Zh. 10B1139(88)   M.R. 89c47083

[144] G. H. Hardy, J. E. Littlewood, G. Pólya: *Inequalities,* Cambridge University Press, London 1934

[145] D. Haudidier: *Théorie spectrale non linéaire et opérateurs de Nemitski,* Preprint Univ. Brest 1976

[146] S. Hechscher: *Variable Orlicz spaces,* Proc. Acad. Amsterdam A 64 (1961), 229-241
R.Zh. 4B330(62)

[147] R. A. Hunt: *On $L(p, q)$ spaces,* Enseignement Math. 12 (1966), 249-276
Zbl. 181.403   R.Zh. 9B507(68)   M.R. 36♯6921

[148] A. D. Ioffe: *Monotone superpositions in Orlicz spaces* (Russian), Doklady Akad. Nauk SSSR 201, 4 (1971), 784-786 [= Soviet Math., Doklady 12, 6 (1971), 1719-1722]
Zbl. 245.46040   R.Zh. 4B1028(72)   M.R. 47♯7529

[149] A. D. Ioffe, V. M. Tihomirov: *Duality of convex functions and extremal problems* (Russian), Uspehi Mat. Nauk 23, 6 (1968), 57-116 [= Russian Math. Surveys 23, 6 (1968), 53-124]
Zbl. 167.422   R.Zh. 8B484(69)   M.R. 44♯5797

[150] Ju. N. Jelenskij: *The Nemytskij operator in cones of spaces of continuous functions* (Russian), Perm' 1982 (VINITI No. 6437-82)
R.Zh. 4B1020(83)

[151] Ju. N. Jelenskij, N. Ju. Kaljuzhnaja, S. V. Netsvetajeva: *Acting of the Nemytskij operator in cones of the space of continuous functions* (Russian), Perm' 1987 (VINITI No. 8055-V87)
R.Zh. 3B898(88)

[152] N. A. Jerzakova: *On a measure of noncompactness* (Russian), Kujbyshev Gos. Univ. Priblizh. Kach. Metody Issled. Diff. Uravn. 1 (1982), 58-60
R.Zh. 12B1190

[153] N. A. Jerzakova: *On measures of noncompactness in Banach spaces* (Russian), Kand. Diss., Univ. Voronezh 1983

[154] M. Josephy: *Composing functions of bounded variation*, Proc. Amer. Math. Soc. 83, 2 (1981), 354-356
Zbl. 475.26005   R.Zh. 3B74(85)   M.R. 83c26009

[155] A. S. Kalitvin: *The superposition operator in spaces with mixed quasi-norms* (Russian), Leningrad 1984 (VINITI No. 6883-84)
R.Zh. 2B1099(85)

[156] L. V. Kantorovich, G. P. Akilov: *Functional analysis* (Russian), Nauka, Moskva 1977 [Engl. transl.: Pergamon Press, Oxford 1982]
Zbl. 357.46020   R.Zh. 10B679(78)   M.R. 56♮16402

[157] L. V. Kantorovich, B. Z. Vulih, A. G. Pinsker: *Functional analysis in semiordered spaces* (Russian), Gostehizdat, Moskva 1950

[158] K. Karták: *A generalization of the Carathéodory theory of differential equations*, Czechoslov. Math. J. 17 (1967), 482-514
Zbl. 189.371   R.Zh. 8B171(68)   M.R. 36♮4050

[159] K. Karták: *On Carathéodory operators*, Czechoslov. Math. J. 17 (1967), 515-519
Zbl. 189.399   R.Zh. 11B183(68)   M.R. 36♮4051

[160] N. V. Kirpotina: *On the continuity of the Nemytskij operator in Banach spaces* (Russian), Metody Reor. Diff. Uravn. Prilozh. Moskva 339 (1975), 76-81
R.Zh. 7B956(76)

[161] W. Kozłowski: *Nonlinear operators in Banach function spaces*, Comm. Math. Prace Mat. 22 (1980), 85-103
Zbl. 471. 47035   R.Zh. 9B835(81)   M.R. 82k47083

[162] W. Kozłowski: *A note on the continuity of nonlinear operators*, Coll. Math. Soc. János Bolyai 35 (1980), 745-750
Zbl. 537.46033   M.R. 86g47084

[163] W. Kozłowski: *Modular function spaces*, M. Dekker, New York 1988
Zbl. 662.46023

[164] M. A. Krasnosel'skij: *Continuity conditions for some nonlinear operators* (Russian), Ukrain. Mat. Zhurn. 2, 3 (1950), 70-86

[165] M. A. Krasnosel'skij: *On the continuity of the operator $Fu(x) = f(x, u(x))$* (Russian), Doklady Akad. Nauk SSSR 77, 2 (1951), 185-188

[166] M. A. Krasnosel'skij: *Topological methods in the theory of nonlinear integral equations* (Russian), Gostehizdat, Moskva 1956 [Engl. transl.: Macmillan, New York 1964]
Zbl. 70.330   R.Zh. 3310(57)   M.R. 28♯2414

[167] M. A. Krasnosel'skij: *On quasilinear operator equations* (Russian), Doklady Akad. Nauk SSSR 214, 4 (1974), 761-764 [= Soviet Math., Doklady 15, 1 (1974), 237-241]
Zbl. 294.47040   R.Zh. 6B943(74)   M.R. 49♯7859

[168] M. A. Krasnosel'skij: *Personal communication* 1983

[169] M. A. Krasnosel'skij, L. A. Ladyzhenskij: *Conditions for the complete continuity of the P. S. Uryson operator* (Russian), Trudy Moskov. Mat. Obshch. 3 (1954), 307-320

[170] M. A. Krasnosel'skij, A. V. Pokrovskij: *Regular solutions of equations with discontinuous nonlinearities* (Russian), Doklady Akad. Nauk SSSR 226, 3 (1976), 506-509 [= Soviet Math., Doklady 17, 1 (1976), 128-132]
Zbl. 341.45013   R.Zh. 5B352(76)   M.R. 58♯30559

[171] M. A. Krasnosel'skij, A. V. Pokrovskij: *On a discontinuous superposition operator* (Russian), Uspehi Mat. Nauk 32, 1 (1977), 169-170
Zbl. 383.47037   R.Zh. 7B976(77)   M.R. 56♯568

[172] M. A. Krasnosel'skij, A. V. Pokrovskij: *Continuity points of monotone operators* (Russian), Comm. Math., Tom. Spec. Honor. L. Orlicz II (1979), 205-216
Zbl. 443.47055   R.Zh. 3B793(80)   M.R. 80k47061

[173] M. A. Krasnosel'skij, A. V. Pokrovskij: *Equations with discontinuous nonlinearities* (Russian), Doklady Akad. Nauk SSSR 248, 5 (1979), 1056-1059 [= Soviet Math., Doklady 20, 5 (1979), 1117-1120]
Zbl. 468.45004   R.Zh. 3B797(80)   M.R. 81c45010

[174] M. A. Krasnosel'skij, A. V. Pokrovskij: *Systems with hysteresis* (Russian), Nauka, Moskva 1983 [Engl. transl.: Springer, Berlin- Heidelberg-New York-Tokyo 1989]
Zbl. 665.47038   M.R. 86e93005

[175] M. A. Krasnosel'skij, Ja. B. Rutitskij: *Differentiability of nonlinear integral operators in Orlicz spaces* (Russian), Doklady Akad. Nauk SSSR 89, 4 (1953), 601-604

[176] M. A. Krasnosel'skij, Ja. B. Rutitskij: *On some nonlinear operators in Orlicz spaces* (Russian), Doklady Akad. Nauk SSSR 117, 3 (1957), 363-366

[177] M. A. Krasnosel'skij, Ja. B. Rutitskij: *Orlicz spaces and nonlinear integral equations* (Russian), Trudy Moskov. Mat. Obshch. 7 (1958), 63-120

[178] M. A. Krasnosel'skij, Ja. B. Rutitskij: *Convex functions and Orlicz spaces* (Russian), Fizmatgiz, Moskva 1958 [Engl. transl.: Noordhoff, Groningen 1961]
Zbl. 95.91   R.Zh. 2B405(61)   M.R. 21♯5144

[179] M. A. Krasnosel'skij, Ja. B. Rutitskij, R. M. Sultanov: *On a nonlinear operator which acts in a space of abstract functions* (Russian), Izvestija Akad. Nauk Azerbajdzh. SSR, Ser. Fiz.-Teh. Mat. Nauk 3 (1959), 15-21

[180] M. A. Krasnosel'skij, G. M. Vajnikko, P. P. Zabrejko, Ja. B. Rutitskij, V. Ja. Stetsenko: *Approximate solutions of operator equations* (Russian), Nauka, Moskva 1969 [Engl. transl.: Noordhoff, Groningen 1972]
Zbl. 194.179   R.Zh. 4B797(70)   M.R. 41♯4271

[181] M.A. Krasnosel'skij, P.P. Zabrejko: *Geometrical methods of nonlinear analysis* (Russian), Nauka, Moskva 1975 [Engl. transl.: Springer, New York 1984]
Zbl. 326.47052   R.Zh. 7B945(76)   M.R. 58♯17976

[182] M. A. Krasnosel'skij, P. P. Zabrejko, Je. I. Pustyl'nik, P. Je. Sobolevskij: *Integral operators in spaces of summable functions* (Russian), Nauka, Moskva 1966 [Engl. transl.: Noordhoff, Leyden 1976]
Zbl. 145.397   R.Zh. 3B507(67)   M.R. 34♯6568

[183] S. G. Krejn, Ju. I. Petunin, Je. M. Semjonov: *Hyperscales of Banach structures* (Russian), Doklady Akad. Nauk SSSR 170, 2 (1966), 265-267 [= Soviet Math., Doklady 7, 5 (1966), 1185-1188]
Zbl. 168.108   R.Zh. 1B340(67)   M.R. 34♯3299

[184] S. G. Krejn, Ju. I. Petunin, Je. M. Semjonov: *Interpolation of linear operators* (Russian), Nauka, Moskva 1978 [Engl. transl.: Transl. Math. Monogr. 54, Amer. Math. Soc., Providence 1982]
Zbl. 499.46044   R.Zh. 12B996(78)   M.R. 81f46086

[185] A. Kufner, O. John, S. Fučik: *Function spaces*, Noordhoff, Leyden 1977
Zbl. 364.46022   R.Zh. 11B777(78)   M.R. 58♯2189

[186] K. Kuratowski: *Topology*, Academic Press, New York - London - Warszawa 1966
Zbl. 158.408   R.Zh. 4A369(68)   M.R. 36♯840

[187] L. A. Ladyzhenskij: *Conditions for the complete continuity of the P. S. Uryson integral operator in the space of continuous functions* (Russian), Doklady Akad. Nauk SSSR 96, 6 (1954), 1105-1108

[188] V. I. Levchenko, I. V. Shragin: *The Nemytskij operator acting from the space of continuous functions into an Orlicz-Nakano space* (Russian), Mat. Issled. 3, 3 (1968), 91-100
Zbl. 241.47044   R.Zh. 7B572(69)   M.R. 42♯3626

[189] K. Lichawski, J. Matkowski, J. Miś: *Locally defined operators in the space of differentiable functions*, Preprint Univ. Bielsko - Biała 1988

[190] J. L. Lions, E. Magenes: *Problèmes aux limites non homogénes et applications III*, Dunod, Paris 1970
Zbl. 197.67   R.Zh. 1B381(71)   M.R. 45♯975

[191] J. S. Lipiński: *On the measurability of functions of two variables*, Bull. Acad. Polon. Sci., Sér. Sci. Math. Astron. Phys. 20 (1972), 131-135
Zbl. 228.28009   R.Zh. 8B44(72)   M.R. 46♯9274

[192] F. Liu: *Luzin property*, Presses Univ. Montréal 95 (1985), 156-165
Zbl. 596.28006   M.R. 86m28003

[193] A. G. Ljamin: *On the acting problem for the Nemytskij operator in the space of functions of bounded variation* (Russian), 11th School Theory Oper. Function Spaces, Chel'jabinsk 1986, 63-64

[194] M. Loève: *Probability theory*, Van Nostrand, Princeton 1955

[195] G. G. Lorentz: *On the theory of the spaces Λ*, Pacific J. Math. 1 (1950), 411-429

[196] G. G. Lorentz: *Bernstein polynomials*, Univ. Toronto Press, Toronto 1953

[197] G.Ja. Lozanovskij: *Reflexive spaces which generalize reflexive Orlicz spaces* (Russian), Doklady Akad. Nauk SSSR 163, 3 (1965), 573-576 [= Soviet Math., Doklady 6,2 (1965), 968-971]
Zbl. 139.302   R.Zh. 11B373(65)   M.R. 33♯539

[198] G.Ja. Lozanovskij: *On Banach structures of Calderón type* (Russian), Doklady Akad. Nauk SSSR 172, 5 (1967), 1018-1020 [= Soviet Math., Doklady 8, 1 (1967), 224-227]
Zbl. 156.366   R.Zh. 7B393(67)   M.R. 34♯8155

[199] G.Ja. Lozanovskij: *Banach structures and their transforms* (Russian), Doct. Diss., Univ. Leningrad 1973

[200] R. Lucchetti, F. Patrone: *On Nemytskii's operator and its application to the lower semi-continuity of integral fractionals*, Indiana Univ. Math. J. 29 (1980), 703-713
Zbl. 476.47049   R.Zh. 6B695(81)   M.R. 82i47104

[201] A. Lunardi: *An extension of Schauder's theorem to little-Hölder continuous functions*, Bull. Unione Mat. Ital. (6) 3-C (1984), 25-34
Zbl. 552.35041   R.Zh. 1B467(85)   M.R. 86b35039

[202] W. A. J. Luxemburg, A. C. Zaanen: *Riesz spaces I*, North-Holland Publ. Co., Amsterdam 1971
Zbl. 231.46014   R.Zh. 2B661(74)   M.R. 58♯23483

[203] P. Mänz: *On the continuity of some nonlinear operators* (Russian), Kand. Diss., Univ. Voronezh 1967

[204] P. Mänz: *On the continuity of the superposition operator acting in normed product spaces* (Russian), Litovsk. Mat. Sbornik 7, 2 (1967), 289-296
Zbl. 177.199   R.Zh. 9B527(68)

[205] P. Mänz: *Der Kompositionsoperator in Räumen abstrakter Funktionen*, Math. Nachr. 67 (1975), 327-335
Zbl. 312.46047   M.R. 52♯6520

[206] A. S. Makarov: *Continuity criteria for the superposition operator* (Russian), Prilozh. Funk. Anal. Priblizh. Vychisl. Kazan. Univ. (1974), 80-83
Zbl. 338.47034   R.Zh. 1B1109(75)   M.R. 58♯30454

[207] L. Maligranda: *Indices and interpolation*, Dissertationes Math., Warszawa 234 (1985), 1-46
Zbl. 566.46038   M.R. 87k46059

[208] M. Marcus, V. J. Mizel: *Absolute continuity on tracks and mappings of Sobolev spaces*, Arch. Rat. Mech. Analysis 45 (1972), 294-320
Zbl. 236.46033   R.Zh. 1B475(73)   M.R. 49♯3529

[209] M. Marcus, V. J. Mizel: *Nemytskij operators on Sobolev spaces*, Arch. Rat. Mech. Analysis 51 (1973), 347-370
Zbl. 266.46029   R.Zh. 4B851(74)   M.R. 50♯978

[210] M. Marcus, V. J. Mizel: *Continuity of certain Nemitsky operators on Sobolev spaces and the chain rule*, J. Analyse Math. 28 (1975), 303-334
Zbl. 328.46028   R.Zh. 8B1005(76)   M.R. 58♯2512

[211] M. Marcus, V. J. Mizel: *Representation theorems for nonlinear disjointly additive functionals and operators on Sobolev spaces*, Trans. Amer. Math. Soc. 228 (1977), 1-45
Zbl. 351.46021   R.Zh. 4B633(78)   M.R. 56♯12871

[212] M. Marcus, V. J. Mizel: *Extension theorems of Hahn-Banach type for nonlinear disjointly additive functionals and operators in Lebesgue spaces*, J. Funct. Anal. 24 (1977), 303-335
Zbl. 342.46021   R.Zh. 1B575(78)   M.R. 57♯1102

[213] M. Marcus, V. J. Mizel: *Extension theorems for nonlinear disjointly additive functionals and operators on Lebesgue spaces with applications*, Bull. Amer. Math. Soc. 82, 1 (1976), 115-117
Zbl. 336.46051   R.Zh. 5B39(77)   M.R. 53♯6305

[214] M. Marcus, V. J. Mizel: *Superposition mappings which operate on Sobolev spaces*, Nonlinear Anal., Theory, Methods Appl. 2, 2 (1978), 257-258
Zbl. 387.46035   R.Zh. 9B681(78)

[215] M. Marcus, V. J. Mizel: *Every superposition operator mapping one Sobolev space into another is continuous*, J. Funct. Anal. 33 (1979), 217-229
Zbl. 418.46024   R.Zh. 3B791(80)   M.R. 80h47039

[216] M. Marcus, V. J. Mizel: *Complete characterization of functions which act, via superposition, on Sobolev spaces*, Trans. Amer. Math. Soc. 251 (1979), 187-218
Zbl. 417.46035   R.Zh. 1B932(80)   M.R. 80j46055

[217] M. Marcus, V. J. Mizel: *A characterization of first order nonlinear partial differential operators on Sobolev spaces*, J. Funct. Anal. 38 (1980), 118-138
Zbl. 444.47048   R.Zh. 1B1132(81)   M.R. 81j47040

[218] A. Matkowska: *On the characterization of Lipschitzian operators of substitution in the class of Hölder's functions*, Zeszyty Nauk. Politech. Łódz. Mat. 17 (1984), 81-85
Zbl. 599.46032   R.Zh. 10B1273(85)   M.R. 86k46034

[219] J. Matkowski: *Functional equations and Nemytskij operators*, Funkc. Ekvacioj Ser. Int. 25 (1982), 127-132
Zbl. 504.39008   M.R. 84i45008

[220] J. Matkowski: *Form of Lipschitz operators of substitution in Banach spaces of differentiable functions*, Zeszyty Nauk. Politech. Łódz. Mat. 17 (1984), 5-10
Zbl. 599.46031   R.Zh. 11B1220(85)   M.R. 87a47097

[221] J. Matkowski: *Personal communication* 1987

[222] J. Matkowski: *On Nemytskii Lipschitzian operator*, Acta Univ. Carolinae 28, 2 (1987), 79-82
Zbl. 639.47052   M.R. 89g47095

[223] J. Matkowski: *On Nemytskii operator*, Math. Japonica 33, 1 (1988), 81-86
Zbl. 644.47061   R.Zh. 9B1085(88)   M.R. 89g47094

[224] J. Matkowski, J. Miś: *On a characterization of Lipschitzian operators of substitution in the space BV < a, b >*, Math. Nachr. 117 (1984), 155-159
Zbl. 566.47033   R.Zh. 2B1097(85)   M.R. 85i47075

[225] I. V. Misjurkejev: *On the continuity of some nonlinear operator* (Russian), Voronezh. Gos. Univ. Trudy Sem. Funk. Anal. 6 (1958), 93

[226] B. S. Mitjagin: *An interpolation theorem for modular spaces* (Russian), Mat. Sbornik 66, 4 (1965), 473-482
Zbl. 142.107   R.Zh. 12B488   M.R. 31♯1562

[227] V. J. Mizel: *Characterization of nonlinear transformations possessing kernels*, Canadian J. Math. 22 (1970), 449-471
Zbl. 203.143   M.R. 41♯7495

[228] V. J. Mizel: Personal communication 1984

[229] V. B. Mosejenkov: *Composition of functions in Sobolev spaces* (Russian), Ukrain. Mat. Zhurn. 34, 3 (1982), 384-388 [= Ukrain. Math. J. 34, 3 (1982), 316-319]
Zbl. 545.46020   R.Zh. 10B23(82)   M.R. 84d46040

[230] J. Moser: *A rapidly convergent iteration method and non-linear differential equations I*, Ann. Scuola Norm. Sup. Pisa 20, 2 (1966), 265-315
Zbl. 144.182   R.Zh. 6B336(67)   M.R. 33♯7667

[231] J. Moser: *A rapidly convergent iteration method and non-linear differential equations II*, Ann. Scuola Norm. Sup. Pisa 20, 3 (1966), 499-535
Zbl. 144.182   R.Zh. 8B320(67)   M.R. 34♯6280

[232] W. Müller, G. Bruckner: *On local Hölder continuity and monotonicity of superposition operators*, Theory Nonlin. Oper. Constr. Aspects, Proc. Int. Summer School Berlin (1977), 355-362
Zbl. 381.47027   R.Zh. 7B943(78)   M.R. 58♯2469

[233] H. Sh. Muhtarov: *On the properties of the operator $Fu = f(u(x))$ in the space H_φ* (Russian), Sbornik Nauchn. Rabot Mat. Kaf. Dagestan. Univ. (1967), 145-150

[234] J. Musielak: *Orlicz spaces and modular spaces*, Lect. Notes Math. 1034 (1983), 1-222
Zbl. 557.46020   R.Zh. 7B703(84)   M.R. 85m46028

[235] I. P. Natanson: *Theory of functions of a real variable* (Russian), GITTL, Moskva 1950 [Deutsche Übersetzung.: Akad.-Verlag, Berlin 1969]

[236] V. I. Nazarov: *The superposition operator in Roumieu spaces of infinitely differentiable functions* (Russian), Izvestija Akad. Nauk BSSR 5 (1984), 22-28
Zbl. 558.46021   R.Zh. 4B1164(85)   M.R. 86d47071

[237] V. I. Nazarov: *A nonlinear differential equation of first order in Roumieu spaces* (Russian), Doklady Akad. Nauk BSSR 28, 9 (1984), 780-783
Zbl. 565.47032   R.Zh. 2B1123(85)   M.R. 86b34007

[238] V. V. Nemytskij: *Existence and uniqueness theorems for nonlinear integral equations* (Russian), Mat. Sbornik 41, 3 (1934), 421-452

[239] V. V. Nemytskij: *On a certain class of nonlinear integral equations* (Russian), Mat. Sbornik 41, 4 (1934), 655-658

[240] V. V. Nemytskij: *The fixed point method in analysis* (Russian), Uspehi Mat. Nauk 1 (1936), 141-174

[241] V. V. Nemytskij: *The structure of the spectrum of nonlinear completely continuous operators* (Russian), Mat. Sbornik 33, 3 (1953); Erratum: Mat. Sbornik 35, 1 (1954)

[242] H. T. Nguyên: *The superposition operator in Orlicz spaces of vector-valued functions* (Russian), Doklady Akad. Nauk BSSR 31,3 (1987), 197-200
Zbl. 622.47064   R.Zh. 7B1021(87)   M.R. 88g47134

[243] H. T. Nguyên: *Orlicz spaces of vector functions and their application to nonlinear integral equations* (Russian), Kand. Diss., Univ. Minsk 1987

[244] Z. Nowak: *Composition operators on Lipschitz spaces and the absolute convergence of Fourier and Taylor series for superpositions of functions*, Arch. Math. 41 (1983), 454-458
Zbl. 514.46016   R.Zh. 5B57(84)   M.R. 85e42006

[245] R. Nugari: *Continuity and differentiability properties of the Nemitskii operator in Hölder spaces*, Glasgow Math. J. 30 (1988), 59-65
Zbl. 637.47035   R.Zh. 11B1158(88)   M.R. 89c47084

[246] T. K. Nurekenov: *Necessary and sufficient conditions for Uryson and Nemytskij operators to satisfy a Lipschitz condition* (Russian), Alma-Ata 1981 (VINITI No. 1459-81)
R.Zh. 7B957(81)

[247] T. K. Nurekenov: *Necessary and sufficient conditions for Uryson operators to satisfy a Lipschitz condition* (Russian), Izvestija Akad. Nauk Kazach. SSR, Ser. Fiz.-Mat. 3 (1983), 79-82
Zbl. 519.45012   R.Zh. 7B957(81)   M.R. 84m45041

[248] T. K. Nurekenov: *Personal communication* 1983

[249] P. M. Obradovich: *On the continuity of the superposition operator* (Russian), Voronezh. Gos. Univ. Probl. Mat. Anal. Slozhn. Sistem 2 (1968), 78-80
R.Zh. 7B573(69)   M.R. 43♯7983

[250] R. O'Neil: *Convolution operators and $L(p,q)$ spaces*, Duke Math. J. 30, 1 (1963), 129-142
Zbl. 178.477   R.Zh. 10B389(63)   M.R. 26♯4193

[251] W. Orlicz: *Über eine gewisse Klasse von Räumen vom Typus B*, Bull. Intern. Acad. Polon. Sér. A 8 (1932), 207-220

[252] W. Orlicz: *Über Räume (L^M)*, Bull. Intern. Acad. Polon. Sér. A 12 (1936), 93-107

[253] M. Otelbajev, G. A. Suvorchenkova: *A necessary and sufficient condition of boundedness and continuity for a class of Uryson operators* (Russian), Sibir. Mat. Zhurn. 20, 2 (1979), 428-432 [= Siber. Math. J. 20, 2 (1979), 307-310]
Zbl. 432.45017   R.Zh. 8B941(79)   M.R. 80g47080

[254] R. S. Palais: *Foundations of global non-linear analysis*, Benjamin, New York - Amsterdam 1968
Zbl. 164.111   M.R. 40♯2130

[255] X. Pan: *Carathéodory operator on $C(X)$* (Chinese), J. Shandong Univ. 2 (1985), 1-5
Zbl. 632.47047

[256] K. R. Parthasarathy: *Probability measures on metric spaces*, Academic Press, New York 1967
Zbl. 153.191   M.R. 37♯2271

[257] J. P. Penot: *Continuité et differentiabilité des opérateurs de Nemitski*, Publ. Math. Univ. Pau 1976, VIII1 - VIII46

[258] J.-C. Peralba: *Topologie cônique faible sur les cônes d'Orlicz: Inf-compacité, sous-différentiel et application à un théorème de surjectivité*, C. R. Acad. Sci. Paris 283 (1976), 771-774
Zbl. 402.46019   R.Zh. 5B350(77)   M.R. 55♯1149

[259] J.-C. Peralba: *Topologie cônique faible sur les cônes d'Orlicz: Inf-compacité, sous-différentiel et application à la surjectivité de l'opérateur de Nemickii*, Travaux Sém. Anal. Conv. 6, 11 (1976), 13 p.
Zbl. 356.46034   M.R. 58♯30115

[260] R. Płuciennik: *On some properties of the superposition operator in generalized Orlicz spaces of vector-valued functions*, Comm. Math. Prace Mat. 25 (1985), 321-337
Zbl. 608.47068   R.Zh. 11B985(86)   M.R. 87i46069

[261] R. Płuciennik: *Boundedness of the superposition operator in generalized Orlicz spaces of vector-valued functions*, Bull. Pol. Acad. Sci. Math. 33, 9/10 (1985), 531-540
Zbl. 587.46027   R.Zh. 5B1033(87)   M.R. 87g46062

[262] R. Płuciennik: *The superposition operator in Musielak-Orlicz spaces of vector-valued functions*, Rend. Circolo Mat. Palermo II-14 (1987), 411-417
Zbl. 637.47036   R.Zh. 6B1217(88)   M.R. 89e47097

[263] R. Płuciennik: *Representation of additive functionals on Musielak–Orlicz spaces of vector-valued functions*, Kodai Math. J. 10, 1 (1987), 49-54
Zbl. 618.46034   R.Zh. 10B981(87)   M.R. 88b46064

[264] A. V. Ponosov: *Some problems in the theory of probabilistic topological spaces* (Russian), Perm'. Gos. Polit. Inst., Krajevye Zad. (1983), 97-100
R.Zh. 5B887(84)

[265] A. V. Ponosov: *Carathéodory operators and Nemytskij operators* (Russian), Perm '1984 (VINITI No. 5141-84)
R.Zh. 11B1042(84)

[266] A. V. Ponosov: *On the theory of locally defined operators* (Russian), Funkc. Diff. Uravn., Perm' (1985), 72-82
Zbl. 618.47056   R.Zh. 5B1195(86)   M.R. 89b47095

[267] A. V. Ponosov: *On the Nemytskij conjecture* (Russian), Doklady Akad. Nauk SSSR 289, 6 (1986), 1308-1311 [= Soviet Math., Doklady 34, 1 (1987), 231-233]
Zbl. 618.47055   R.Zh. 12B1270(86)   M.R. 88b47092

[268] A. I. Povolotskij, S. M. Redlih: *On the solvability of Hammerstein type equations* (Russian), Funk. Anal. 6 (1976), 132-141
Zbl. 407.45007   R.Zh. 8B803(77)   M.R. 58♮30568

[269] R. Randrananja: *Théorie des opérateurs dans certains F-espaces*, Doct. Diss., Univ. Madagascar 1988

[270] R. Randrananja: *Sur les h-opérateurs*, C. R. Acad. Sci. Paris 306 (1988), 667-669
Zbl. 643.47065   R.Zh. 1B1296(89)   M.R. 89c47085

[271] B. Ricceri: *Su due caratterizzazioni della proprietà di Scorza-Dragoni*, Matematiche 35, 1 (1980), 149-154
Zbl. 527.28006   R.Zh. 11B1281(83)   M.R. 84f28011

[272] B. Ricceri, A. Villani: *Separability and Scorza-Dragoni's property,* Matematiche 37, 1 (1982), 156-161
Zbl. 581.28004   R.Zh. 12A540(85)   M.R. 86h28005

[273] J. Robert: *Continuité d'un opérateur non linéaire sur certains espaces de suites,* C. R. Acad. Sci. Paris 259 (1964), 1287-1290
Zbl. 196.446   R.Zh. 5B516(65)   M.R. 29♯3875

[274] S. Roman: *The formula of Faà di Bruno,* Amer. Math. Monthly 87 (1980), 805-809
Zbl. 513.05009   R.Zh. 8B10(81)   M.R. 82d26003

[275] C. Roumieu: *Sur quelques extensions de la notion de distribution,* Ann. Sci. École Norm. Sup. Paris 77 (1960), 41-121
Zbl. 104.334   R.Zh. 9B410(61)   M.R. 22♯12377

[276] Ja. B. Rutitskij: *On a certain generalized Orlicz sequence space* (Russian), Nauchn. Trudy Voronezh. Inzhen.-Stroitel'-skogo Inst. 1 (1952), 271-286

[277] Ja. B. Rutitskij: *On a nonlinear operator in Orlicz spaces* (Ukrainian), Dopovidi Akad. Nauk Ukr. SSR 3 (1952), 161-163

[278] Ja. B. Rutitskij: *Application of Orlicz spaces for the investigation of certain functionals in* L^2 (Russian), Doklady Akad. Nauk SSSR 105 (1955), 1147-1150

[279] Ja. B. Rutitskij: *On integral operators in Orlicz spaces* (Russian), Doklady Akad. Nauk SSSR 145, 5 (1962), 1000-1003 [= Soviet Math., Doklady 3,2 (1962), 1149-1152]
Zbl. 171.353   R.Zh. 6B368(63)   M.R. 25♯5393

[280] B. N. Sadovskij: *Limit-compact and condensing operators* (Russian), Uspehi Mat. Nauk 27, 1 (1972), 81-146 [= Russian Math. Surveys 27,1 (1972), 85-155]
Zbl. 243.47033   R.Zh. 6B774(72)   M.R. 55♯1161

[281] M.-F. Sainte-Beuve: *On the extension of von Neumann-Aumann's theorem,* J. Funct. Anal. 17 (1974), 112-129
Zbl. 286.28005   R.Zh. 2B742(75)   M.R. 51♯10564

[282] G. Scorza-Dragoni: *Un teorema sulle funzioni continue rispetto ad una e misurabili rispetto ad un'altra variabile,* Rend. Sem. Mat. Univ. Padova 17 (1948), 102-106

[283] H. H. Schaefer: *Aspects of Banach lattices*, Math. Assoc. Amer. Studies 21 (1980), 158-227
Zbl. 494.46021   R.Zh. 9B610(82)   M.R. 82g46048

[284] Je. M. Semjonov: *On a scale of spaces having the interpolation property* (Russian), Doklady Akad. Nauk SSSR 148, 5 (1963), 1038-1041 [= Soviet Math., Doklady 4, 1 (1963), 235 - 239]
Zbl. 194.148   R.Zh. 8B376(63)   M.R. 26♯2870

[285] Je. M. Semjonov: *On imbedding theorems for Banach spaces of measurable functions* (Russian), Doklady Akad. Nauk SSSR 156, 6 (1964), 1292-1295 [= Soviet Math., Doklady 5, 1 (1964), 831-834]
Zbl. 136.109   R.Zh. 11B391(64)   M.R 30♯3368

[286] G. Ja. Shilov: *Differentiability of functions in linear spaces* (Russian), (with an appendix by P.P. Zabrejko), Jaroslav. Gos. Univ. (1978), 6-120
R.Zh. 9B829(80)   M.R. 82b26010

[287] T. Shimogaki: *A generalization of Vajnbergs theorem I*, Proc. Japan. Acad. 34, 8 (1958), 518-523

[288] T. Shimogaki: *A generalization of Vajnbergs theorem II*, Proc. Japan. Acad. 34, 10 (1958), 676-680

[289] I. V. Shragin: *On the weak continuity of the Nemytskij operator* (Russian), Uchen. Zapiski Moskov. Obl. Ped. Inst. 57 (1957), 73-79

[290] I. V. Shragin: *On some operators in generalized Orlicz spaces* (Russian), Doklady Akad. Nauk SSSR 117, 1 (1957), 40-43

[291] I. V. Shragin: *On some nonlinear operator* (Russian), Nauchn. Doklady Vyssh. Shkoly 2 (1958), 103-105

[292] I. V. Shragin: *The Nemytskij operator from C into L_M* (Russian), Uchen. Zapiski Moskov. Obl. Ped. Inst. 77, 5 (1959), 161-168

[293] I. V. Shragin: *On the weak continuity of the Nemytskij operator in generalized Orlicz spaces* (Russian), Uchen. Zapiski Moskov. Obl. Ped. Inst. 77, 5 (1959), 169-179

[294] I. V. Shragin: *On the measurability of some functions* (Russian), Uchen. Zapiski Moskov. Obl. Ped. Inst. 77, 5 (1959), 181-186

[295] I. V. Shragin: *On a nonlinear operator in Orlicz spaces* (Russian), Uspehi Mat. Nauk 14, 4 (1959), 233-235

[296] I. V. Shragin: *On the continuity of the Nemytskij operator in Orlicz spaces* (Russian), Uchen. Zapiski Moskov. Obl. Ped. Inst. 70 (1959), 49-51

[297] I. V. Shragin: *On the continuity of the Nemytskij operator in Orlicz spaces* (Russian), Doklady Akad. Nauk SSSR 140, 3 (1961), 543-545 [= Soviet Math., Doklady 2, 2 (1961), 1246-1248]
Zbl. 161.350   R.Zh. 8B456(64)   M.R. 28♯463

[298] I. V. Shragin: *On the continuity of the Nemytskij operator* (Russian), Trudy 5-j Vsjesojuznoj Konf. Funk. Anal. Akad. Nauk Azerbajdzh. SSR Baku (1961), 272-277
R.Zh. 7B416(62)

[299] I. V. Shragin: *On the boundedness of the Nemytskij operator in Orlicz spaces* (Russian), Kishin. Gos. Univ. Uchen. Zapiski 50 (1962), 119-121
R.Zh. 6B413(63)   M.R. 35♯5986

[300] I. V. Shragin: *Some properties of the Nemytskij operator in Orlicz spaces* (Russian), Mat. Sbornik 65, 3 (1964), 324-337 [= Math. USSR Sbornik
Zbl. 199.205   R.Zh. 4B540(65)   M.R. 30♯1394

[301] I. V. Shragin: *On the continuity of the Nemytskij operator in Orlicz spaces* (Russian), Kishin. Gos. Univ. Uchen. Zapiski 70 (1964), 49-51
R.Zh. 11B435(64)   M.R. 32♯2871

[302] I. V. Shragin: *The Hammerstein equation in the space of continuous functions* (Russian), Kishin. Gos. Univ. Uchen. Zapiski 82 (1965), 40-43
R.Zh. 9B399(66)   M.R. 34♯1891

[303] I. V. Shragin: *Measurability of the supremum and of its realizing function* (Russian), Mat. Issled. 3, 1 (1968), 215-220
R.Zh. 1B69(69)   M.R. 41♯5583

[304] I. V. Shragin: *Measurability of the superposition of discontinuous functions* (Russian), Trudy Tambov. Inst. Him. Mashinostroenija 3 (1969), 7-9
Zbl. 242.28005   R.Zh. 7B77(70)   M.R. 42♯6180

[305] I. V. Shragin: *The Nemytskij operator in spaces which are generated by gen-functions* (Russian), Doklady Akad. Nauk SSSR 189, 1 (1969), 63-66 [= Soviet Math., Doklady 10, 6 (1969), 1372-1375]
Zbl. 197.404   R.Zh. 6B747(70)   M.R. 40♮7837

[306] I. V. Shragin: *Conditions for the measurability of superpositions* (Russian), Doklady Akad. Nauk SSSR 197, 2 (1971), 295-298 [= Soviet Math., Doklady 12, 2 (1971) 465-470]; Erratum: Doklady Akad. Nauk SSSR 200, 1 (1971), vii
Zbl. 222.26014   R.Zh. 7B88(71)   M.R. 42♮3561

[307] I. V. Shragin: *The superposition operator in co-ordinate spaces of Orlicz type* (Russian), Trudy Tambov. Inst. Him. Mashinostroenija 6 (1971), 80-88
          R.Zh. 10B679(71)

[308] I. V. Shragin: *The superposition operator in modular function spaces* (Russian), Studia Math. 43 (1972), 61-75
Zbl. 278.46029   R.Zh. 11B68(72)   M.R. 47♮915

[309] I. V. Shragin: *The Nemytskij operator that is generated by a non-Carathéodory function* (Russian), Uspehi Mat. Nauk 27, 3 (1972), 217-218
Zbl. 249.47070   R.Zh. 11B963(72)   M.R. 52♮11687

[310] I. V. Shragin: *Superpositional measurability* (Russian), Izvestija Vyssh. Uchebn. Zaved. Mat. 1 (1975), 82-89
Zbl. 333.60006   R.Zh. 10B793(75)   M.R. 51♮10565

[311] I. V. Shragin: *Co-ordinate spaces and the superposition operator in them* (Russian), Perm' 1975 (VINITI No. 31-75)
          R.Zh. 5B902(75)

[312] I. V. Shragin: *Abstract Nemytskij operators are locally defined operators* (Russian), Doklady Akad. Nauk SSSR 227, 1 (1976), 47-49 [= Soviet Math., Doklady 17, 2 (1976), 354-357]
Zbl. 338.47035   R.Zh. 7B955(76)   M.R. 54♮1039

[313] I. V. Shragin: *Imbedding conditions for classes of sequences and some consequences* (Russian), Mat. Zametki 20, 5 (1976), 681-692 [= Math. Notes 20, 5 (1976), 942-948]
Zbl. 349.46011   R.Zh. 3B370(77)   M.R. 56♮1022

[314] I. V. Shragin: *On the continuity of locally defined operators* (Russian), Doklady Akad. Nauk SSSR 232, 2 (1977), 292-295 [= Soviet Math., Doklady 18, 1 (1977), 75-78]
Zbl.372.46032   R.Zh. 5B567(77)   M.R. 56♮6495

[315] I. V. Shragin: *Conditions for convergence of superpositions* (Russian), Differentsial'nye Uravn. (1977), 1900-1901 [= Differential Equ. 13, 10 (1977), 1326-1327]
Zbl. 365.28004   R.Zh. 3B793(78)   M.R. 57♯16526

[316] I. V. Shragin: *The necessity of the Carathéodory conditions for the continuity of the Nemytskij operator* (Russian), Perm'. Gos. Polit. Inst., Funkc.-Diff. Krajev. Zad. Mat. Fiz. (1978), 128-134
R.Zh. 8B661(78)

[317] I. V. Shragin: *On the Carathéodory conditions* (Russian), Uspehi Mat. Nauk 34, 3 (1979), 219-220 [= Russian Math. Surveys 34, 3 (1970), 220-221]
Zbl. 417.28003   R.Zh. 11B988(79)   M.R. 80k28009

[318] I. V. Shragin: *On an application of the Luzin theorem, the Tietze-Uryson theorem, and a measurable selection theorem* (Russian), Perm'. Gos. Polit. Inst., Krajevye Zad. (1979), 171-175
R.Zh. 5B971(80)

[319] I. V. Shragin: (A, B)-*continuity of the Nemytskij operator* (Russian), Perm'. Gos. Polit. Inst., Krajevye Zad. (1980), 180-185
R.Zh. 6B1020(81)

[320] I. V. Shragin: *Orlicz spaces which are generated by functions with vector argument* (Russian), Funkc. Diff. Uravn., Perm' (1985), 64-69
R.Zh. 5B964(86)

[321] N. V. Shuman: *Personal communication* 1985

[322] E. Sinestrari: *Continuous interpolation spaces and spatial regularity in nonlinear Volterra integrodifferential equations*, J. Integral Equ. 5 (1983), 287-308
Zbl. 519.45013   R.Zh. 4B531(84)   M.R. 85h45029

[323] M. S. Skaff: *Vector valued Orlicz spaces I*, Pacific J. Math. 28 (1969), 193-206
Zbl. 176.110   R.Zh. 11B497(69)   M.R. 54♯3395a

[324] M. S. Skaff: *Vector valued Orlicz spaces II*, Pacific J. Math. 28 (1969), 413-430
Zbl. 176.110   R.Zh. 11B498(69)   M.R. 54♯3395b

[325] W. Ślęzak: *Sur les fonctions sup-mesurables*, Problemy Mat. 3 (1982), 29-35
M.R. 85e28006

[326] S. L. Sobolev: *Applications of functional analysis in mathematical physics* (Russian), Izd. Leningr. Univ., Leningrad 1950 [Engl. transl.: Transl. Math. Monogr. 7, Amer. Math. Soc., Providence 1963]

[327] Je. P. Sobolevskij: *The superposition operator in Hölder spaces* (Russian), Voronezh 1984 (VINITI No. 3765-84)
R.Zh. 11B1041(84)

[328] Je. P. Sobolevskij: *The superposition operator in Hölder spaces* (Russian), Voronezh 1985 (VINITI No. 8802-V)
R.Zh. 4B1171(86)

[329] Je. P. Sobolevskij: *On a superposition operator* (Russian), Voronezh 1985 (VINITI No. 8803-V)
R.Zh. 4B1170(86)

[330] F. Szigeti: *On Niemitzki operators in Sobol'ev spaces*, Z. Angew. Math. Mech. 63, 5 (1983), T 332
Zbl. 549.47014   R.Zh. 10B873(83)   M.R. 85a47056

[331] F. Szigeti: *The composition of functions which belong to Sobol'ev spaces* (Russian), Diff. Uravn. Prilozh., Izdat. Mosk. Gos. Univ. (1984), 44-46
M.R. 88e34003

[332] F. Szigeti: *Multivariable composition of Sobol'ev functions*, Acta Sci. Math. 48 (1985), 469-476
Zbl. 595.46038   R.Zh. 7B101(86)   M.R. 87f46058

[333] F. Szigeti: *Composition of Sobol'ev functions and applications I*, Ann. Univ. Sci. Budapest Eötvös Sect. Math. (to appear)

[334] F. Szigeti: *Necessary conditions for certain Sobol'ev spaces*, Acta Math. Hung. 47 (1986), 387-390
Zbl. 626.46023   R.Zh. 4B80(87)   M.R. 87k46077

[335] F. Szigeti: *Composition of Sobol'ev functions and applications*, Notas Mat. Univ. Andes. 86 (1987), 1-25
R.Zh. 1B87(89)   M.R. 89g46071

[336] P.L. Ul'janov: *Absolute convergence of Fourier-Haar series for superpositions of functions* (Russian), Anal. Math. 4 (1978), 225-236
Zbl. 428.42011   R.Zh. 5B48(79)   M.R. 80b42012

[337] P.L. Ul'janov: *Some results on series in the Haar system* (Russian), Doklady Akad. Nauk SSSR 262, 3 (1982), 542-545 [= Soviet Math., Doklady 25, 1 (1982), 87-90]
Zbl. 507.42014   R.Zh. 6B52(82)   M.R. 83d42026

[338] P.L. Ul'janov: *On an algebra of functions and Fourier coefficients* (Russian), Doklady Akad. Nauk SSSR 269, 5 (1983), 1054-1056 [= Soviet Math., Doklady 27, 2 (1982), 462-464]
Zbl. 527.42015   R.Zh. 12B77(83)   M.R. 84j42039

[339] P.L. Ul'janov: *Compositions of functions and Fourier coefficients* (Russian), Trudy Mat. Inst. Steklova 172 (1985), 338-348 [= Proc. Steklov Inst. Math. 3 (1987), 367-378]
Zbl. 575.42005   M.R. 87h42005

[340] H. D. Ursell: *Some methods of proving measurability*, Fundamenta Math. 32 (1939), 311-330

[341] M. M. Vajnberg: *Existence of eigenfunctions for a system of nonlinear integral equations* (Russian), Doklady Akad. Nauk SSSR 61, 6 (1948), 965-968

[342] M. M. Vajnberg: *On the continuity of some special operators* (Russian), Doklady Akad. Nauk SSSR 73, 2 (1950), 253-255

[343] M. M. Vajnberg: *Existence theorems for systems of nonlinear integral equations* (Russian), Uchen. Zapiski Moskov. Obl. Ped. Inst. 18, 2 (1951)

[344] M. M. Vajnberg: *On some variational principles in the theory of operator equations* (Russian), Uspehi Mat. Nauk 7, 2 (1952),

[345] M. M. Vajnberg: *On the structure of a certain operator* (Russian), Doklady Akad. Nauk SSSR 92, 2 (1953), 213-216

[346] M. M. Vajnberg: *The operator of V. V. Nemytskij* (Russian), Ukrain. Mat. Zhurn. 7, 4 (1955), 363-378

[347] M. M. Vajnberg: *Variational methods in the study of nonlinear operators* (Russian), Gostehizdat, Moskva 1956 [Engl. transl.: Holden-Day, San Francisco- London-Amsterdam 1964]
Zbl. 122.355   R.Zh. 5737(57)   M.R. 31#638

[348] M. M. Vajnberg: *On a nonlinear operator in Orlicz spaces* (Russian), Studia Math. 17 (1958), 85-95

[349] M. M. Vajnberg, I. V. Shragin: *The Nemytskij operator and its potential in Orlicz spaces* (Russian), Doklady Akad. Nauk SSSR 120, 5 (1958), 941-944

[350] M. M. Vajnberg, I. V. Shragin: *Nonlinear operators and Hammerstein equations in Orlicz spaces* (Russian), Doklady Akad. Nauk SSSR 128, 1 (1959), 9-12

[351] M. M. Vajnberg, I. V. Shragin: *The Nemytskij operator in generalized Orlicz spaces* (Russian), Uchen. Zapiski Moskov. Obl. Ped. Inst. 77, 5 (1959), 145-159

[352] T. Valent: *Sulla differenziabilità dell'operatore di Nemytsky*, Atti Accad. Naz. Lincei, VIII. Ser., Rend. Cl. Sci. Fis. Mat. Nat. 65 (1978), 15-26
Zbl. 424.35084   R.Zh. 8B811(80)   M.R. 82f47080

[353] T. Valent: *Osservazioni sulla linearizzazione di un operatore differenziale*, Atti Accad. Naz. Lincei, VIII. Ser., Rend. Cl. Sci. Fis. Mat. Nat. 65 (1978), 27-37
Zbl. 424.35085   R.Zh. 8B812(80)   M.R. 81j47053

[354] T. Valent: *Local theorems of existence and uniqueness in finite elastostatics*, Proc. IUTAM Symp. Finite Elasticity, Lehigh Univ. 1981, 401-421
Zbl. 512.73038   M.R. 84b73019

[355] T. Valent: *A property of multiplication in Sobolev spaces: some applications*, Rend. Sem. Mat. Univ. Padova 74 (1985), 63-73
Zbl. 587.46037   M.R. 87c46041

[356] T. Valent: *Boundary value problems of finite elasticity - local theorems on existence, uniqueness and analytic dependence on data*, Springer Tracts Nat. Philos. 31 (1987), 1-191
Zbl. 648.73019   R.Zh. 10B617(88)

[357] T. Valent, G. Zampieri: *Sulla differenziabilità di un operatore legato a una classe di sistemi differenziali quasi-lineari*, Rend. Sem. Mat. Univ. Padova 57 (1977),311-322
Zbl. 402.35027   M.R. 82f35053

[358] A. Vanderbauwhede: *Center manifolds, normal forms and elementary bifurcations*, Preprint Univ. Gent 1987

[359] A. Vanderbauwhede, S. A. van Gils: *Center manifolds and contractions on a scale of Banach spaces*, J. Funct. Anal. 72 (1987), 209-224
Zbl. 621.47050   R.Zh. 1B1121(88)   M.R. 88d58085

[360] R. Vaudène: *Quelques propriétés de l'opérateur de Nemickii dans les espaces d'Orlicz*, C. R. Acad. Sci. Paris 283 (1976), 767-770
Zbl. 342.46018   R.Zh 5B352(77)   M.R. 55♯1148

[361] R. Vaudène: *Quelques propriétés de l'opèrateur de Nemickii dans les espaces d'Orlicz*, Travaux Sém. Anal. Conv. 6, 8 (1976), 17 p.
Zbl. 409.46025   M.R. 58♯30112

[362] I. Vrkoč: *The representation of Carathéodory operators*, Czechoslov. Math. J. 19 (1969), 99-109
Zbl. 175.147   R.Zh. 12B754(69)   M.R. 39♯1719

[363] S. W. Wang: *Differentiability of the Nemytskij operator* (Russian), Doklady Akad. Nauk SSSR 150, 6 (1963), 1198-1201 [= Soviet Math., Doklady 4, 1 (1963), 834-837]
Zbl. 161.350   R.Zh. 1B489(64)   M.R. 27♯1794

[364] S. W. Wang: *On the products of Orlicz spaces*, Bull. Acad. Polon. Sci., Sér. Sci. Math. Astron. Phys. 11 (1963), 19-22
Zbl. 107.91   R.Zh. 6B444   M.R. 29♯6300

[365] S. W. Wang: *Some properties of the Nemytskij operator* (Chinese), J. Nanjing Univ. Natur. Sci. Ed. 2 (1981), 161-173
Zbl. 466.45009   R.Zh. 6B995(82)   M.R. 83d47071

[366] S. Yamamuro: *On the theory of some nonlinear operators*, Yokohama Math. J. 10 (1962), 11-17
Zbl. 188.456   M.R. 27♯6132

[367] S. Yamamuro: *A note on the boundedness property of nonlinear operators*, Yokohama Math. J. 10 (1962), 19-23
Zbl. 188.456   R.Zh. 8B5120(64)   M.R. 27♯5126

[368] A. C. Zaanen: *Note on a certain class of Banach spaces*, Indag. Math. 11 (1949), 148-158

[369] A. C. Zaanen: *Linear analysis*, North-Holland Publ., Amsterdam 1953

[370] A. C. Zaanen: *Ries spaces II*, North-Holland Publ. Comp., Amsterdam 1983
Zbl. 519.46001   R.Zh. 10B763(84)   M.R. 86b46001

[371] P. P. Zabrejko: *Nonlinear integral operators* (Russian), Voronezh. Gos. Univ. Trudy Sem. Funk. Anal. 8 (1966), 1-148
                    R.Zh. 3B508(67)   M.R. 52♯15120

[372] P. P. Zabrejko: *On the differentiability of nonlinear operators in L_p spaces* (Russian), Doklady Akad. Nauk SSSR 166, 5 (1966), 1039-1042 [= Soviet Math., Doklady 166, 5 (1966), 224-228]
Zbl. 161.351   R.Zh. 8B488(66)   M.R. 33♯1772

[373] P. P. Zabrejko: *On the theory of integral operators in ideal function spaces* (Russian), Doct. Diss., Univ. Voronezh 1968

[374] P. P. Zabrejko: *Schaefer's method in the theory of Hammerstein integral equations* (Russian), Mat. Sbornik 84, 3 (1971), 456-475 [= Math. USSR Sbornik 13, 3 (1971), 451-471]
Zbl. 251.45009   R.Zh. 6B458(71)   M.R. 55♯6149

[375] P. P. Zabrejko: *Ideal function spaces I* (Russian), Jaroslav. Gos. Univ. Vestnik 8 (1974), 12-52
R.Zh. 2B479(75)   M.R. 57♯7139

[376] P. P. Zabrejko: *On the theory of integral operators I* (Russian), Jaroslav. Gos. Univ. Kach. Priblizh. Metody Issled. Oper. Uravn. 6 (1981), 53-61
R.Zh. 4B1039(82)   M.R. 87k47114

[377] P. P. Zabrejko: *On the theory of integral operators II* (Russian), Jaroslav. Gos. Univ. Kach. Priblizh. Metody Issled. Oper. Uravn. 7 (1982), 80-89
R.Zh. 3B1159(83)   M.R. 89a47049a

[378] P. P. Zabrejko: *On the theory of integral operators III* (Russian), Jaroslav. Gos. Univ. Kach. Priblizh. Metody Issled. Oper. Uravn. 9 (1984), 8-15
R.Zh. 5B1036(85)   M.R. 89a47049b

[379] P. P. Zabrejko: *Ideal spaces of vector functions* (Russian), Doklady Akad. Nauk BSSR 31, 4 (1987), 298-301
Zbl. 624.46017   R.Zh. 9B843(87)   M.R. 88i46049

[380] P. P. Zabrejko, A. I. Kosheljov, M. A. Krasnosel'skij, S. G. Mihlin, L. S. Rakovshchik, V. Ja. Stetsenko: *Integral equations* (Russian), Nauka, Moskva 1968 [Engl. transl.: Noordhoff, Leyden 1975]
Zbl. 159.410   R.Zh. 10B382(68)

[381] P. P. Zabrejko, M. A. Krasnosel'skij: *On the $\mathcal{L}$-characteristic of linear and nonlinear operators* (Russian), Uspehi Mat. Nauk 19, 5 (1964), 187-189
R.Zh. 5B517(65)

[382] P. P. Zabrejko, H. T. Nguyễn: *On the theory of Orlicz spaces of vector functions* (Russian), Doklady Akad. Nauk BSSR 31, 2 (1987), 116-119
Zbl. 626.46013   R.Zh. 7B823(87)   M.R. 88c46045

[383] P. P. Zabrejko, H. T. Nguyễn: *Linear integral operators in ideal spaces of vector functions* (Russian), Doklady Akad. Nauk BSSR 32, 7 (1988), 587-590
Zbl. 653.46035   R.Zh. 12B792(88)   M.R. 89e47076

[384] P. P. Zabrejko, P. M. Obradovich: *On the theory of Banach spaces of vector functions* (Russian), Voronezh. Gos. Univ. Trudy Sem. Funk. Anal. 10 (1968), 12-21
                        R.Zh. 7B450(68)   M.R. 58♮2238

[385] P. P. Zabrejko, A. I. Povolotskij: *Existence and uniqueness theorems for solutions of Hammerstein equations* (Russian), Doklady Akad. Nauk SSSR 176, 4 (1967), 759-762 [= Soviet Math., Doklady 8, 5 (1967), 1178-1181]
Zbl. 165.136   R.Zh. 3B363(68)   M.R. 36♮4293

[386] P. P. Zabrejko, A. I. Povolotskij: *Eigenvectors of the Hammerstein operator* (Russian), Doklady Akad. Nauk SSSR 183, 4 (1968), 758-761 [= Soviet Math., Doklady 9, 6 (1968), 1439-1442]
Zbl. 194.450   R.Zh. 4B323(69)   M.R. 39♮833

[387] P. P. Zabrejko, A. I. Povolotskij: *Bifurcation points of Hammerstein equations* (Russian), Doklady Akad. Nauk SSSR 194,3 (1970), 496-499 [= Soviet Math., Doklady 11,5 (1970), 1220-1223]
Zbl. 214.116   R.Zh. 2B479(71)   M.R. 42♮3517

[388] P. P. Zabrejko, A. I. Povolotskij: *On the theory of Hammerstein equations* (Russian), Ukrain. Mat. Zhurn. 22, 2 (1970), 150-162 [= Ukrain. Math. J. 22, 2 (1970), 127-138]
Zbl. 193.397   R.Zh. 7B428(70)   M.R. 41♮7398

[389] P. P. Zabrejko, A. I. Povolotskij: *Eigenfunctions of the Hammerstein operator* (Russian), Differentsial'nye Uravn. 7, 7 (1971), 1294-1304 [= Differential Equ. 7, 7 (1971), 982-990]
Zbl. 236.45004   R.Zh. 2B479(71)   M.R. 44♮4603

[390] P. P. Zabrejko, A. I. Povolotskij: *On bifurcation points of the Hammerstein equation* (Russian), Izvestija Vyssh. Uchebn. Zaved. Mat. 6 (109) (1971), 43-53
Zbl. 266.45006   R.Zh. 11B539(71)   M.R. 45♮4233

[391] P. P. Zabrejko, A. I. Povolotskij: *The Hammerstein operator and Orlicz spaces* (Russian), Jaroslav. Gos. Univ. Kach. Priblizh. Metody Issled. Oper. Uravn. 2 (1977), 39-51
Zbl. 406.45009   R.Zh. 10B1047(78)

[392] P. P. Zabrejko, Je. I. Pustyl'nik: *On the continuity and complete continuity of nonlinear integral operators in L_p spaces* (Russian), Uspehi Mat. Nauk 19, 2 (1964), 204-205
R.Zh. 12B422(65)

[393] A. Zhou: *On the continuity and boundedness of W.W. Nemytskij operators in Banach function spaces* (Chinese), Nature J. 11 (1988), 395-396
R.Zh. 3B1103(89)

[394] I. V. Zorin, I. V. Shragin: *On the imbedding of classes of measurable vector functions* (Russian), Perm'. Gos. Polit. Inst., Krajevye Zad. (1979), 176-182
R.Zh. 5B970(80)

[395] W. Zygmunt: *Product measurability and Scorza–Dragoni's type property*, Rend. Sem. Mat. Univ. Padova 79 (1988), 301-304
Zbl. 649.28012   R.Zh. 2B902(89)

List of Symbols

AC	(space of absolutely continuous functions)	229
$\alpha(N)$	(Hausdorff measure of noncompactness)	57
BV	(space of functions of bounded variation)	173
$B_r(X)$	(closed ball)	40
$\mathcal{B}(\Omega)$	(Borel subsets)	10
$\beta(N)$	(special Sadovskij functional)	58
C	(space of continuous functions)	163
C^k	(space of differentiable functions)	205
C^∞	(space of smooth functions)	207
C_k^α	(binomial coefficient)	207
$co\,N$	(convex hull)	21
D_s	(partial derivative)	208
D_u	(partial derivative)	208
D^α	(partial derivative)	228
$\mathcal{D}(F)$	(domain of definition)	48
Δ	(special set)	44
Δ_h^i	(iterated difference)	232
$\delta(G)$	(special set)	59
$\delta(s)$	(special function)	44
E_M	(Orlicz space)	121
$\eta(N)$	(special Sadovskij functional)	58
$F[X]$	(direct transform)	87
$F^{-1}[Y]$	(inverse transform)	88
$F'(x)$	(Fréchet derivative)	70
$F'(\infty)$	(asymptotic derivative)	74
F^*	(rearranged operator)	157
$f^+(s,u)$	(right s-regularization)	180
$f^-(s,u)$	(left s-regularization)	175
$f_+(s,u)$	(right u-regularization)	32
$f_-(s,u)$	(left u-regularization)	32
$f \preceq g$	(special ordering)	12
$f \simeq g$	(sup-equivalence)	12
$\Phi(x,D)$	(integral functional)	23
Φx	(integral functional)	37

$\lambda(D)$	(probability measure) 7
$\lambda(x, h)$	(distribution function) 19
$\lambda_c(x, h)$	(distribution function) 19
$M(= L_\infty)$	(Lebesgue space) 42
$M(u_0)$	(ideal space) 42
M_ϕ	(Marcinkiewicz space) 151
M_p	(Marcinkiewicz space) 153
$M(s, u)$	(Young function) 119
$\tilde{M}(s, u)$	(associated Young function) 122
$M_*(s, u)$	(Sobolev–Young conjugate function) 251
$M \sqsubseteq N$	(special ordering) 123
$M \sqsubset N$	(special ordering) 122
$M[x]$	(Orlicz modular) 120
$\mathcal{M}$	(algebra) 7
$\mathcal{M}(\mathcal{B})$	(special algebra) 18
$\mathcal{M} \otimes \mathcal{B}$	(product algebra) 17
$\overline{\mathcal{M} \otimes \mathcal{B}}$	(completion) 31
$\mu(D)$	(measure) 7
$\mu(x, h)$	(distribution function) 142
Ω'	(accumulation points) 167
Ω_c	(continuous part) 8
Ω_d	(discrete part) 8
$\omega(N)$	(special Sadovskij functional) 58
$\omega(x, \sigma)$	(modulus of continuity) 164
$\omega_F(r, \delta)$	(modulus of continuity) 101
P_D	(multiplication operator) 9
P_ω	(averaging operator) 69
$\tilde{p}$	(associate number) 90
p_*	(Sobolev conjugate number) 251
Π	(special set) 45
$\pi(N)$	(special Sadovskij functional) 57
$\psi(N)$	(Sadovskij functional) 57
$R[x_1, x_2]$	(random interval) 68
$R_\mu(L)$	(space of smooth functions) 217
$R_\mu(0)$	(Beurling class) 222
$R_\mu(\infty)$	(Roumieu class) 222
$\rho(x, y)$	(metric) 8
S	(space of measurable functions) 8

S_0	(space of simple functions) 9
S^0	(space of completely measurable functions) 141
S-lim	(limit in measure) 65
$\operatorname{supp} x$	(support of a function) 41
$\operatorname{supp} N$	(support of a set) 41
Σ	(special set) 44
$\Sigma(N)$	(Σ-hull) 44
Σ_M^r	(Orlicz class) 120
σ_λ	(dilatation operator) 145
$\sigma_X(\lambda)$	(dilatation function) 145
θ	(zero function) 8
$\lceil \vartheta \rceil$	(least upper integer) 93
$\lfloor \vartheta \rfloor$	(greatest lower integer) 229
$\operatorname{var}(\mu, \Omega)$	(total variation) 164
$\operatorname{var}(x; a, b)$	(total variation) 173
W_p^k	(Sobolev space) 228
$W_p^{k,0}$	(Sobolev space) 228
W_p^{-k}	(Sobolev space) 228
$W^k L_M$	(Sobolev–Orlicz space) 250
$W^k E_M$	(Sobolev–Orlicz space) 251
$W(r, \tau, \rho)$	(special set) 190
$W_\phi(s, u, r, \delta)$	(special set) 185
X^0	(regular part) 41
$\tilde{X}$	(associate space) 43
X^*	(dual space) 43
X^σ	(rearranged space) 145
X^κ	(power space) 156
x^*	(decreasing rearrangement) 142
x^{**}	(average rearrangement) 143
$\|x\|_X$	(norm) 40
$[x]_\phi$	(special functional) 150
$< x >_\phi$	(special functional) 151
$< x, y >$	(integral pairing) 42
$\xi_F(r)$	(noncompactness function) 58
χ_D	(characteristic function) 9
Y/X	(multiplicator space) 62
Y/X^n	(higher-order multiplicator space) 75

Subject Index

unit 41

V-pair 66, 93, 132, 148
V_0-pair 66, 93, 132, 149
V^n-pair 75, 93, 135, 149
V_0^n-pair 75, 93, 135, 149
Vandermonde determinant 221

W-boundedness 54
weak Orlicz space 154

Young function 119, 137, 250
– , associated 122
Young inequality 122

Printed in the United States
By Bookmasters